The Urban Book Series

The Urban Book Series is a resource for urban studies and geography research worldwide. It provides a unique and innovative resource for the latest developments in the field, nurturing a comprehensive and encompassing publication venue for urban studies, urban geography, planning and regional development.

The series publishes peer-reviewed volumes related to urbanization, sustainability, urban environments, sustainable urbanism, governance, globalization, urban and sustainable development, spatial and area studies, urban management, transport systems, urban infrastructure, urban dynamics, green cities and urban landscapes. It also invites research which documents urbanization processes and urban dynamics on a national, regional and local level, welcoming case studies, as well as comparative and applied research.

The series will appeal to urbanists, geographers, planners, engineers, architects, policy makers, and to all of those interested in a wide-ranging overview of contemporary urban studies and innovations in the field. It accepts monographs, edited volumes and textbooks.

Indexed by Scopus.

Rubén C. Lois-González · Jose A. Rio Fernandes
Editors

Urban Change in the Iberian Peninsula

A 2000–2030 Perspective

Springer

Editors
Rubén C. Lois-González
Facultade de Xeografía e Historia
Universidade de Santiago de Compostela
Santiago de Compostela, Spain

Jose A. Rio Fernandes
Facultade de Letras da
Universidade de Porto/CEGOT
Porto, Portugal

ISSN 2365-757X ISSN 2365-7588 (electronic)
The Urban Book Series
ISBN 978-3-031-59678-0 ISBN 978-3-031-59679-7 (eBook)
https://doi.org/10.1007/978-3-031-59679-7

This Springer imprint is published by the registered company Springer Nature Switzerland AG
The registered company address is: Gewerbestrasse 11, 6330 Cham, Switzerland

Introduction

Some consider cities to be reflections of the world, where issues such as poverty and inequality, pollution and disease, violence and insecurity are concentrated. Others, however, prefer to affirm that cities have paved the way for hope in an increasingly urban world, capable of extending our longevity on the planet, reinforcing average levels of well-being and reducing poverty. Additionally, cities also define our individual and social existence because we either live in cities or engage in a significant portion of our activities in them that allow us to be part of the modern world.

For all these reasons, cities deserve comprehensive analysis and reflection. In this case, our focus is on the Iberian Peninsula, a territory in southern Europe that has undergone intense urbanisation since the second half of the 20th century. With this objective in mind, a set of contributions centred on Spain and Portugal has been developed, addressing similar urban issues and city dynamics. Indeed, various reasons led to the creation of this collective work, which acknowledges the diversity of urban situations observed in two different countries. Thus, in addition to considering each context, we recognise the existence of urban spaces of unequal size and characteristics, which often pose challenges in adapting the concept of the city.

The classic study of the city has always differentiated between an approach focussed on analysing the main nucleus of a territory, forming part of an urban network or system. This approach is opposed to the analysis of the inner space of the city, its morphology, structure, social diversity and land use. Furthermore, present-day cities are tangible realities but also projected as images for mass consumption, promotional or iconic purposes. Everything revolves around cities, which stand before us, the simple residents in large, complex locations. Places where we coexist with thousands or even millions of people, whether in concentrated and continuous spatial structures that have characterised cities in history or in dispersed spatial arrangements articulated in a reticular diffusion of urbanisation processes and practises.

Theoretical approaches to the urban world would be incomplete without acknowledging that planning, urban design and urbanism are also ways of understanding the city, seeing how it was erected and planned in the past until it reached the reality we know today. Furthermore, present-day urban government and governance seek to change our living conditions and material elements of life daily. They also contribute to the progress or the goal of sustainability for the future. In this sense, the processes of approval of general, partial, special plans or new Urban Agendas shape the future, delineating the new sectors that will be incorporated into the city in the short term or making it more habitable. All this in an era where issues such as ecological requirements in areas such as mobility or the treatment of urban waste will play a central role in understanding the urban environment.

Throughout the book, a path is traced between a theoretical dimension, which illuminates general features while obscuring the individuality of each case, and the understanding of a unique example. In these chosen examples, the greatest risk is to undervalue the understanding of processes that do not have a merely local or even national dimension. To counteract this risk, emphasis has been placed on developing two approaches. The first is based on critical analysis, cultivated by urban social geography since the late 1970s. Through it, general interpretations can be obtained starting from unique cases and framing the city in the understanding of a financialised capitalist system. The second approach insists on studying urban planning, city projects and urban governance. It is a more positivist and functional model, but it also allows for establishing general interpretations and norms that transcend from isolated urban examples.

This work, coordinated by two neighbouring professors and friends from Santiago de Compostela and Porto, brings together 26 works structured into 6 themes, involving around thirty authors. Professionally, most of them are geographers, although urban planners and sociologists are also present. It proposes a reading of cities but also of the world from the cities of the two Iberian countries: Spain and Portugal. Additionally, three essential dimensions for spatial and territorial analysis are present and we highlight them in this brief presentation.

An Outsider Perspective

Chicago, Paris, London, as well as more recent urban dynamics on the West Coast of the United States, in China and in Central and Northern Europe, constitute the basis of much of the classic urban studies. They are the protagonists of the books and articles we read from reference publishers (often translated), whether the treatises of Peter Hall, Henri Lefebvre, David Harvey and Edward Soja or the textbooks that have influenced generations of urban geography students, such as those by Harold Carter, Paul Claval or Michael Pacione.

Naturally, each national geography, Portuguese or Spanish, was also capable of producing its own references and sometimes (in some cases and moments of history) of giving special value to somewhat parochial and egocentric views in the name of specificity on the global stage. In others, these interpretations arose in response to a lack of translations and difficulty in accessing and understanding what came from other countries. The reference books were mostly produced in English, which partly explains the rise of authors from languages with Latin roots, especially the French, who also retained their influence.

It is also worth considering that the world can—and should—be viewed from cities outside those considered to be central cities in the economic organisation of the planet. So why don't we study Madrid or cities hosting people from the African continent, like Barcelona or Lisbon to understand the processes of inequality in the world? Or try to understand the changes introduced by tourism in cities (and the difficulty of access to housing) by analysing Porto or Lisbon, when the changes are recent and intense? How can we not consider the cases from Portugal and Spain, where there are cities that have almost entirely dedicated themselves to being consumer cities? After all, perceptions of urban dynamics associated with new challenges of soft and sustainable mobility can be analysed from Madrid or Lisbon. Also, if we want to understand gender inequalities, we can do so from Iberian cities. Yes, Portugal and Spain are part of the world. And they are rich worlds seen from within, from medium-sized cities like Albacete or Aveiro, to metropolises like Madrid. However, cities are above all windows that allow us to perceive the world and how its governance is currently viewed, contributing to a more polycentric and democratic urban geography debate.

Views On and from the City

The six parts of the book are organised into two groups of works from Spain and Portugal. Part I refers to Urbanisation and brings together four contributions focussed on studying population concentration, city growth and urbanisation expansion in general. Part II has inequality as its central theme, addressing contemporary urban problems such as the persistence of residential segregation, difficulty accessing housing or gender inequalities. In the analysis of all these issues, the fundamental role of the state as a mitigator of social contrasts, the importance of a financialised market, the value of real estate and the differences between different parts of the city will always be present.

Part III chapters groups texts that examine the importance of the real estate sector, finance and central economic activities grouped in cities. It is about knowing the financial city, the speculative processes in urban space and, at the same time, the prominence of large cities as undisputed centres of the main companies in a country or region. The texts of Part IV deal with consumers and residents, starting from the brand images projected by the city. In Porto, a novel study will address the floating population that is increasingly relevant in this city, while in Madrid, concepts of consumer and the consumed city will be discussed. Part V focuses on mobility issues, with several works elaborating on specific dynamics. Proposals range from the possibilities opened by the end of the private car city to new forms of urban mobility. Finally, Part VI groups texts that address issues related to governance and urban planning. It begins with a historical approach and continues with the challenges, decentralisation processes and recent advances in urban planning in Spain and Portugal. Finally, specific cases of governance processes in the Lisbon region, Barcelona and a medium-sized city in Castilla-La Mancha, Albacete, are studied.

Beyond Understanding, a Comparison

This book does not carry out a comparative analysis between countries or between cities. It is the result of the union of a set of efforts to articulate the theoretical component with the understanding of different national and local contexts, in what we see as a contribution to better understand cities and the world, because the world is becoming increasingly urban and cities are becoming increasingly essential in the world.

The book is valuable, naturally, for the contribution of each author or group of authors and for the organisation and refinement of the texts they have provided. However, we believe it allows for more than one set of singular readings by providing an interpretation from two complementary perspectives. That is, for each theme, subtheme and case study, it brings visions from Spain and Portugal on urban and national contexts, allowing for the verification and evaluation of differences. We believe that by placing together the texts from each country, it offers the possibility of knowing ourselves better by confronting us with other situations. This is the comparison we seek: not that of mathematical contrasts, impossible in the face of different realities, but the enrichment of understanding each situation. In this way, we would like to contribute to two important objectives: understanding the cities of each of the two countries. Discovering cities not only from national authors who write about them or in their comparison or contextualisation with readings that come from works written by authors based in the United Kingdom, the United States, the Netherlands, France or Germany, but in a richer way, based on the enormous tradition of urban studies in the Iberian Peninsula. For Fernando Pessoa, reading is dreaming hand in hand with others.

We, the organisers of this book, hope that readers dream of visiting cities again—and the world that reaches us through them—and that, with us, they learn to see them from various perspectives through the authors who wrote the texts we have gathered, to whom we are very grateful.

Santiago de Compostela, Spain Rubén C. Lois-González
Porto, Portugal Jose A. Rio Fernandes

Contents

Part I
Urbanization

Chapter 1
Population Change, Migration and Mobility Patterns in Portugal

Maria Lucinda Fonseca

Abstract This chapter explores the most recent trends in migration, population change and settlement patterns in Portugal. The trend towards urban concentration, the bipolarisation of the urban network and the depopulation and economic decline of large areas of the country, developed since the middle of last century, are the result of the rural exodus, emigration and the sharp decline in the birth rate. In the last decade, particularly in recent years, there have been signs of revitalisation of some rural areas, due to the combined effects of the attraction of labour immigrants to work in agriculture, the return of Portuguese emigrants to their home regions and changes in the pattern of residential mobility of the Portuguese population, due to the widening of commuting belts and to a lesser extent some counter-urbanisation movements, especially after the COVID-19 pandemic. The study is based on secondary data from the Population Census and registers of the foreign population. Although these migratory flows have not significantly changed the settlement patterns and the regional distribution of the population, they can have a positive impact on the demographic and economic dynamics of some areas which, for several decades, have been affected by depopulation and ageing.

Keywords Migration · Population change · Urban concentration · Rural areas · Portugal

1.1 Introduction

The geographic distribution of the Portuguese population is characterised by the historical trend towards greater density of occupation in coastal regions, both on the western coast and on the Algarve (south), the bipolarisation of the urban network

---

M. L. Fonseca (✉)
Centre of Geographical Studies, Institute of Geography and Spatial Planning, University of Lisbon, Lisbon, Portugal
e-mail: fonseca-maria@edu.ulisboa.pt

Associate Laboratory Terra, Lisbon, Portugal

R. C. Lois-González and J. A. Rio Fernandes (eds.), *Urban Change in the Iberian Peninsula*, The Urban Book Series, https://doi.org/10.1007/978-3-031-59679-7_1

around the cities of Lisbon and Porto and the persistence of the contrasts between the settlement system in the North and South of the country inherited from the Middle Ages, with a predominance of small villages and towns in the North and Centre regions and the most concentrated settlement in the territories south of the Tagus river, where urban centres relevant at a regional scale exist since the Middle Ages and Arab urbanising influence.

The post-Industrial Revolution urbanisation process in Portugal was quite late, compared to other European countries. In 1960, 43.6% of the active population worked in the primary sector, 65.7% of the inhabitants lived in localities with less than two thousand inhabitants, and places with ten thousand or more residents concentrated only 22.6% of the total population of the country.

The sixties, seventies and eighties of the last century correspond to a period marked by profound demographic, economic, social and political transformations. Accession to EFTA in 1960, the great wave of emigration to Europe between 1960 and 1973, the emergence of the colonial war in 1961, the effort to industrialise and improve the country's infrastructure, with successive Development Plans, the development of tourism, the end of the dictatorship and the decolonisation of African territories (Cape Verde, Guinea-Bissau, Angola and Mozambique) in the 1970s, and the membership to the European Economic Community in 1986, contributed decisively to ending the long period of isolation and underdevelopment of a traditional and predominantly rural society. These changes gave rise to a new phase of Portugal's repositioning in Europe and in the world and of opening to the values of modern life and urban ways of life, with important repercussions on demographic and family structures, population distribution patterns, the economic base and functioning of labour markets, the living conditions of the population and the asymmetries of regional development (Ferrão 2005).

The trend towards the bipolarisation of the urban network and the demographic decline of large areas of the North, Centre and Alentejo regions are the result of a double migratory process that has developed since the middle of the last century: (i) the rural exodus, mainly to the metropolitan areas of Lisbon and Porto, but also to medium-sized cities, in particular for district capitals; and emigration, particularly to Western European countries (Fonseca 1990, 2022; Gaspar et al. 1989; Fonseca et al. 2017; Ferrão 2018). In other words, the rural exodus and emigration (mainly of the younger population), combined with the sharp decline in the birth rate, depopulated and aged the interior of the country, leading to a progressive demographic hollowing out and a worsening of the already weak economic dynamism of some regions. On the other hand, the displacement of the population to the coast led to a rapid growth of the main cities located there and the emergence of new urban centres, around Lisbon and Porto. The increase in immigration, from the mid-eighties and especially in the second half of the 1990s and the beginning of the new millennium, reinforced the concentration of the population in the two metropolitan areas and in the Algarve (Fonseca 2019). However, in the last decade, particularly in recent years, there are signs of revitalisation of some rural areas, due to the combined effects of the attraction of labour immigrants to work in agriculture, the return of Portuguese emigrants to their home regions and changes in the pattern of residential mobility of the Portuguese population, due to the widening of commuting belts and to a lesser extent some counter-urbanisation movements, especially after the COVID-19 pandemic.

The hypothesis of a "rural revival" or "urban exodus" during the early stages of the COVID-19 pandemic has been widely discussed in the media and the upward trend of internal migrations from dense cities to small towns and rural areas during the pandemic crisis has been documented in the literature by studies carried out in different countries: Spain (Gonzalez-Leonardo et al. 2022a, b), Japan (Fielding and Ishikawa 2021), United Kingdom (Rowe et al. 2023), Sweden (Vogiazides and Kawalerowicz 2022) and Australia (Borsellino et al. 2022). In Germany, there was also a decline in net migration rates in the largest German cities (Stawarz et al. 2022).

In the Portuguese case, although these migratory flows have not significantly changed the pattern of regional distribution of the population, they can have a positive impact on the demographic and economic dynamics of some areas which, for several decades, have been affected by a sharp demographic decline and ageing.

This chapter explores the evolving migration patterns and demographic dynamics in Portugal from historical urban concentration to an emerging trend of rural revival in several municipalities, in areas close to major cities and also in sparsely populated spaces that have experienced considerable increases in net migration—internal, international or both.

The study is based on data from the Population Census on the distribution of the Portuguese population, by size classes of the places of residence, as well as on the evolution of the population and migrations, internal and international. The text is structured as follows: the first section offers a brief overview of the urbanisation pattern and demographic change in Portugal. Section two analyses the main trends of international migration, and section three examines the latest trends of intermunicipal mobility patterns, focusing our attention on migration to rural areas. Finally, in the conclusion, we discuss the key findings and some policy implications.

1.2 Urbanisation and Demographic Dynamics in Portugal

Lisbon and Porto are the only cities that until 1981 had more than 100,000 inhabitants, the former being the only one with more than 500,000. The phenomenon of urbanisation and population concentration observed since the 1960s is clearly illustrated by the persistent reduction of inhabitants in localities with fewer than 2000 residents and by the growth of urban centres with populations between 10,000 and 20,000 and between 20,000 and 100,000 inhabitants (Table 1.1). The pace of urbanisation reflects the demographic dynamics of the country, with the growth of cities being less intense in periods of strong emigration and demographic decline, such as the 1960s–1970s, and more intense between 1970 and 1981, due to the arrival of returnees from former African colonies. In the following decade, despite the stagnation of the country's population, resulting from the drastic reduction in natural population growth, there was an expansion of the suburban areas of Lisbon and Porto and cities with more than 20,000 and less than 100,000 inhabitants increased (Table 1.1).

As Roca and Pimentel (2003), point out, from the nineties of the last century onwards, two opposing tendencies coexisted, largely resulting from the configuration of internal migrations and the increase in immigration: reinforcement of urban concentration in the coastal regions and urban deconcentration (peri-urbanisation), especially in the regions of Lisbon and Porto, with these cities losing population and urbanisation spreading to the transitional rural space surrounding the two metropolitan areas. In the interior, the main dynamic was a continuing concentration in the main urban centres and the decline of the population residing in rural areas.

In 2021, the Metropolitan Areas of Lisbon and Porto together accounted for 46.7% of the population and only 5.7% of the area of mainland Portugal. Therefore, they are the regions of the country with the highest population densities: 951.9 and 850.6 inhabitants/km^2, respectively, i.e. 8.5 and 7.6 times higher than the national average (112.2 inhabitants/km^2). Thus, of the twenty-two municipalities with a population density greater than one thousand inhabitants/km^2, eleven were in the Lisbon Metropolitan Area, eight in the Porto Metropolitan Area and only three (Entroncamento, Funchal and Braga) were located outside these areas (Fig. 1.1).

Figure 1.2 also shows the imbalance in the occupation of the national territory. About seventy per cent of the municipalities (213 out of 308) have population densities below 140 inhabitants/km^2, 93 (29.2%) less than 30 inhabitants/km^2 and only 60 (19.5%) above 250 inhabitants/km^2. The less populated areas are situated in the border territories of the North and Central regions and in the interior of Alentejo, while the most densely populated areas correspond to the coastal strip north of Setúbal and the Algarve coast.

In the last decade, migration balances have not been sufficient to offset the effects of the fall in the birth rate and the negative natural growth rates since 2007, so the country registered, between 2011 and 2021, a population decline of – 2.1%, reversing

Table 1.1 Rate of change of the population of Portugal, by size of localities, 1960–2021

Years	Rate of change of population by size of population centre (%)						
	Total	< 2000	2000–4999	5000–9999	10,000–19,999	2000–100,000	100,000 or more inhabitants
1960–1970	– 2.5	n.a	n.a	n.a	20.7	46	– 2.8
1970–1981	13.5	n.a	n.a	n.a	18.9	69.2	5.5
1981–1991	0.3	– 8.8	5.9	26.6	36.7	14.1	– 4.1
1991–2001	5.0	– 8.4	10.7	30.3	17.3	19.3	21.8
2001–2011	2.0	– 11.8	0.7	18.7	25.7	10.8	13.2
2011–2021	– 2.1	– 2.4	10.9	1.2	– 7.7	– 7.8	– 0.1

n.a.—not available

Source INE—Population Census, with own treatment

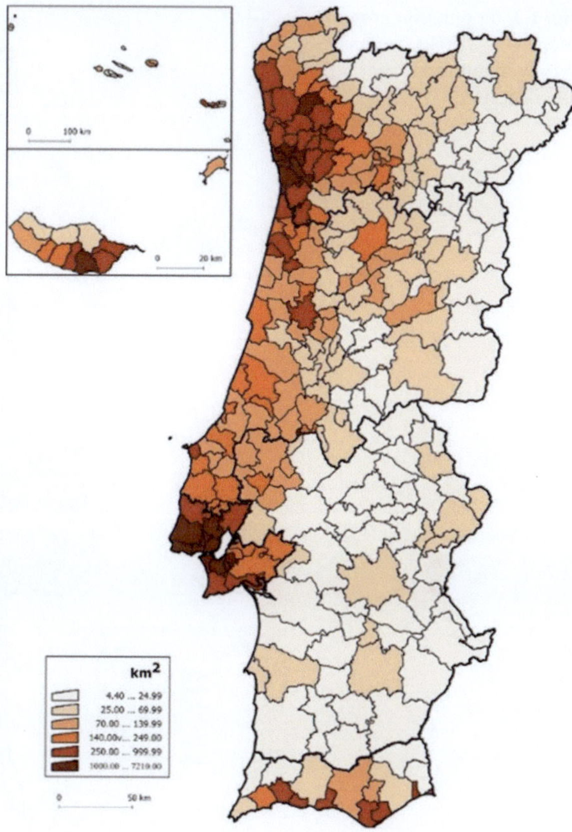

Fig. 1.1 Population density in Portuguese municipalities: average number of residents per square kilometre, 2021. *Source* INE, with authors' calculations

the growth trend observed from the seventies of the last century. Regarding geographical distribution, the concentration of the population in predominantly urban areas increased (+ 0.23% than in 2011), while moderately urban and rural areas recorded declines of – 5.09% and – 11.26% respectively (INE 2023). 253 municipalities have lost resident population (83.8%), with a special incidence in the North, Centre and Alentejo regions (Fig. 1.2).

It must be noted that the population decline is not restricted to the interior of the country, but also covers vast areas of the coast. Only two regions—Lisbon Metropolitan Area and Algarve registered a population growth. The largest declines occurred in Alentejo (7.0% less) and the Autonomous Region of Madeira (6.4% less). However, in the first of these regions, contrary to this trend, the municipality of Odemira stands out, registering the highest population growth rate in Portugal, due to the attraction of immigrant workers for agriculture and also retired sun-seekers and other lifestyle migrants. In the peripheral regions, only Viseu and Sernancelhe registered a slight demographic increase. Despite this, as can be seen in Table 1.1, it is interesting to note that while the decrease in population extends to small and

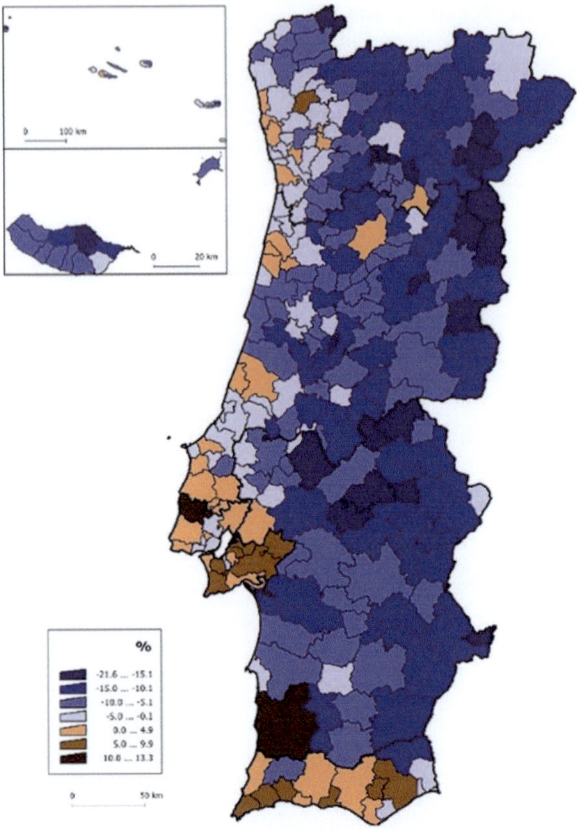

Fig. 1.2 Population growth rate, 2011–21 (%). *Source* INE, with authors' calculations

medium-sized cities (between 10,000 and 100,000 inhabitants), the smaller settlements, between 2000 and 5000 residents, after the stagnation observed between 2001 and 2011, show signs of some dynamism, registering a population increase of 10.9%. These transformations indicate changes in the geographical mobility patterns of the population, so we will analyse below the recent trends of migration in Portugal and its contribution to the revitalisation of some rural areas.

1.3 Regional Immigration Patterns in Portugal: From Metropolisation to Dispersion

The evolution and regional distribution of the Portuguese population have been, throughout history, profoundly marked by international migrations. Traditionally a country of emigration, from the mid-eighties of the last century, there was an increase in migratory flows to Portugal, namely from the former African colonies and Brazil, giving rise to the formation of immigrant communities with a significant numerical expression and a diverse sociodemographic composition, especially in the Lisbon Metropolitan Area and the Algarve (Fonseca et al. 2005; Góis and Marques 2018).

It was especially after integration into the European Economic Community in 1986, that the post-colonial cycle of immigration, characterised by the predominance of migratory flows from the former African colonies, gave way to a phase of high growth and diversification of the migrants' countries of origin. Thus, between 1980 and 1990, the number of foreign citizens with residence permits in Portugal doubled, from 50,750 to 107,707.

Growth in immigration continued in the 1990s, growing by 6.8% annually between 1990 and 2000. In addition to the maintenance of traditional flows, there was an expansion and diversification of the areas of international recruitment of labour, namely from Eastern European countries, with particular emphasis on Ukraine. This evolution is the result of the combined effects of a favourable economic situation observed during this period and also of more structural factors, namely the increase in the average level of education of the Portuguese population and the consequent increase in their expectations of professional mobility towards more qualified professions, the maintenance of labour-intensive sectors marked by strong seasonality, such as tourism and construction, as well as Portugal's external image, in the context of the EU, as a country where it was easy to enter and work, even if in irregular conditions (Fonseca 2009).

Thus, in the year 2000, there were immigrants from more than a hundred nationalities, with the citizens of the former African colonies, Brazil and European countries continuing to be clearly dominant, a fact that suggests a strong association between immigration to Portugal, the colonial past and the dominant historical, cultural and economic links. It should also be noted that, compared to other European countries, the percentage of foreign population was still quite small, representing only about 2% of the total number of residents (Baganha et al. 2004).

From 2004 onwards, at the end of the cycle of major public works, in the context of economic recession and greater effectiveness of the mechanisms for controlling irregular immigration and the employment of undocumented workers, there was a reversal of the trend of growth in immigration. Despite this, the stock of foreign population with residence permits in the country continued to increase, as many undocumented immigrants who obtained a residence permit under the extraordinary regularisation regime of 2001 were able to obtain a residence permit a few years later.

The emergence of the economic and financial crisis of 2008 marks the beginning of a new phase of international migration in Portugal, characterised by a decrease in immigration, an increase in emigration and a change in the profiles of the foreign population residing in Portugal, with emphasis on the growth in the number of international students, researchers, highly qualified professionals, self-employed workers, investors and retirees and the decrease in labour immigration (Oliveira and Gomes 2018; Fonseca 2019). The decrease in immigration and the return of immigrants to their country of origin or remigration to another destination, combined with the acquisition of Portuguese nationality by an increasing number of foreigners, especially after the amendment of the nationality law in 2006, were reflected in the decrease, as of 2010, in the stock of foreign resident population with residence permits in Portugal.

The recovery of the Portuguese economy, which began in 2014, was accompanied by the continued increase in immigration from 2016 onwards, and in 2022, according to the records of the Foreigners and Borders Service (SEF), the number of foreigners with legal resident status in the country amounted to 781,247, which is equivalent to 7.5% of the resident population in Portugal. It should also be noted that in addition to documented immigrants, there is also a high number of foreign citizens in an irregular situation, given that, as stated by Fonseca et al. 2005; Machado 1997; Sampaio 2017, there is a high proportion of immigrants who, throughout their stay in Portugal, have already gone through periods of irregularity, due to the difficulties of obtaining the necessary documentation before migration, or to the long waiting times either to regularise their situation after entering Portugal, either to renew residence permits in the country.

In 2022, the immigrants registered by the Portuguese authorities were distributed among about one hundred and ninety nationalities, the ten biggest groups coming from Brazil, the United Kingdom, Cape Verde, India, Italy, Angola, France, Ukraine, Nepal and Guinea-Bissau.

The trend towards greater geographical diversity of the origins of immigrants residing in Portugal is associated with different motivations for migration and different demographic, socio-economic and cultural profiles of migrants. As previously mentioned, in addition to economic and family motivations, there has also been an increase in the attraction of international students, retirees and other immigrants looking for a new lifestyle (Iorio and Fonseca 2018; Alves 2021; Montzuma and McGarrigle 2019; McGarrigle 2021; Sampaio 2022; Sardinha 2013, 2015).

The attraction of privileged and highly skilled immigrants is due, to a large extent, to the perception of Portugal as a safe country, with immigrant-friendly policies, quality of life and lower cost of living, as well as tax advantages for non-habitual residents and, in the particular case of the United Kingdom, also the effects of BREXIT (Montezuma and McGarrigle 2019; McGarrigle 2021). In addition, it is also worth mentioning the effect of the programme for granting residence permits for investment activity, commonly known as *golden visas*, created in 2012 to attract foreign investment.[1] In the case of French citizens, one of the groups that registered the greatest growth, the attraction of Lisbon and other Portuguese regions, to the detriment of Maghreb countries, such as Morocco or Tunisia, where many bought properties for holidays or for residence after retirement, is also due to the effects of the "Arab Spring". More recently, with the accelerating effect of the pandemic COVID-19 on

[1] The Residence Permit Regime for Investment Activity (ARI), created in 2012, was amended in 2023 by Law No. 56/2023, of 6 October, which revoked the granting of residence permits for real estate investment activities. This change was justified by its effects on the increase in real estate speculation, mainly in the Algarve and in the metropolitan areas of Lisbon and Porto.

the digital transition and the expansion of teleworking, Portugal has also become attractive to digital nomads.

In 2022, 63% of foreigners residing in Portugal with legal resident status lived in the Lisbon Metropolitan Area and the Algarve. However, the most recent migratory waves present a greater geographical dispersion, and the Lisbon Metropolitan Area has been losing relative importance. In addition, in territories with low population density, even low numbers of immigrants can make a relevant contribution to mitigating the effects of ageing and demographic decline, given that, in 2021, 61.1% of foreigners with a residence permit in Portugal were between 20 and 49 years old, while the Portuguese population of the same age group represented only 35.4% of the total.

The number of immigrants with legal resident status, per hundred inhabitants, by municipality, in 2011 and 2022 (Figs. 1.3 and 1.4), shows, more clearly, the differences in the impact of immigration, also allowing the identification of some relevant changes between the two dates. In the North Region, both in 2011 and 2022, the share of the foreign population is, in all municipalities, lower than the national average. Despite this, some municipalities on the border with Galicia (Valença, Monção and Vila Nova de Cerveira), the Metropolitan Area of Porto (namely the city of Porto), Braga and some municipalities of Trás-os-Montes and Douro stand out with a proportion of foreign population higher than the average of the region.

Between 2011 and 2022, the growth of the foreign population in all regions (NUT II) of the country is highlighted, especially in the North Region. Although this region continues to be the one with the lowest proportion of foreign population, it is noteworthy that, for the first time, in some municipalities of Alto Minho (Valença and Vila Nova de Cerveira), Braga, Porto and Bragança the percentage of inhabitants of foreign nationality is higher than the national average.

In both 2011 and 2022, the Algarve is the region where the share of immigrant population is most expressive. In the last year, all the municipalities in this region, with the exception of Alcoutim, have values well above the national average (Fig. 1.4). The highest values correspond to Vila do Bispo (42.2%), Albufeira (39.5%), Aljezur (36.7%) and Lagos (36.3%).

The Lisbon Metropolitan Area is, after the Algarve, the region with the highest proportion of foreign population in 2022 (13.2%). Compared to 2011, stands out the high growth in the number of immigrants, namely in the city of Lisbon and in the suburban municipalities of Cascais, Odivelas and Amadora.

In the Centre Region (Região Centro) the foreign population is more pronounced in the sub-region of Oeste, especially in the municipalities of Caldas da Rainha, Óbidos, Nazaré and Torres Vedras and in the sub-region of Leiria, namely in Pedrogão Grande, Alvaiázere and Marinha Grande. Aveiro and some rural municipalities situated near Coimbra (Arganil, Góis and Penela) also have values above the national average.

In the interior of the central region, between 2011 and 2022, the percentage of immigrants in the total population increased from 1.9 to 5.2%. At the municipal level, Penamacor, Idanha-a - Nova and Vila Velha de Ródão stand out as the proportion of

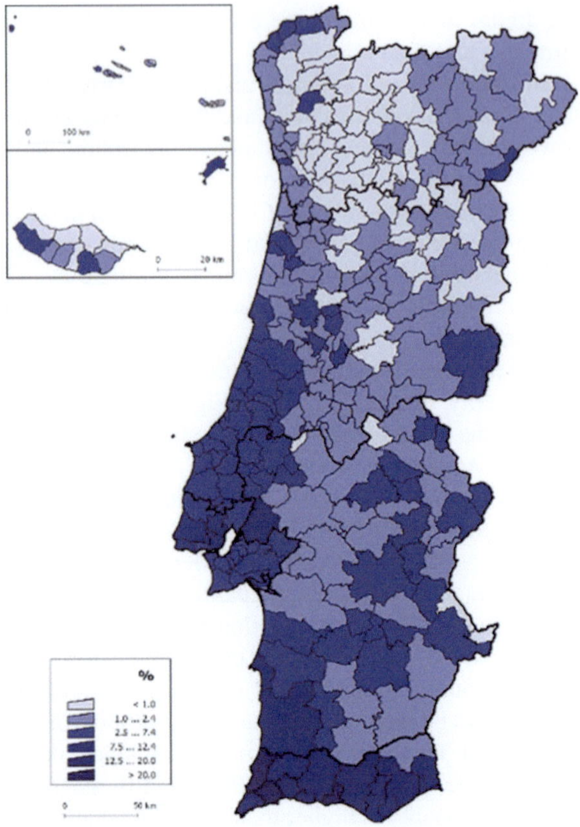

Fig. 1.3 Foreign resident population per 100 inhabitants by municipality of residence, 2011. *Source* SEF and INE, with authors' calculations

foreign residents is higher than the regional average. On the other hand, in Viseu-Dão-Lafões, Beiras and Serra da Estrela, the foreign population with residence permits represents less than 5% of the total inhabitants in all municipalities in this area, with the exception of Fundão (Fig. 1.4).

In Alentejo, the growth of the immigrant population near the coast stands out, especially in Odemira, being the second municipality with the highest proportion of foreign nationals (39.5%) in Portugal in 2022. In Ferreira do Alentejo, Cuba and Sines, the percentage of foreign population is also higher than the country's average.

In the Autonomous Regions, Lajes das Flores, in the Azores (8.8%), and Calheta and Porto Santo in Madeira, with 10.7% and 8.6% respectively, are the municipalities with the highest proportion of immigrants.

In summary, it can be concluded that the regional distribution of the foreign population, in 2022, compared to 2011, although still very concentrated in the Lisbon Metropolitan Area and in the Algarve, presents a greater dispersion throughout the national territory, highlighting the growth of immigrants in the transition spaces, on the outskirts of the urban and industrial regions of the coast and some cities in the interior, as well as in low-density rural areas, especially in the Alentejo, which are

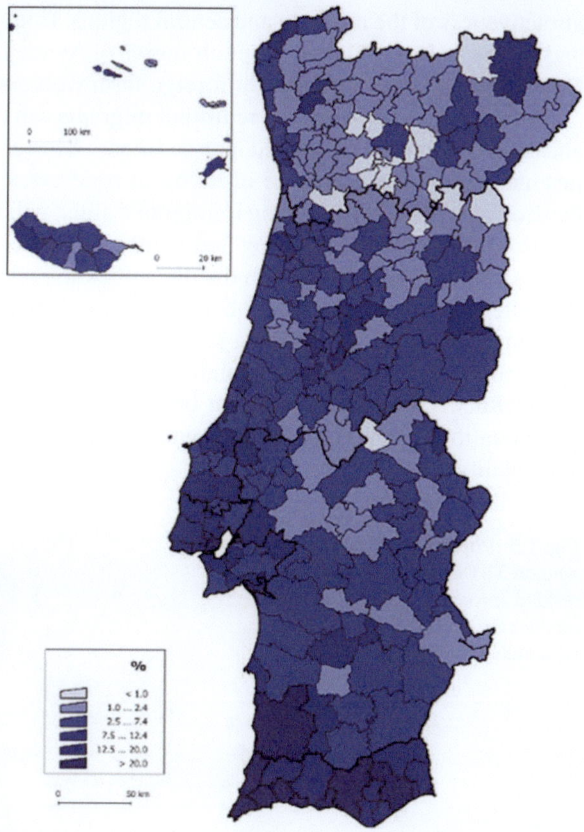

Fig. 1.4 Foreign resident population per 100 inhabitants by municipality of residence, 2021. *Source* SEF and INE, with authors' calculations

strongly affected by demographic decline and ageing, but where there have been large investments in agriculture, which increasingly resort to the recruitment of foreign workers to cope with the shortage of local labour (Fonseca et al. 2021; Carvalho 2021). There is also an increase in privileged immigrants in some areas, attracted by the tranquillity, climate and quality of life offered by rural environments, remote work opportunities and also attractive investment possibilities. Another relevant aspect is related to the effects of local policies to attract and integrate them, as is the case in the municipality of Fundão.

International migrations to Portugal, in addition to the foreign population, also include Portuguese emigrants. In the absence of records of these movements, the data collected by the 2021 Population Census regarding the place of residence on 31 December 2019 is a good indicator of the recent attraction of international immigrants, both foreign citizens and Portuguese returnees.

As can be seen in Fig. 1.5, the population that lived outside the country in the year prior to the census is more representative in the Algarve and in the Lisbon Metropolitan Area, regions where, as we have seen above, the municipalities with the highest proportions of foreigners are located. In addition, there is also a greater attractiveness of the northern and central regions, from where most of the emigration had its origin. In the Alentejo, the sub-region of Alentejo Litoral stands out, especially Odemira, due to the attraction of foreign farm workers and life-style immigrants.

The foreign population and returning migrants can contribute to the demographic and economic revitalisation of depressed areas. A lower cost of living, family or social ties and previous experiences of living in rural areas motivate some to migrate to these areas. However, attracting immigrants alone will not solve the problem of most of the country's areas in ageing and demographic decline. Rural development will necessarily have to be based on the diversification of the socio-economic structure, the increase in the productivity of agriculture and the creation of support services for residents, with a view to creating jobs and improving individual and family incomes and well-being that allow the maintenance of the population and the occupation of the territory. Otherwise, it will not be possible to stem the exodus of the younger, more qualified population or to sustain a high negative balance between mortality and natality rates.

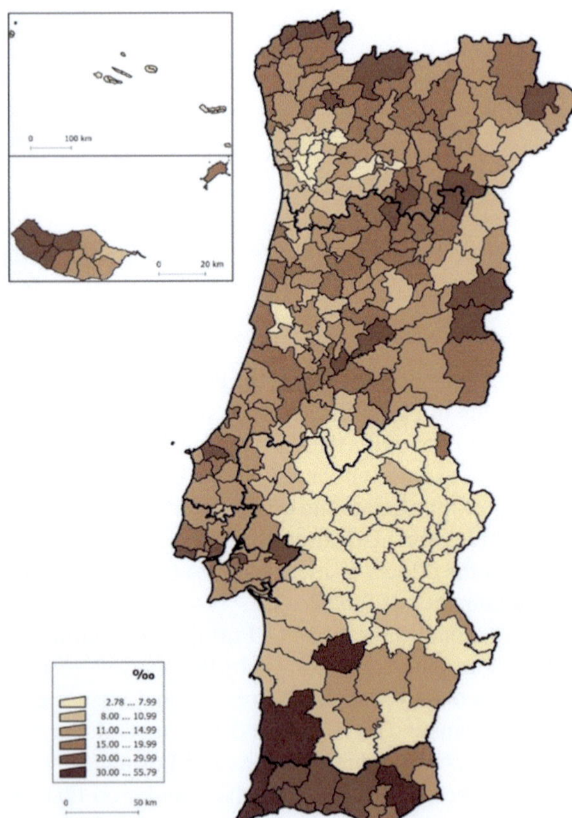

Fig. 1.5 Population in 2021, who on 31 December 2019 resided abroad, per 1000 inhabitants. *Source* INE, with authors' calculations

1.4 Recent Trends of Residential Mobility and Migration: Urban Deconcentration and Rural Revival?

In Portugal, there is no information that allows an accurate characterisation of the internal movements of the resident population.

In this chapter, we use data collected in the 2021 census, regarding the place of residence one year before the census (December 31, 2019). It should be noted that this information does not allow us to know whether there has been any other change of residence, in the country or even abroad, between the reference times considered, constituting only a generic indicator of the patterns of internal geographical mobility and the attraction of the population residing abroad, since it simply allows to quantify, for each municipality, the number of residents in 2021 who in 2019 lived in another country.

In Portugal, like in other Southern and Eastern European countries, the residential mobility of the population is relatively low (Bell et al. 2015). At the time of the 2021 census, only 13.7% of the population lived in a different dwelling from the one they lived in on December 31, 2019, and 90.6% lived in the same municipality where they lived in 2019. Only 7% moved to another municipality, 146,469 (1.4%) lived abroad and 98,007 (1%) had not yet been born. From these figures we may note several things, one being that according to data from the 2011 population census, referred to by Fonseca et al. (2022), the percentage of the population that one year before the census was carried out, lived in another municipality is much lower than that observed in 2020 (2.6%).

As mentioned earlier, internal migrations have reinforced the concentration of the population in urban areas, certainly due to the provision of better opportunities for employment, education and access to services. However, recent changes, such as the expansion of remote work during the pandemic, the inflated cost of housing in large cities, changing lifestyle preferences and public policy initiatives for rural development must explain the growing movement towards rural areas.

In 2020, all regions (NUT III) of the country, apart from the metropolitan areas of Lisbon and Porto and the Algarve, recorded positive net-internal migration rates. The largest population gains due to internal migrations occurred in insular, ultra-peripheral territories, especially on the island of Corvo, in the Autonomous Region of the Azores, and in peri-urban areas close to the main cities, where many residents had secondary residences, especially north and east of Lisbon Metropolitan Area.

Sixty-two municipalities (20.1%) recorded negative internal migration balances, of which about one third belonged to the metropolitan areas of Lisbon and Porto and the Algarve (Fig. 1.6). Negative migratory balances are also observed in municipalities on the coast and in the interior, where the main cities in the region are located (Braga, Aveiro, Coimbra, Covilhã, Évora and Beja). On the other hand, the surrounding territories registered positive balances.

These results reveal that the intermunicipal residential mobility of the population tends to be greater in metropolitan areas and in the largest urban agglomerations on the coast, where the phenomenon of urban deconcentration (peri-urbanisation) has

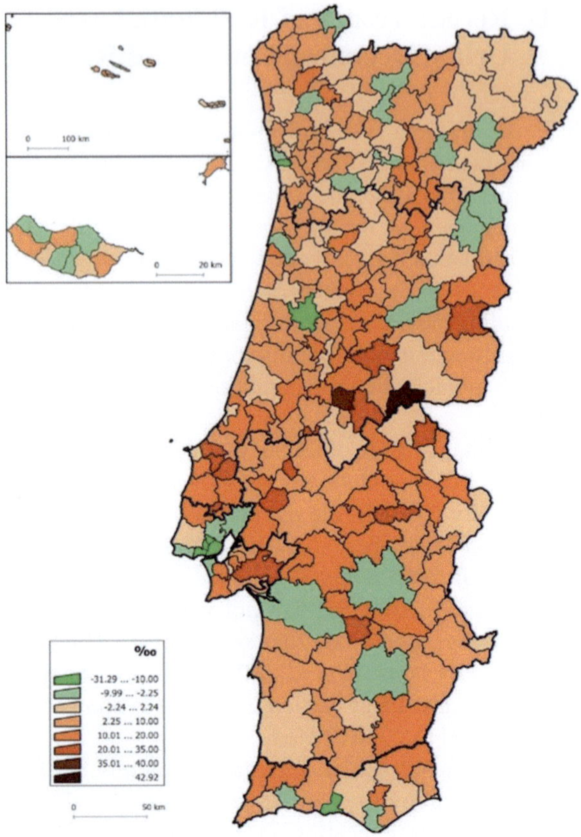

Fig. 1.6 Intermunicipal net migration rate in the year prior to the 2021 census (‰). *Source* INE, with authors' calculations

gained some expression, thus increasing the distance between places of residence and work/study (Fonseca et al. 2022). At the same time, there is also an increase in internal geographical mobility for some rural areas in the interior of the country, especially in the vicinity of the main urban centres of these regions. Although population gains of the most marginal territories are still low and insufficient to compensate for the negative natural balance, they are important to mitigate the risk of depopulation of these areas.

1.5 Discussion of Results and Concluding Remarks

This study contributes to the knowledge of the most recent trends in demographic change and immigration and internal geographical mobility of the Portuguese population, and their effects on the urban system and the regional demographic dynamics.

Despite the limitations of the statistical information, data from population census and records of the stock of foreign population with residence permits in Portugal, made it possible to verify that in recent years the population distribution in Portugal has undergone significant changes. Although the trend towards concentration in the urban areas of the western coast and the Algarve remains, there is also evidence of an increase in the attractiveness of some rural areas in the interior of the North and Centre of the country and in the Alentejo region, affected by demographic decline since the middle of the last century.

Several factors may have contributed to this shift: investments in transportation and infrastructure, technological advancements, changing attitudes towards urban living, the desire for a closer connection to nature and a slower pace of life, lower costs of living, better housing conditions and policy initiatives for rural development.

Improvements in accessibility, due to investments in transportation and infrastructure, have made it easier for people to live outside major urban centres, while still being able to access employment opportunities and services. The emergence of the COVID-19 pandemic has also contributed to stimulating the residential mobility of skilled professionals and young families to small towns and rural areas, due to increased opportunities to work remotely and the perception of rural areas as safer and healthier places than large cities. In this context, the maintenance of ties with the place of origin, both of internal migrants and the Portuguese diaspora, are also important factors in decisions to return to the homeland or to maintain secondary residences in these places.

It is still too early to assess whether these trends are only conjunctural or whether they will continue. In any case, these changes highlight the increasing diversity of factors that interfere in the decision-making processes of migration and residential mobility. In addition to economic opportunities, quality of life and connections to the place are increasingly important aspects to be considered in migratory aspirations and in the choice of a place to live. Therefore, if it is to attract residents to less populated areas, it seems that public policies and strategies should enhance territorial identities, strengthen relations with the Portuguese diaspora and internal migrants, and foster the formation of rural communities.

Regarding the settlement of the foreign population, whether labour or lifestyle immigrants, it is important to create conditions that facilitate their reception and integration, in order to prevent possible situations of conflict, xenophobia and racism, establishing trust relationships and promoting a sense of belonging to the place.

Despite the positive aspects of rural migration, there are major limitations and challenges resulting from infrastructural deficits, lack of employment opportunities and difficulties in accessing services, notably in health and education. The progressive and already long process of demographic and economic decline in several areas of the country, combined with the closure of public services, such as courts, schools, health centres, hospitals, post offices and tax offices, due to very low population thresholds, makes it difficult to retain the younger and more qualified population and reinforces the marginalisation of these territories.

As stated by Ferrão (2018), public policies may play a relevant role in regulating the interactions between endogenous and exogenous factors that have led to the progressive marginalisation and peripheralisation of rural areas, doubly affected by demographic ageing and decline, but their capacity to reverse their negative demographic dynamics is limited, mainly due to the lack of articulation of the different development and territorial cohesion policy instruments and the practises of the various actors. Therefore, the definition of strategies must be participatory to meet the aspirations of local communities. In addition, it is necessary to make the application of the available financial support instruments more flexible, otherwise, they will not be effective and not adjusted to the specific needs of each territory.

Acknowledgements This work is part of the Re-Place (Reframing Non-Metropolitan Left Behind Places Through Mobility and Alternative Development) project, co-funded by the European Union within the Horizon Europe Programme. Grant agreement ID: 101094087.

References

Alves EP (2021) Angolan and Cape Verdean students in Portuguese higher education: motivations and socio-spatial trajectories. Ph.D. thesis in Human Geography (Migration Studies), University of Lisbon

Baganha MI, Marques JC, Góis P (2004) The unforeseen wave: migration from Eastern Europe to Portugal. In: Baganha MI, Fonseca ML (eds) New waves: migration from Eastern to Southern Europe. Fundação Luso-Americana, Lisboa, pp 23–39

Bell M, Charles-Edwards E, Ueffing P, Stillwell J, Kupiszewski M, Kupiszewska D (2015) Internal migration and development: comparing migration intensities around the world. Popul Dev Rev 41(1):33–58. https://doi.org/10.1111/j.1728-4457.2015.00025.x

Borsellino R, Bernard A, Charles-Edwards E, Corcoran J (2022) A regional renaissance? The shifting geography of internal migration under COVID-19. Aust Geogr 1–19. https://doi.org/10.1080/00049182.2022.2074622

Carvalho JM (2021) A imigração e a agricultura no Alentejo no século XXI. In: Revista Migrações, Observatório das Migrações, dezembro 2021, n.° 17, ACM, Lisboa, pp 87–104

Ferrão J (2005) Dinâmicas Demográficas: Uma Visão Panorâmica. In: Ferrão J, Salgueiro TB (eds) *Geografia de Portugal*, vol 2—Sociedade, Paisagens e Cidades. Círculo de Leitores, pp 50–71

Ferrão J (2018) Despovoamento em áreas rurais: entre a inevitabilidade e a capacidade de transformação. Cultivar. Cadernos de Análise e Prospetiva. Março 2018, nº 11, pp 13–19

Fielding T, Ishikawa Y (2021) COVID-19 and migration: a research note on the effects of COVID-19 on internal migration rates and patterns in Japan. Popul Space Place 27(6):2499. https://doi.org/10.1002/psp.2499

Fonseca ML (1990) População e Território – do País à Área Metropolitana. Centro de Estudos Geográficos, Lisboa

Fonseca ML, Ormond M, Malheiros J, Patrício M, Martins F (2005) Reunificação Familiar e Imigração em Portugal. Observatório da Imigração, ACIME, Lisboa. http://www.oi.acidi.gov.pt/docs/Estudos%20OI/Estudo_OI_15.pdf

Fonseca ML (2009) Imigração, diversidade e novas paisagens étnicas e culturais. In: Lages M, Matos AT (eds) Portugal: percursos de interculturalidade. ACIDI, Lisboa, pp 49–96
Fonseca ML (2022) Urbanización, migraciones y concentración de la población en Portugal. In: Fernandes RJA, González RCL, Sposito MEB (eds) El mundo visto de las ciudades. Tirant Humanidades, pp 51–58
Fonseca ML (2019) Migrações, envelhecimento demográfico, urbanização e despovoamento das áreas rurais. In: Jacinto R (ed) Transversalidades: Fotografia sem Fronteiras 2019. Centro de Estudos Ibéricos, Guarda, pp 196–206
Fonseca ML, Abreu D, Esteves A (2017) Ageing and migration: some reflexions on the effects of the economic and financial crisis on demographic trends in Portuguese regions. In: Fonseca M, Fratesi U (eds) Regional upgrading in Southern Europe. Spatial disparities and human capital. Advances in Spatial Science. The Regional Science Series. Springer, Cham, pp 265–298
Fonseca ML, Esteves A, Moreno L (2021) Migration and the reconfiguration of rural places: the accommodation of difference in Odemira, Portugal. Popul Space Place 27(8):e2445. https://doi.org/10.1002/psp.2445
Fonseca ML, Esteves A, Silva S, Abreu D (2022) Padrões de distribuição geográfica e migrações internas da população estrangeira residente em Portugal (2005–2011). In: Livro de Homenagem à Professora Fernanda Cravidão. Imprensa da Universidade, Coimbra
Gaspar J, Abreu D, Ferrão J, Jensen-Butler C (1989) Portugal, os Próximos 20 Anos. VI volume. Ocupação e Organização do Espaço. Uma Perspectiva. Fundação Calouste Gulbenkian, Lisboa
Góis P, Marques JP (2018) Retrato de um Portugal migrante: a evolução da emigração, da imigração e do seu estudo nos últimos 40 anos. e-cadernos CES [Online]. URL: http://journals.openedition.org/eces/3307. https://doi.org/10.4000/eces.3307
González-Leonardo M, Rowe F, Fresolone-Caparrós A (2022a) Rural revival? The rise in internal migration to rural areas during the COVID-19 pandemic. Who moved and Where? J. Rural Stud 96:332–342
González-Leonardo, M., López-Gay A, Newsham N, Recano J, Rowe F (2022b) Understanding patterns of internal migration during the COVID-19 pandemic in Spain. Popul Space Place 28(6):e2578. https://doi.org/10.1002/psp.2578
INE - Instituto Nacional de Estatística - Censos (2021) O que nos dizem os Censos sobre dinâmicas territoriais. INE, Lisboa, 2023. Available at https://www.ine.pt/xurl/pub/66320870. e-ISBN: 978-989-25-0636-4; ISBN: 978-989-25-0674-6
Iorio JC, Fonseca ML (2018) Estudantes brasileiros no ensino superior português: construção do projeto migratório e intenções de mobilidade futura. Finisterra 53(109):3–20. https://doi.org/10.18055/Finis14556
Machado FL (1997) Contornos e especificidades da imigração em Portugal. Sociologia, Problemas e Práticas 24:9–44
McGarrigle J (2021) The locational choice of urban lifestyle migrants in Lisbon: beyond tourism imaginaries. In: Dominguez-Mujica J, McGarrigle J, Parreño JM (eds) International residential mobilities: from lifestyle migrations to tourism gentrification. Series Geographies of Tourism and Global Change. Springer Nature, pp 105–124. https://doi.org/10.1007/978-3-030-77466-0_7.
Montezuma J, McGarrigle J (2019) What motivates international homebuyers? Investor to lifestyle 'migrants' in a touristic city. Tour Geogr Int J Tour Space Place Environ 21(2):214–234
Oliveira CR, Gomes N (2018) Indicadores de Integração de Imigrantes. Relatório Estatístico Anual 2018. Coleção Imigração em Números do Observatório das Migrações. ACM, Lisboa
Roca MN, Pimentel D (2003) Causas prováveis das migrações internas em Portugal na década de noventa. GeoInova. Revista Do Departamento De Geografia e Planeamento Regional 8:91–116
Rowe F, Calafiore A, Arribas-Bel D, Samardzhiev K, Fleischmann M (2023). Urban exodus? Understanding human mobility in Britain during the COVID-19 pandemic using Meta-Facebook data. Popul Space Place 29(1):e2637. https://doi.org/10.1002/psp.2637 (Epub 2022 Dec 7. PMID: 36718419; PMCID: PMC9877951)

Sampaio D (2022) Migration, diversity and inequality in later life: ageing at a crossroads. Palgrave Macmillan. https://doi.org/10.1007/978-3-031-10894-5
Sampaio V (2017) Os processos de regularização de imigrantes realizados em Portugal: que avaliação? Tese de Doutoramento em Políticas Públicas apresentada no ISCTE-IUL – Instituto Universitário de Lisboa, Lisboa, Portugal
Sardinha J (2013) Lifestyle migrants in central Portugal. Strategies of settlement and socialisation In: Janoschka M, Haas H (eds) Contested spatialities, lifestyle migration and residential tourism. Routledge, London
Sardinha J (2015) Idyllic seekers and liminal beings: lifestyle migrants in the rural regions of Central Portugal. In: Torkington K, Sardinha J, David I (eds) Practising the good life: lifestyle migration in practices. Cambridge Scholars Publishing, Cambridge, pp 33–51
Stawarz N, Rosenbaum-Feldbrügge M, Sander N, Sulak H, Knobloch V (2022) The impact of the COVID-19 pandemic on internal migration in Germany: a descriptive analysis. Popul Space Place e66. https://doi.org/10.1002/psp.2566
Vogiazides L, Kawalerowicz J (2022) Urban exodus in covid times: who moves out of the inner city of Stockholm and where do they go? Stockholm Research Reports in Demography, vol. 2022. Stockholm University, p 4. https://su.figshare.com/articles/preprint/Urban_exodus_in_Covid_times_Who_moves_out_of_the_inner_city_of_Stockholm_and_where_do_they_go_/19248190.

Chapter 2
Regional Concentration and Urban Expansion in Spain. Fragmentation and Intra-Urban Inequality

Jesús M. González-Pérez

Abstract Parallel to the Portuguese example, in Spain cities and the main metropolises have grown rapidly since 1950. This growth translates once again into a process of littoral concentration and, to a large extent, into a polycentric urban system. On the one hand, two cities and their metropolitan regions define the top level of the Spanish territorial hierarchy, Madrid and Barcelona. Another five cities (Valencia, Sevilla, Málaga, Bilbao and Zaragoza), constitute a second step with prominence to centralise regional urban networks in the country, of entity and internal cohesion. Below, there is still a proliferation of regional metropolises (Valladolid, Vigo, Gijón, Alicante, Murcia, etc.) and a large number of cities that act as both deeply rooted economic-provincial centres and administrative capital. All this panorama defines a very complex urban system, which is understood in terms of the strong role that the region and the regional (expressed in Autonomous Communities with powers) acquire in Spain. In turn, one of the main problems in the recent urbanisation process has been higher social inequalities within cities. We first look at the distribution and level of concentration of Spain's urban population through a dual external-internal analysis. We then approach intra-urban inequalities based on income and housing indicators.

Keywords Urban system · Metropolitanisation · Urban hierarchy · Urban inequality · Urban fragmentation

2.1 Introduction

Population in Spain aligns with global demographic dynamics linked to increased urbanisation and population growth in cities. The effects of the crisis in industrial cities (1970s) and the start of the era of post-industrial cities (1980) have brought about dual-scale consequences: the new role of cities within a territory that redefines

J. M. González-Pérez (✉)
Department of Geography, University of the Balearic Islands, Palma, Spain
e-mail: jesus.gonzalez@uib.es

R. C. Lois-González and J. A. Rio Fernandes (eds.), *Urban Change in the Iberian Peninsula*, The Urban Book Series, https://doi.org/10.1007/978-3-031-59679-7_2

urban hierarchy and development pillars, and the emergence of intra-urban and intra-metropolitan demographic processes that impact the very form and structure of cities.

Although the diffuse city model has partially modified the very idea of cities since the 1980s in Spain, population concentration has traditionally been linked to urban centres. On the one hand, although the total population of a city is a partial way to gauge urban nature, it is size—either in terms of demographics or density—which acts as a distinguishing feature in comparison to other settlements. On the other hand, all big cities act as a central location for something, whether this be retail, public services, job creation in industry, etc. (Lois et al. 2012). Thus, population density is a distinguishing feature of a city for both external (as an urban system) and internal (as an urbanised area) analysis.

Higher social inequalities within cities have been one of the main problems in recent urbanisation processes. Inequality and fragmentation are inherent realities of post-Fordist capitalism that are changing cities in myriad ways: they foster the urban concentration of capital, employment and culture; make urban space subject to tourist development; and contribute to social exclusion, differentiation and marginalisation processes by creating social divides (due to income, age, race and geography) and defining fractures within the urban fabric.

The city that emerged post-1973 sprawls with social inequalities (González-Pérez et al. 2018). The three milestones of 1973, 2007 and 2020 mark the post-Fordist city and are tied to higher urban inequality, capitalist crises and changes in the economic cycle:

1. Cities after the 1973 crisis and general social inequality. According to Hammett (2001), most urban inequality between rich and poor appeared in the 1980s. Terms such as dual cities (Mollenkopf and Castells 1991) and divided cities (Fainstein et al. 1992) have been used since the urban crisis of the 1980s to analyse inequality in late-capitalist cities.
2. The 2007–08 global financial crisis led to economic crisis and neoliberal austerity policies that widened the gap between social classes, creating new urban spaces characterised by segregation and exclusion. The initial crisis and subsequent austerity policies collectively spurred a new research stage into urban inequalities. In this sense, the fragile post-crisis recovery period (2015) was particularly imbalanced and socially polarising. Indeed, authors such as Harvey (2012) and Peck (2012) point out that the financial crisis was particularly urban.
3. The COVID-19 crisis in 2020 was, once again, an essentially urban crisis that highlighted the weak foundations and vast social and urban inequalities underpinning the supposed economic recovery (González-Pérez and Piñeiro-Mantiñán 2020; Montoya et al. 2021).

In this vein, the dual aim of this chapter is to firstly analyse urban population distribution and concentration in Spain from a dual external-internal analysis perspective, to then look at intra-urban inequalities in selected Spanish cities based mainly on income and housing indicators.

2.2 City and Territory

2.2.1 Urban Hierarchy and Settlement System

Spanish population growth between 1960 and 2022 is mainly due to growth in municipalities with over 10,000 inhabitants (a population threshold that defines urban centres in Spain) in opposition to the demographic decline in rural areas. Municipalities with over 10,000 inhabitants only comprise 9.26% of all municipalities in Spain yet, as of 2021, were home to 79.76% of the entire Spanish population. There are 63 municipalities with over 100,000 inhabitants, representing 39.9% of the population. In 1960, 418 municipalities (4.54% of the total) had over 10,000 inhabitants and were home to 56.87% of the population. Only 26 municipalities housed over 100,000 inhabitants in that year, representing 27.74% of the Spanish population.

At the end of the 1930s, Jefferson (1939) interpreted the double-headed nature of the Spanish system as a consequence of the historical and political origins of Spain. The weak spread of urban growth inland from the coast is a key economic factor, with the continual appreciation of coastal subsystems (Juaristi 1995). At the start of the 1990s, the Spanish urban system was defined as a half-ring structure with an urbanised coast and a scantly urbanised inland area focussed on Madrid (Ferrer 1992). Without changing this two-pronged approach, the effects from establishing autonomous regions included a consolidated decentralised spatial and functional structure, weakening the links between the periphery and the centre and increasing ties between sub-centres (Precedo 1999). Be that as it may, Spanish cities have undergone the most radical and quickest transformations in their history since the mid-twentieth century (Nel.lo 2017). These are underpinned by three main features: the population shift towards urban centres, the formation of a large cluster of metropolitan areas and the establishment of activity hubs in the country (Nel.lo 2004).

At present, old rural and urban spaces have been fully integrated and the urbanisation process has established an interdependent and fully consolidated global network (Nel.lo 2017) (González-Pérez et al. 2018). Madrid and Barcelona, the two unrivalled cities at the forefront of the urban system in Spain, have populations in excess of one million. Valencia (with nearly 800,000) is the main counterweight in this dual pole system, followed by three other cities with populations of over 500,000 inhabitants (Sevilla, Zaragoza and Málaga). Alongside Bilbao (around 350,000 inhabitants), these six cities make up the seven main metropolitan areas in Spain. This system is rounded off by a large network of municipalities, 57 in total (if we include Bilbao), with between 100,000 and 500,000 inhabitants, comprising regional and provincial capitals (including Murcia, Palma, Valladolid and A Coruña), important industrial cities with significant weight in regional structures (such as Vigo, Gijón, Elche or Cartagena) and major urban centres in the metropolitan area of major cities (e.g. Hospitalet, Badalona, Móstoles, Getafe, Dos Hermanas or Telde). Nonetheless, it is what one could term small cities—municipalities with between 10,000 and 100,000 inhabitants—which house most of the population. Alongside mid-sized cities (100,000–500,000), these centres are driving population growth.

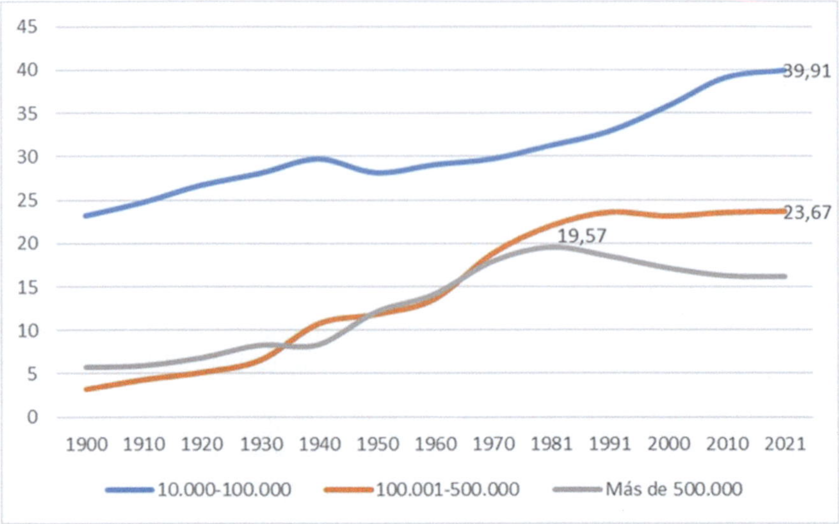

Fig. 2.1 Population change in urban municipalities (over 10,000 inhabitants) by municipality size (1900–2021). *Source* produced by the authors based on population census and the ongoing register of inhabitants (Instituto Nacional de Estadística, INE)

Meanwhile, the largest cities (over 500,000 inhabitants) have been experiencing slower growth and losing demographic weight since 1981, aligning with the start of the post-industrial urban system (Fig. 2.1).

2.2.2 *Urban Imbalance: Coastalisation and Hubs for Spatial Growth*

The urbanisation process in 21st-century Spain includes three main socio-demographic dynamics: the importance of metropolitanisation (both in large and mid-sized cities), the urban focus of development poles (the Mediterranean, Madrid, the Ebro, Guadalquivir and Atlantic) and the *coastalisation* of how the main urban centres are distributed, fostering coastal urbanisation over inland 'desertification' (González-Pérez 2022). Of the 63 municipalities with over 100,000 inhabitants in 2021, 37 are non-central cities in metropolitan areas and 26 are located in coastal provinces. Despite widening polycentrism, four inland provinces in mainland Spain (Ávila, Palencia, Segovia and Soria) currently only have one municipality with over 10,000 inhabitants (Fig. 2.2).

The Atlas Estadístico de las Áreas Urbanas de España (Statistical Atlas of Urban Areas in Spain) produced by the Ministries of Transport and Sustainable Mobility, and Housing and Urban Agenda, has defined and categorised urban areas for the last two decades. The latest edition (Ministerio de Vivienda y Agenda Urbana 2022)

Fig. 2.2 Municipalities with over 10,000 inhabitants in 2021. *Source* Instituto Geográfico Nacional (IGN). Atlas Nacional de España (https://interactivo-atlasnacional.ign.es/index.php#bbox=-1517988,5467589,2179252,1322079&c=indicator&i=r_16_t.r_16_t&s=2021&t=A02&view=map7)

included two categories: large and small urban areas. As a general rule, large areas have a municipality with at least 50,000 inhabitants, while municipalities included in multi-municipal urban areas have at least 1,000 inhabitants. In total, there are 86 areas with over 50,000 inhabitants (either single or multi-municipal) covering a total of 755 municipalities housing over 32 million inhabitants (2021). In other words, 9.6% of the land and 9.3% of all municipalities are home to 69% of the Spanish population. Of the 86 areas, 20 include a single municipality while the remaining 66 comprise several municipal areas. The latter represents true Spanish conurbations.

With regard to the resident population in these urban areas by autonomous region, we note three main aspects. Firstly, clear spatial concentration on the coast and in the main development poles. In Madrid, nearly 95% of residents live in one of the large urban areas, followed by Catalonia and Asturias with around 80% of the local population. By comparison, the urban areas of Extremadura and Castille-La Mancha are home to less than 40% of the local population. Secondly, tourist areas on the Mediterranean coast and in the Canaries are highly represented. Thirdly, and as highlighted in red in Fig. 2.3, the resident population in these areas has decreased in eight of the 17 autonomous regions compared to figures from 2017. Although this may be due to cyclical processes since the figures are not particularly significant, the possibility of new trends should be monitored and analysed. This process can be seen across different types of areas, especially in autonomous regions with the highest urban population: Madrid (−0.25%) and Catalonia (−0.15%) (Fig. 2.3).

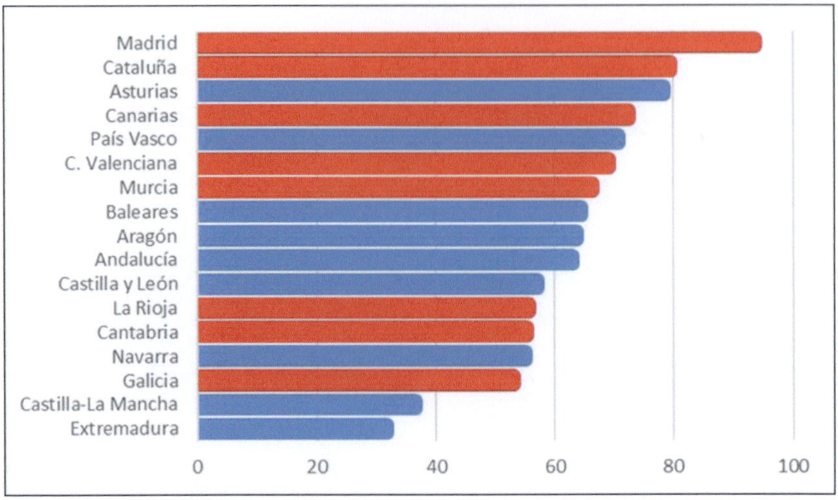

Fig. 2.3 Resident population in urban areas by autonomous region in 2021. *Source* Produced by the authors based on the Atlas Estadístico de las Áreas Urbanas (Ministerio de Vivienda y Agenda Urbana 2022)

2.3 Metropolitanisation: Urban Growth and Sprawl

Another way to approach the analysis of population concentration is an internal analysis of urban spaces. Urban sprawl or deconcentration refers to the trend in cities to spread into wide agglomerations where current building, inhabitant and economic growth are mostly higher than in what would normally be considered the main city (Lois et al. 2012).

With the exception of Ceuta and Melilla, excluded from this analysis since the municipality and autonomous city limits match, population density in five urban areas exceeds 2000 inhabitants/km^2: Madrid and four areas on the tourist coast. Arrecife is at the top of the list. This small city is the capital of the island of Lanzarote in the Canaries. With regard to private dwelling density, four areas exceed 1000 homes/km^2. All of these areas are found on the tourist coast due to the ponderance of second or holiday residences (González-Pérez 2022). The clear leader here is Torrevieja, a large urban area with the highest growth in housing stock over the 1981–2011 period (Table 2.1).

The increasing relevance of urban peripheries obliges us to reinterpret current cities, since metropolitan municipalities (peri-urban, urban–rural) record the highest growth. Figure 2.4 shows population evolution between 1996 and 2021 in the municipalities of the 13 biggest urban areas at present. The central circle is the main city and the external ring includes all other municipalities comprising the respective urban areas. With the exception of Asturias, urban peripheries have seen general demographic growth. For example, a mere 0.42% in Bilbao to over 10% in other areas. Nevertheless, this does not represent a population decline in the main municipality

Table 2.1 Urban areas with the highest population and private dwelling density in 2021

Urban area	Inhabitants/km^2 (2021)	Urban area	Homes/km^2 (2021)
Arrecife	2805.9	Torrevieja	1385.3
Valencia	2513.7	Valencia	1253
Madrid	2165.4	Arrecife	1115.2
Las Palmas de Gran Canaria	2070.7	Roquetas de Mar	1041.9
Bilbao	1786.5	Salamanca	991.4
Roquetas de Mar	1724.7	Las Palmas de Gran Canaria	949.5
Salamanca	1623.1	Madrid	929.2
Barcelona	1591.8	Bilbao	846.5
Santander - Torrelavega	1255.4	Barcelona	716.3
Málaga	1241.3	Sant Feliú de Guixols	687.3

Source produced by the authors based on the Atlas Estadístico de las Áreas Urbanas (Ministerio de Vivienda y Agenda Urbana 2022)

in the urban area. Except for Sevilla, Bilbao and the cities in the Bay of Cadiz (Cádiz and Jerez), the eleven remaining areas have experienced population growth. Indeed, three have seen growth of over 10%: Zaragoza (the main city in the Ebro hub with an extensive area that soaks up some peri-urban growth) and two in the Mediterranean tourist region (Palma and Murcia) (Fig. 2.4).

2.4 The Interior Space of Cities. Urban Fragmentation and Inequality

In addition to suburbanisation, phenomena such as residential segregation or social inequalities increased within the framework of industrial restructuring in the 1980s and the development of metropolisation, globalisation and neoliberal policies (González-Pérez et al. 2023). The end result is a new urban geography characterised by greater fragmentation in urban structure (De Mattos 2002).

Although the term urban fragmentation is multi-faceted (Szupiany 2018) and not founded on a particularly large developed theoretical framework, it has been used alongside other terms in recent decades due to the similarities between them, including dual city (Castells 1989), social division of space (Duhau 2013), cracked city (Marcuse and van Kempen 2002) and fractal city (Soja 2008). Processes creating fragmentation dynamics have existed since the 1990s: metropolisation and urban sprawl, gentrification, the creation of gated apartment complexes, globalisation and differential access to technology (Prèvôt-Schapira and Cattaneo Pineda 2008). A consensus has emerged in recent years on the appearance of a new pattern of segregation where urban fragmentation underpins the structure of today's cities, based on

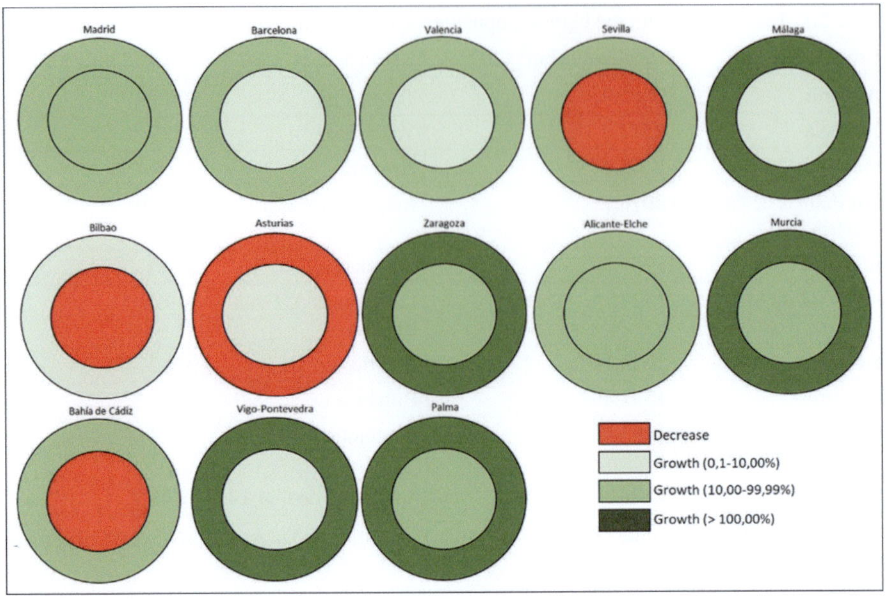

Fig. 2.4 Demographic change in the main urban areas in Spain: Central Municipality and other municipalities (1996–2021). *Source* Produced by the authors based on the Atlas Estadístico de las Áreas Urbanas[1] (Ministerio Fomento 2000, 2018; Ministerio de Vivienda y Agenda Urbana 2022)

the multiplication of separate parts with limited small-scale permeability and framed around large-scale nodes (Burgess 2018).

The 2008 financial crisis was especially urban. While Spain was still healing from scars in 2020, the COVID-19 pandemic generated a new crisis highlighting the weak structures underpinning the supposed economic recovery, as well as the major social inequalities that lay behind it. The 2020 health crisis was also fundamentally urban, not only due to its impact on health but also to the consequences it had for less advantaged social classes. In this vein, higher intra-urban inequalities, the vulnerability affecting many areas of cities and the higher number of individuals at risk of social exclusion are just a few shared features from the last two crises (González-Pérez and Piñeira-Mantiñán 2020).

The intra-urban contrasts in income level and the increasing difficulties in accessing housing are the main variables when drawing up current urban social

[1] The following urban areas have increased the number of municipalities they comprise: Madrid (from 28 in 2000 to 52 in 2022), Barcelona (from 164 to 165), Valencia (from 44 to 45), Málaga (from 7 to 8), Zaragoza (from 2 to 15), Murcia (from 3 to 10), Vigo-Pontevedra (from 6 to 14) and Palma (from 2 to 8). This helps to explain why many have recorded growth above 100%. What we term central municipality comprises more than one municipality in four urban areas: Asturias (Gijón and Oviedo), Alicante-Elche (Alicante and Elche), Bahía de Cádiz (Jerez de la Frontera and Cádiz) and Vigo-Pontevedra (Vigo and Pontevedra).

maps. Nonetheless, they are also useful in detecting the emergence of new patterns of inequality in Spanish cities.

On one hand, income level inequality is measured by calculating the income gap between the highest and lowest net earnings per household on the 2021 census, for the period covering 2015–2021 in the top five Spanish cities. On the other, the Gini coefficient is used for the same cities over the same seven-year period.

Madrid has the highest income gap across all years, with a maximum difference of nearly 80,000 EUR between the highest and lowest income brackets in 2018. With the exception of some small fluctuations, the gap continuously increased in all cities up to 2018. This helps to demonstrate how the supposed general economic recovery of so-termed post-crisis period was weaker than initially thought and based on inequality dynamics. However, a trend change is observed since 2018 with a positive reduction in the differences, although one should bear in mind that the gap is wider in 2021 than in 2015 for the five cities in question (Table 2.2).

This negative performance can be contrasted with two more positive variables. On the one hand, excluding Barcelona, growth in average net income per household is higher in the lowest income bracket in the census than in the highest. On the other hand, while only Zaragoza has a better Gini Index than the Spanish average in 2021 (33), this marks a positive trend and it has progressively declined across the five cities. Nevertheless, this decline is less marked in the two main cities, Madrid and Barcelona, which saw a reduction somewhat above one percentage point between 2015 and 2021. The index level for Sevilla fell by just under three points, while Zaragoza and Valencia recorded a drop of over three points. Similar to what was seen with the income gap, Madrid recorded the most negative result for the entire period in question (Fig. 2.5).

These trends in income level contrast however with one of the main urban issues in Spain leading to major contrasts and intra-urban inequalities: access to housing

Table 2.2 Income gap (EUR) in the five municipalities with the highest number of inhabitants in 2021

	Madrid	Barcelona	Valencia	Sevilla	Zaragoza
2015	68,747	61,736	67,696	55,189	66,235
2016	72,074	64,568	69,981	50,954	69,663
2017	73,919	66,981	71,997	58,856	69,771
2018	79,846	71,279	77,399	62,323	76,746
2019	77,874	69,869	75,159	63,136	75,376
2020	76,085	70,869	73,768	61,974	72,273
2021	74,119	66,313	72,129	63,501	71,030
2015–2021 (highest income bracket) (%)	10.77	10.77	28.02	12.02	10.77
2015–2021 (lowest income bracket) (%)	28.64	9.53	34.86	13.40	27.62

Source produced by the authors based on the Atlas de distribución de renta de los hogares

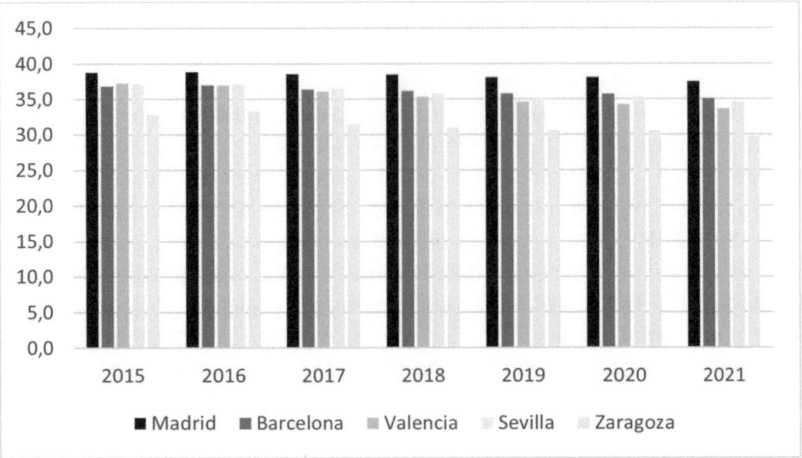

Fig. 2.5 Evolution of the Gini Coefficient (2015–2021) in the five cities with the highest population. *Source* Produced by the authors based on the Atlas de distribución de renta de los hogares

both in terms of owning and renting. This is a particularly serious problem for the most disadvantaged social groups and poorest neighbourhoods.

Below is a comparison of the latest data released for income level (2021) with the evolution of the housing index (2015–2021) in three Spanish cities with the highest home rental prices (Madrid, Barcelona and Palma). The latter city is the capital of the Balearic Islands and is known for its focus on tourism and the complex real estate market attracting investment capital from overseas (González-Pérez 2001, 2017, 2019). The net average income level per household in Palma is 38,016 EUR (2021). Two of the five districts are below the average. The area with the lowest income is precisely where the home rental index rose the most over the 2015–2021 period. Moreover, the neighbourhood with the third lowest income level recorded the second-highest jump in the rental index. In Barcelona, six of the city's ten neighbourhoods have an income level below the average (41,454 EUR). Nevertheless, the four areas with the highest growth in the rental index (above 115) are among the six areas with the lowest income, all below 40,000 EUR. District 8 has the second-lowest income level in the city but recorded the highest growth in rental prices. A total of 12 urban areas in Madrid have an income level below the city average (43,953 EUR). The eight areas with the highest increase in the rental index (above 113) recorded an income level below the average. The rental index increased the most in districts 18 and 13 which are respectively the areas with the fifth lowest and lowest income levels in the city (Table 2.3).

In short, housing is a key element explaining the social map of the city and represents a main factor in urban inequality. Thus, despite some improvement in income level in the most disadvantaged neighbourhoods, they still have to deal with a problematic rental market that particularly penalises the most vulnerable areas.

Table 2.3 Ratio between net average household income (2021) and the evolution of the home rental index (2015–2021). Census districts in Madrid, Barcelona and Palma (Majorca)

	Madrid				
	Income	% rent		Income	% rent
D. 1	38.360	110,204	D. 12	30.023	112,907
D. 2	46.699	113,043	D. 13	28.681	114,298
D. 3	57.448	112,430	D. 14	38.421	113,836
D. 4	60.217	111,375	D. 15	41.311	112,067
D. 5	69.670	111,963	D. 16	54.316	112,195
D. 6	39.991	113,030	D. 17	30.673	112,388
D. 7	56.281	112,122	D. 18	35.278	115,371
D. 8	55.125	112,465	D. 19	37.141	113,480
D. 9	63.205	110,989	D. 20	39.321	113,440
D. 10	33.790	113,465	D. 21	51.400	111,166
D. 11	32.050	113,768			
Barcelona					
D. 1	28.528	110,184	D. 6	42.530	113,325
D. 2	44.947	113,635	D. 7	36.839	115,376
D. 3	36.071	115,072	D. 8	31.169	115,999
D. 4	54.767	112,006	D. 9	37.669	113,998
D. 5	67.917	112,264	D. 10	39.199	115,059
Palma					
D. 1	46.243	116,243	D. 4	33.623	118,554
D. 2	41.392	117,236	D. 5	36.503	115,547
D. 3	38.846	118,059			

Source produced by the authors based on the Atlas de distribución de renta de los hogares and the Sistema estatal de índices de alquiler de vivienda

This situation represents an extraordinarily huge theoretical effort to access housing in many cities and in view of the data above, particularly impacts the most disadvantaged neighbourhoods. For example, the areas in Madrid where the highest percentage of wages are used to pay rent are Usera, Centro, Puente de Vallecas and Carabanchel. Usera recorded the highest price increase while the percentage of salary required to cover housing rental stood at 32.3%.

This situation is equally concerning when looking at access to property ownership. This is measured by the theoretical effort of access to housing. The income level a family needs to fully commit to purchasing a home in Spain is at its highest for the last ten years. The recommended level should not exceed 30% of income, a threshold which began to disappear from the third quarter of 2022, when the European Central Bank (ECB) kickstarted major interest rate hikes in an attempt to curb inflation. Critical accessibility reached a rate of 45%. According to statistics from the Bank of Spain, Spanish households bear an annual theoretical level of 38.9% (Q2 2023). This is the highest level since 2011.

In turn, according to data from Tinsa (a property valuation company), three provincial capitals recorded an annual theoretical effort ratio above 45% in the first quarter of 2023. Firstly, Palma is the most stressed market in terms of accessibility, with a ratio of 53.1%. This means the city's population needs to use over half their salary to pay the first year of a mortgage covering 80% of the value of an average home there. The city is followed by Barcelona (47.4%), Pamplona (45.6%), Madrid (44.2%) and Málaga (42%). A full 16.5 years' of salary is required in the Balearic Islands to pay for a home. The Region of Madrid (8.6) and Catalonia (8.5) are a long distance behind.

Foreign investors in holiday spots play a major role in generating stress and effort for residents. This is due to there being a luxury market with high prices and the speculative investment value characterising many of these transactions. For this reason, the average effort required in tourist areas in the province can even be higher than in the capital city. This is true for the Balearic Islands, where average effort stands at 58.5%, as well as Málaga, Girona, Alicante, Tenerife and Gran Canaria.

In light of this situation, the Housing Act of 12/2023 aims to contain prices and increase the supply of rental flats at reasonable rents, providing autonomous regions with the ability to declare municipalities, cities or territorial areas where there is a particular risk of insufficient housing stock for the local population 'stressed residential market zones'. This decision is made by the autonomous regions that are able to declare areas as stressed when at least one of the following two requirements is met: a) the average cost of a mortgage or lease (including expenses) exceeds 30% of the average household income in the area, and b) the purchase or rental price in the last five years has recorded an aggregate increase of at least 3% above aggregate CPI growth in the autonomous region. A study from Atlas Real Estate Analytics (Fragua) analysing data from over 11,000 postcodes in Spain concluded that at least 2298 areas in the country meet one or both requirements to be deemed stressed markets, in accordance with the new 2023 Housing Act. The province of Barcelona is home to 198 stressed zones, followed by Madrid (179), the Balearic Islands (138), Valencia (116) and Málaga (111).

At an intra-urban level, nearly all neighbourhoods in Madrid can be deemed stressed zones. Three low-income working-class neighbourhoods in the most disadvantaged area of southern Madrid stand out: Usera (recording the highest price rises across the region with a salary percentage required to pay rent in excess of 32%), Puente de Vallecas (where families require 31.2% of their salary on average to pay the rent) and Carabanchel (30.3%). In spite of this, the conservative regional government (led by the PP or Partido Popular) has no plans to apply the regulation.

All districts in Barcelona are deemed stressed zones. Contrary to the approach in Madrid, Barcelona City Council governed by the Socialist Party (PSOE) signed a decree to declare Barcelona a stressed housing market zone, providing approval and a legal basis to the procedure initiated by the Government of Catalonia as the competent administration.

Prices have risen three points above the CPI in all districts in Palma. In this sense, the entire municipality is deemed a stressed zone. Platja de Palma (the main tourist and hotel area in the city), Llevant (home to some of the most disadvantaged neighbourhoods and estates) and Norte (home to working-class areas with old social housing and middle-class suburbs) lead the list of neighbourhoods where the highest average income is required to pay for a home. The new conservative regional government (PP) has agreed to not declare stressed zones in the Balearic Islands, leaving Palma unclassified.

2.5 Conclusion

The urban system in Spain is fully consolidated. Municipalities with over 10,000 inhabitants are home to nearly 80% of the Spanish population as of 2021. This multi-pole system is underpinned by the growth in small and mid-sized cities. Those housing over 500,000 inhabitants have lost demographic weight since 1981. The so-called large urban areas in Spain are concentrated in coastal areas and main development hubs, with a strong focus on tourist areas in the Mediterranean and Canary Islands.

Spanish cities are recognised for high population and building density rates, and municipalities in peripheral urban areas have recorded the highest rates in demographic growth in the last two decades. This is true for most large urban areas in Spain. Nevertheless, despite what may occur in other countries, this does not always involve degrowth or a serious decline in the central role of capital cities in urban areas. Cities are expanding their polycentrism yet sustaining a high level of internal compression with major demographic densities and noted functional diversity.

At an intra-urban level, inequality and fragmentation are two of the most notable features. The recent crises have clearly highlighted the expansion in the urban sprawl of gentrification and impoverishment, which compete for space and represent unequal geographical development. The income gaps between the wealthiest and poorest neighbourhoods have declined in major cities. Nonetheless, there is a new essential element: housing. This topic is not only at the forefront of current urban debates but has also become a major variable in understanding the social map of cities.

Acknowledgements Grant project: "Cities in transition. Urban fragmentation and new socio-spatial patterns of inequality in the post-pandemic context. The case of the urban area of Palma (Mallorca)" (PID2021-122410OB-C31), funded by MCIN/AEI/https://doi.org/10.13039/501100011033 and by "ERDF A way of making Europe".

References

Burgess R (2018) Segregación y fragmentación urbana: algunos comentarios. Revista Iberoamericana de Urbanismo 14:79–86

Castells M (1989) The informational city: information, technology, economic restructuring and the urban-regional process. Blackwell, Oxford and Cambridge

De Mattos C (2002) Transformación de las ciudades latinoamericanas. ¿Impactos de la globalización? Eure 28(85):5–10. https://doi.org/10.4067/S0250-71612002008500001

Duhau E (2013) La división social del espacio metropolitano. Nueva Sociedad 243:79–91. https://www.nuso.org/media/articles/downloads/3917_1.pdf

Fainstein SS, Gordon I, Harloe M (1992) Divided cities: New York and London in the contemporary world (studies in urban and social change). Blackwell Publishers, New Jersey

Ferrer M (1992) Los sistemas urbanos. Síntesis, Madrid

González-Pérez JM et al (2018) Asentamientos humanos. In: J Sancho-Comíns (ed) España en mapas. Una síntesis geográfica. Serie Compendios del Atlas Nacional de España (ANE), Instituto Geográfico Nacional, Centro Nacional de Información Geográfica, Madrid, pp 204–226

González-Pérez JM (2001) Formación de la trama urbana y transformaciones socioeconómicas recientes en la ciudad de Palma de Mallorca (1960–2001). Geographicalia 40:75–100

González-Pérez JM (2017) Les ciutats de les Balears. Processos d´urbanització i urbanisme. Lleonard Muntaner, Palma

González-Pérez JM (2019) The dispute over tourist cities. Tourism gentrification in the historic Centre of Palma (Majorca, Spain). Tour Geogr 22–1:171–191. https://doi.org/10.1080/14616688.2019.1586986

González-Pérez JM, Parreño-Castellano JM, Sánchez-Aguilera D (2023) Fragmentación urbana. Marco conceptual para el análisis de la desigualdad en España. In: Arnáez-Vadillo J et al (eds) Geografía: Cambios, Retos y Adaptación. AGE and Universidad de La Rioja, Madrid, pp 811–820

González-Pérez JM and Piñeira-Mantiñán MJ (2020) La ciudad desigual en Palma (Mallorca): geografía del confinamiento durante la pandemia de la COVID-19. Boletín de la Asociación de Española de Geografía (BAGE) 87. https://doi.org/10.21138/bage.2998

González-Pérez JM (2022) La concentración de población. In JA Rio, RC Lois and ME Beltrão (eds) El mundo visto de las ciudades. Tirant Humanidades, Valencia, pp 31–40

Hammet C (2001) Social segregation and social polarization. In: Paddison R (ed) Handbook of urban studies. SAGE Publications, Thousand Oaks, pp 162–176

Harvey D (2012) The urban roots of financial crises: reclaiming the city fir anti-capitalist struggle. Socialist Register 48:1–35. https://baierle.files.wordpress.com/2012/01/harvey_the-urban-roots-of-financial-crises.pdf

Juaristi J (1995) El sistema urbano español y sus relaciones con el sistema urbano europeo. In: AJ Campesino MÁ, Troitiño, Campos ML (eds) Las ciudades españolas a finales del siglo XX, AGE and Universidad de Castilla-La Mancha, Cuenca, pp 17–28

Lois RC, González-Pérez JM, Escudero L (2012) Los espacios urbanos. El estudio geográfico de la ciudad y la urbanización. Biblioteca Nueva, Madrid

Marcuse P, van Kempen R (2002) States, cities, and the partitioning of urban space: conclusion. In: Marcuse P, van Kempen R (eds) Of states and cities: the partitioning of urban space. Oxford University Press, Oxford, pp 258–268

Ministerio de Fomento (2018) Áreas urbanas en España 2018, Madrid

Ministerio de Fomento (2000) Atlas Estadístico de las Áreas Urbanas en España 2000, Madrid

Ministerio de Ministerio de Vivienda y Agenda Urbana (2022) Áreas urbanas en España 2023. Ministerio de Transportes, Movilidad y Agenda Urbana, Madrid

Mollenkopf JH, Castells M (1991) Dual city: restructuring New York. Russell Sage, New York

Montoya MA, Krstikj K, Rehner J, Lemus-Delgado D (2021) COVID-19 and cities: experiences, responses, and uncertainties (the urban book series). Springer, Berlin

Nel.lo O (2004) ¿Cambio de siglo, cambio de ciclo? Las grandes ciudades españolas en el umbral del siglo XXI. Ciudad y territorio, Estudios Territoriales, XXXVI, 141–142:33–46

Nel.lo O (2017) El proceso de urbanización: motor y expresión de las transformaciones sociales y territoriales. In: J Romero (ed) Geografía Humana de España. Tirant Humanidades, Valencia, pp 289–366

Peck J (2012) Austerity urbanism. City: analysis of urban trends, culture, theory, policy action. City 16:6:626–655. https://doi.org/10.1080/13604813.2012.734071

Precedo A (1999) Ciudad y desarrollo urbano. Síntesis, Madrid

Prèvôt-Schapira MF, Cattaneo-Pineda R (2008) Buenos Aires: la fragmentación en los intersticios de una sociedad polarizada. Eure 34(103):73–92

Soja E (2008) Postmetrópolis. Estudios críticos sobre las ciudades y las regiones. Traficantes de Sueños, Madrid

Szupiany EB (2018) La ciudad fragmentada: una lectura de sus diversas expresiones para la caracterización del modelo latinoamericano. Revista De Estudios Sociales Contemporáneos 19:99–116

Chapter 3
Urbanisation and Territorial Change in Portugal

Teresa Sá Marques, Hélder Santos, and Miguel Saraiva

Abstract Portugal urbanisation process arrived later in Portugal than in most other European countries and was especially intense. It was associated to industrialisation and to a subsequent suburbanisation around the city of Lisbon and diffuse urbanisation process in the Northwest. Road infrastructure and the welfare state supported the urbanisation process whereas the economy became much more tertiary. Early twenty-first century analyses show serious problems caused by: often undisciplined urbanisation, with irrational land fragmentation, strong real estate speculation, poor urban infrastructure and strong pressure on ecological resources. That and a bipolar metropolitanisation that seems to be moving towards a monopolar structure, facing an apparently "dormant" or "forgotten" "rest of the country" poses important challenges, that must leave us to recognise that the "seeds" or "anchors" of change are located in metropolitan areas and that the centre of the public policies must recognise that the present urban paradigm focuses on the need to create more equitable and sustainable spaces.

Keywords Urbanisation · Territorial concentration · Industrialisation · Tertiarisation · Welfare state

3.1 Framework

The dynamic development of societies has led to a strong concentration of people and activities in urban centres and to a profound transformation of rural contexts.

T. S. Marques (✉) · H. Santos · M. Saraiva
University of Porto, Porto, Portugal
e-mail: tmarques@letras.up.pt

H. Santos
e-mail: hcsantos@letras.up.pt

M. Saraiva
e-mail: mmsaraiva@letras.up.pt

R. C. Lois-González and J. A. Rio Fernandes (eds.), *Urban Change in the Iberian Peninsula*, The Urban Book Series, https://doi.org/10.1007/978-3-031-59679-7_3

Compared to other European countries, Portugal experienced a process of urbanisation at a relatively late stage. In the period 1875–1900, some countries in Northern and Western Europe became highly urbanised (Bairoch 1985). For example, by 1910, 75% of England's population already lived in urban areas, and even one considers different statistical methodologies it is relevant stressing that in Portugal the corresponding figure was only 16%). Historically, such intense urbanisation has been accompanied by strong population movements from rural to urban areas (rural–urban migration). This movement, together with the decline in urban mortality rates, has led to an increase in urban growth. Demographic transitions have clearly made urban changes more dynamic (Friedmann 2010; Randolph and Storper 2023; Marques 2004).

According to urban theory, urbanisation is positively related to economic development (Garreau 1991; Scott and Storper 2015; Randolph and Storper 2023). The link between these two processes is justified by agglomeration economies (Jacobs 1969; Marshall 1890). Urban expansion thus goes hand in hand with industrialisation and tertiarisation of the economy, as well as rising productivity and incomes. Industrialisation led to a massive reorganisation of the internal space of each country, with development concentrating in a few areas and, consequently, the emergence of an intra-national pattern of territorial inequalities (Camagni 2001; Scott 1998). In addition, Fordist industrialisation also led to the concentration of capital and labour on urban and metropolitan areas.

Globalisation has intensified post-Fordism, fuelling economic growth processes in metropolitan areas (Scott 1998). Agglomeration processes promote proximity and shared learning networks, stimulating innovation, wealth creation and rising incomes (Ascher 1995; Camagni 2001; Lacour 1998). These economic dynamics will lead to new urbanised areas integrated into metropolitan or metapolitan areas (Ascher 1995; Gaspar 1999; Randolph and Storper 2023). Due to this new phase of urbanisation, most cities, especially the bigger ones, move away from their monocentric and dense model, based mainly on physical proximity, to a model based on road infrastructure and car mobility (Ascher 1999, 2001; Mangin 2004). The city changed in size and shape, expanding its perimeter, supported by road infrastructure, which has led to a proliferation of facilities, commerce and services on the outskirts. As urban spaces become multicentric, urban systems evolve into more polycentric structures. Mobility is central in this process, as it supports urban expansion and allows the emergence of new urban forms, while the central city remains depopulated and in crisis (Friedmann 2010; Scott and Storper 2015).

It's a fact that the first industrialisation and urbanisation process took place in Portugal at the end of the nineteenth century. But an industrialisation process strong enough to change the rural character of the country's economy only took off in the 1960s and 70s, with a new wave of the rural exodus as urban growth lagged behind that of European countries.

Later, the changes brought about by the country's integration in the European Union (1986) and the globalisation of economy accelerated changes in the structure of activities and productivity levels, making the metropolitan areas (especially Lisbon) more attractive. Living standards improved and owning a motorised vehicles became common (Ferrão and Jensen-Butler 1986; Marques 2004; Reis 2018). Concomitantly, European funding promoted the construction of infrastructures that clearly contributed to expanding and fragmenting the urban areas, reinforcing discontinuities and polynuclearities.

However, while urban areas expanded and grew, rural areas experienced intense social and economic losses. They became "forgotten or abandoned" areas where agroforestry and natural resources tried to subsist and continue to characterise the landscape. The diversity of ecosystems was no longer sufficiently valued, allowing the new urbanisation to take hold and territorial inequalities to deepen (Gaspar 1987; Salgueiro 1992; Marques et al. 2021a, 2021b).

The economic and financial crisis of 2008 put the brakes on urbanisation and a social crisis took hold in Portugal (Marques et al. 2016). Later, the pandemic crisis and serious international instability were reflected in an energy crisis with major social consequences. We now live in a highly unstable and uncertain environment because of the powerful and predictable impacts of climate and technological change. Faced with a more socially fragmented and dual territory, major challenges lie ahead in terms of regeneration and the construction of a new, more inclusive and sustainable urbanity, as well as greater dialogue and cooperation between cities and rural areas (Marques and Maia 2022a).

It is, therefore, necessary to analyse not only the demographic and economic dynamics, but also the spatial dynamics of the physical substrates (infrastructure, equipment and housing) and the territorial interrelationships, in order to understand the complex nature of urbanisation in Portugal, where urban models have changed and new challenges have emerged.

3.2 Urban Organisation and Urbanisation Processes

Europe experienced an accelerated process of urbanisation between 1800 and 1910 (Fig. 3.1). In 1950, Portugal's urban population rate was only around 25%, rising to 43% in 1980, 54% in 2000 and 66% in 2020, which being a huge acceleration is slow if compared to other southern European countries such as Spain, France and Italy.

From the 1960s until the economic and financial crisis of 2008, urbanisation in Portugal increased substantially and continuously, supported by the expansion of infrastructural networks and easier access to housing loans at low interest rates, particularly after adoption of the euro (1999), which increased the opportunities for appropriation of urban capital gains (Fig. 3.2a, b; Box 1). This strong constructive pressure, both private and public, took place in a context of favourable urban planning. This allowed for a large supply of developable land. These circumstances led to intense urban expansion, often fragmented, dispersed or linear, supported by high construction rates and little concern for architectural and urban quality (Carmo and Marques 2018; Fernandes and Seixas 2018; Ferrão and Marques 2003).

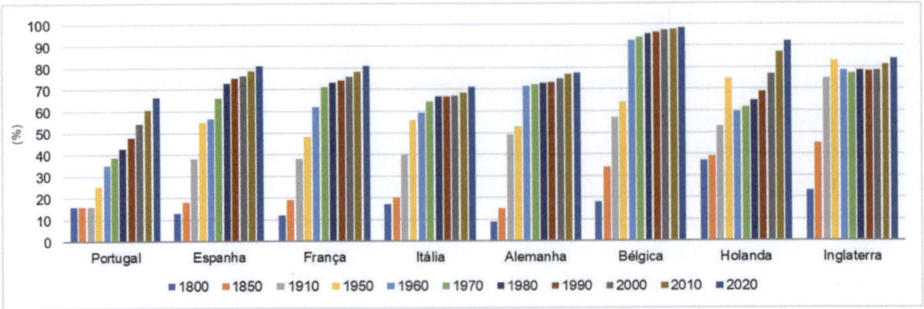

Fig. 3.1 Urban population (1800–2020). *Sources* 1800–1950: Bairoch (1985); 1960–2020: World Bank Open Data

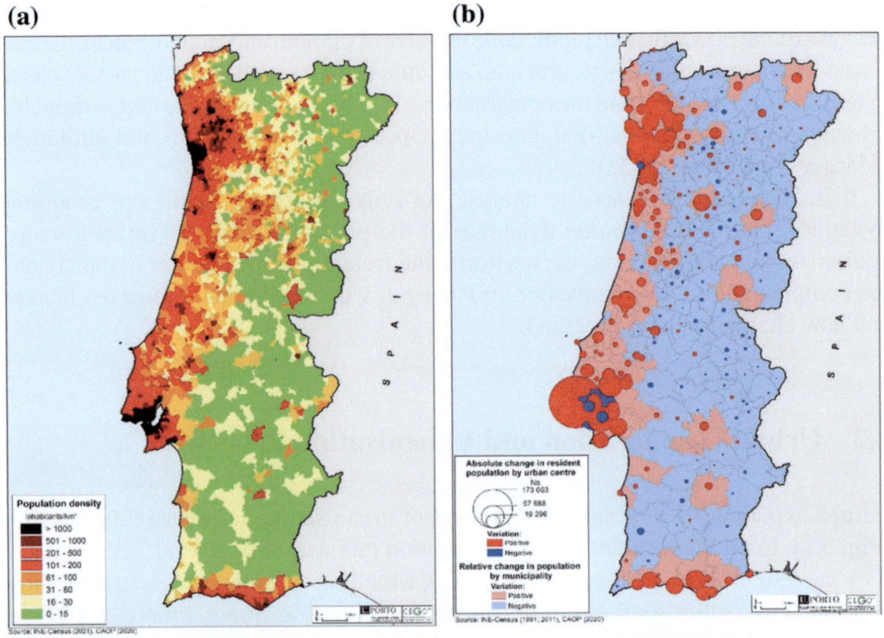

Fig. 3.2 **a** Population density (2021). *Sources* INE, Census (1991, 2011, 2021). **b** Change in resident population (1991–2011). *Sources* INE, Census (1991, 2011, 2021)

Regarding spatial planning, public policies have not always ensured the quality of urban expansion. The growth of under-equipped and often unqualified urban peripheries has not always guaranteed a minimum level of proximity services for residents. The housing supply was also poorly regulated, and the bad quality of construction was reflected in the thermic discomfort and energy inefficiency of the buildings (Carmo et al. 2014; Marques et al. 2019; Matos et al. 2019). In addition, the dynamics of construction have not been limited or structured by the provision of public transport, which has contributed to an increase in commuting times and/or transport costs, due to the intensive use of private transport, creating problems of sustainability and efficiency in urban management and natural resources management (Carmo and Marques 2018; Fernandes and Seixas 2018).

In Portugal, extensive and mono-functional urbanisation has been encouraged, supported by road infrastructure, private transport and a commercial offer where shopping centres played a relevant role (Fernandes 1998). The promotion of spatial specialisation separated residential areas from employment and service areas, reinforcing social and territorial segmentation. We have seen the engineering of the land, the smothering of rivers and streams, the decline in biodiversity and the degradation of water and soil, as well as the increase in resource-consuming practices in production, distribution and consumption, leading to destruction, degradation and waste.

Early twenty-first century analyses show serious problems caused by urban dynamics: often undisciplined urbanisation, irrational land fragmentation, strong real estate speculation, poor urban infrastructure and strong pressure on ecological resources (Carmo and Marques 2018; Marques and Queiroz 2022).

Box 1

Artificialized territories account for 5.1% of the mainland, or around half a million hectares, and reflect a high ratio of urbanized, built-up and infrastructural areas per inhabitant. These territories increased between 1995 and 2015, but at different rates (Carmo and Marques 2018). Between 1995 and 2007 the rate was very high (8000 ha/years), but with the economic recession the rate slowed down (4500 ha/year between 2007 and 2010 and 1300 ha/year between 2010 and 2015). The economic and financial crisis has slowed the property market and public and private investment. Since 2015, the dynamics of land artificialisation have increased again, as real estate investment, especially foreign investment, has increased.

These urbanization processes have coexisted with the loss of population in a large part of the country. Rural areas have become less important demographically: in 1981 they accounted for 57% of the resident population, but by 1991 this had fallen to 18% (by 2021 to 12%). This abrupt demographic decline boosted urbanization in Portugal.

Between 2011 and 2021, these urbanization processes lost momentum, reflecting the impact of the 2008 economic and financial crisis, later followed by the shock caused by the Covid-19 pandemic. In 2021, the Lisbon Metropolitan Area accounted for 28% of the national population, while 17% of the population lived in the Porto Metropolitan Area. Together, the two MAs accounted for 45% of the country's resident population in a bipolarisation that has become more pronounced, with Lisbon being increasingly dominant (Fig. 3.2a, b).

The economic crisis of 2008 has led to a slowdown in urbanisation and construction and an increase of restrictions on access to credit. The changes made to the institutional, regulatory and normative framework for land and land transformation were intended to promote urban regeneration and rehabilitation, particularly in city centre that had previously been in a state of disrepair and abandonment. But it was foreign investment in residential and tourist property that drove new construction and urban regeneration, especially in Lisbon and Porto, leading to huge increases in housing and land prices. Difficulties in accessing housing for younger and lower-income families and populations worsened as a consequence, creating a housing crisis, especially in metropolitan areas and medium-sized cities (Marques and Maia 2022a; Marques et al. 2022).

With the shock of the pandemic (COVID-19), the urban model and sprawling of recent decades is being called into question. That is partly the consequence of the lockdown periods when people felt that the urban scale of sociability and proximity, with cafes and small shops, based on short distances and neighbourly relations—which was characteristic of the traditional city—was missing from most of the urban structures (Falanga and Ferrão 2023; Marques and Maia 2022b). The poor quality and energy inefficiency of housing was also a major constraint for many families during the long periods of confinement. While services and public infrastructure were lacking in the urban periphery, especially in the "dormitory neighbourhoods", the weakening of community ties and neighbourhood and family networks in urban spaces led to greater loneliness and social isolation (Drago 2021; Ferrão et al. 2022; Marques et al. 2021b). In practise, the morphology and quality of the city had been neglected.

3.3 Urbanisation Processes, Industrialisation and Tertiarisation

Urban-industrial processes emerged in Portugal mainly in the 1960s and 1970s, especially with Portugal's accession to EFTA (Gaspar et al. 1998). They rose alongside existing cities, giving rise to processes of urbanisation, or around factories, driving peri-urbanisation processes. Industrialisation and the resulting urbanisation are inextricably linked to the spread of transport, modern communications and the electricity grid (Ferrão 1987; Ferrão and Jensen-Butler 1986; Gama 1987). An example of sub-urbanisation can be seen around Lisbon, with industrialisation processes on the south bank (Barreiro–Almada) or further north (in Vila Franca de Xira), where typical Fordist modes of production based on economies of scale and capital- and labour-intensive industries emerged (Gaspar et al. 1998). Another process, following a diffuse urban-industrial model, took place in the north-west of Portugal (Fonseca 2001; Marques 1988; Domingues and Marques 1987). The initial process of relocation is more intense from Porto to Vale do Ave, eventually spreading to Vale do Cávado. This is an industrial dynamic based on relatively complex socio-economic networks, whose territoriality is structured around diffuse settlements with fragile infrastructures.

The 1970s saw a restructuring of the production process due to the oil crisis, with industrial relocations, intensification of globalisation and, in the specific case of Portugal, a political revolution, with the consequent nationalisation of most sectors of the economy. Portugal developed an urban-industrial model, of the peripherical type in the centre-periphery model, which was developed mainly along the Lisbon-Setúbal and Porto-Braga axes. (Ferrão and Jensen-Butler 1986). Recent changes in the urbanisation process are partially a cause of the strengthening of tertiary education and an increasing functional orientation of cities towards the provision of services. This momentum accelerated particularly in the mid-1980s, with Portugal's accession to the EEC and the subsequent support for the modernisation of the production system and services. In this period, the urban-industrial system began to spread (Gaspar et al. 1998). It is present from Lisbon to Porto, in the Northwest and in most cities, although in some small and medium-sized cities with a low density context, tertiarisation is mainly driven by the provision of public services (administration, education, health) (Gaspar et al. 1998) and the peripheralisation of commercial activities (Fernandes 1998; Marques 2003, 2004).

The 1990s saw an industrial crisis emerging. The loss of industrial employment in Portugal is associated with the process of deindustrialisation of the main industrial centres in the metropolitan areas, although there has been a strengthening of the western coastal strip and some industrial expansion in previously peripheral municipalities (Vale 2005). Tertiary development continues to intensify and the metropolitan centres are losing jobs, while the peripheries are becoming more "central" (Fernandes 1998; Domingues 1993).

The turn of the century saw a series of policies aimed at the revitalisation of the urban-industrial model, the promotion of technological innovation through science and technology parks (e.g. Taguspark), functional conversion of urban spaces associated with major events such as Expo 98 and Porto European Capital of Culture (2001), investment in the qualification of urban space and the strengthening of centrality and attractiveness (Francisco 2018).

Over the last twenty years, the two metropolitan regions have become more concentrated, with the coastal urban-industrial area between Setúbal and Viana do Castelo polarised around a diverse group of medium-sized towns, while the urbanisation of the Algarve coast has increasingly specialised in tourism (Fig. 3.3a; Box 2). Meanwhile, a vast area lost population and economic activity and was abandoned. The emergence of the knowledge-based economy (Lundvall and Johnson 1994), driven by economic innovation processes, has highlighted the importance of geographical proximity and agglomeration. Once again, the take-up of these processes in Portugal has been slow, leaving the country in a peripheral position in relation to the countries at the centre of the EU (Santos 2022; Carvalho and Vale 2018; Vale and Caldeira 2007; Fontes 2005). However, the geography of innovation includes a number of activities and actors with a distinctly urban profile, so that this type of activity will be concentrated in the main metropolitan areas (Balland et al. 2020; Ribeiro et al. 2015,

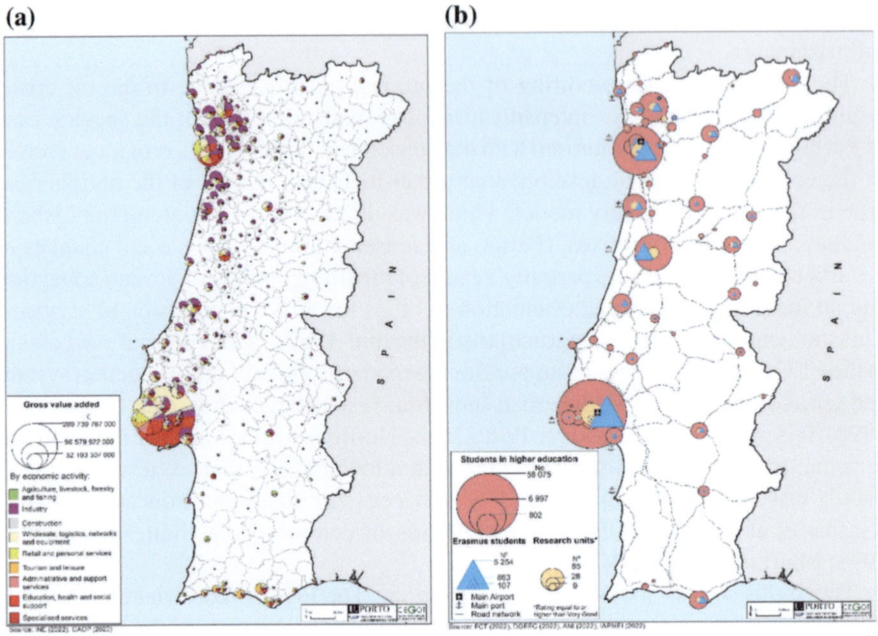

Fig. 3.3 **a** Gross Value Added (GVA) (2022). *Sources* INE (2022), FCT (2022), DGEEC (2022), ANI (2022), IAPMEI (2022). **b** Higher education and research units (2022). *Sources:* NE (2022), FCT (2022), DGEEC (2022), ANI (2022), IAPMEI (2022)

2016; Florida 2005), although there may be some eccentric exceptions (Carvalho and Vale 2018).

Box 2

In Portugal, agriculture played an important role until the country joined the EEC. The decline of agriculture, and later of industry, will interact with the growth of trade. In 1980, Portugal was the country in the European Union (EU-15) with the highest percentage of rural population (World Development Indicators 2002). However, between 1980 and 2000, the agricultural population fell from 71 to 34% of the total population. Employment in manufacturing grew until 1991 and then fell sharply until 1998, while tertiary activities (retail and wholesale trade, personal and social services, and financial and business services) grew steadily.

In 2001, the percentage of the working population in the primary sector (10.8%) and in the secondary sector (35%) showed that the country was lagging behind in tertiarization (59.9%) compared to a Europe (EU-15) that was already highly tertiarised (67% of the active workers, according to the Eurostat Yearbook 2002). In Portugal, the tertiary sector increased sharply until 2011 (70.5%) and at a slower pace until 2021 (to 72.2%).

In recent years, the trend towards tertiarization has been very strong in the metropolitan areas of Lisbon and Porto and in the Algarve (Fig. 3.3a), particularly in terms of Gross Value Added (GVA) by economic activity. Nevertheless, industrial activity is still very important, especially in the north-west of Portugal, but also along the coast between Setúbal and Porto. From the point of view of the knowledge economy driven by innovation, the geography is associated with human capital and institutional concentration in the two metropolitan areas (Fig. 3.3b) and in a small group of medium-sized cities.

3.4 Road Infrastructural Investments as a Catalyst for Urbanisation and Territorial Concentration

In the twentieth century, particularly in the post-war years, the massification of the automobile, along with road investments and low rents or home ownership outside city centres, accelerated the expansion of urban areas, and then suburbanization (Nechyba and Walsh 2004). Even today, commuting patterns, particularly outbound, are still deemed to considerably influence urban sprawl (Weilenmann et al. 2017). On one hand, by augmenting mobility and increasing infrastructural investment, metropolitan regions gain increased accessibility to larger areas and activities (Black 2010; Louro et al. 2021; Vulevic et al. 2020). However, the other side of the coin is that such a dispersion favours private car, leading to lower public transport usage; increases in trip distance, congestion, accidents, fuel consumption and traffic pollution; and results in an unequal provision of services and goods that may be associated with territorial segregation and vulnerabilities (Monteiro et al. 2023; Nechyba and Walsh 2004). Such contrasts are also visible in Portugal, where car-based commuting is dominant and supports the development of distinct functional areas and regions, uneven in size and configuration (Marques et al. 2020).

The urbanisation processes associated to the (often European-funded) improvement of transport and communication networks, resulted in significant transformations that shaped the territorial behaviour and localisation of people and companies. That resulted in the widening of the employment basins, but also in an increase of the articulation of and between urban centres, fostering multi-polarities. Therefore, new territorial configurations and interactions emerged expanding beyond administrative boundaries. The western coastline became closer to the interior, the North created more connections to the South, and small and medium-sized cities interacted more intensely between themselves and with both metropolitan and rural areas, diluting contrasts and territorial representations (Pacheco 2001).

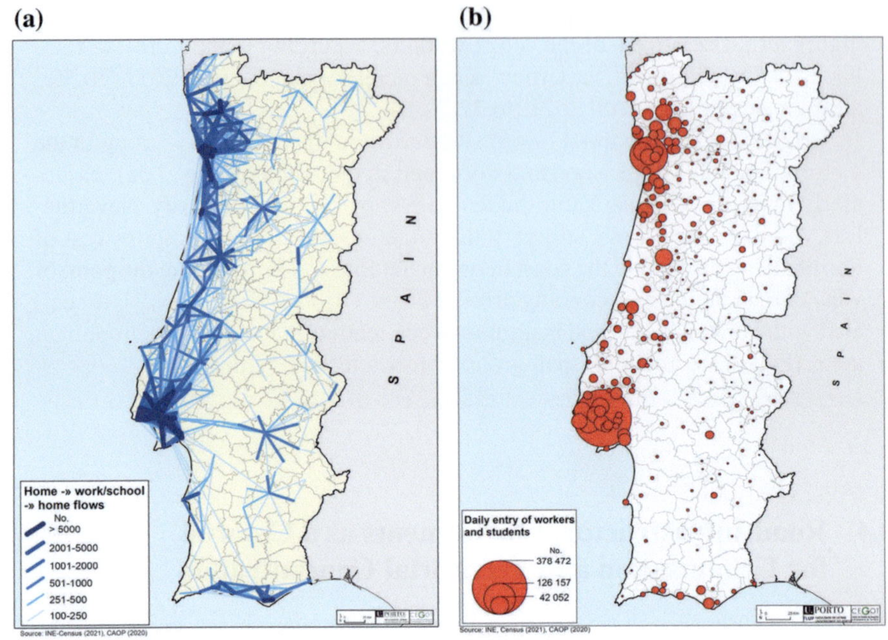

Fig. 3.4 **a** Commuter movements (2021); *Source* INE, Census (2021). **b** Daily attractiveness (2021). *Source* INE, Census (2021)

Considering the relatively small size of Portugal and the distances between urban centres, proximity plays a key role in the functional inter-urban and urban–rural relationships. They are perceptible in the commuting patterns of the employed and student population (Fig. 3.4a), establishing clear territorial organisations oriented towards the most polarising urban areas (Fig. 3.4b), that contrast with the rest of the country, and places with little polarising capacity, apparently disconnected, subsisting through small local internal economies and dynamics.

Portugal increased significantly in terms of freeways in the past 30 years (see Box 3), becoming the second country in Europe, after Spain, in terms of kilometres of freeway per inhabitant. According to the 2021 Census, 67% of the population travelled by car in daily commuting, including 32% that moved to other municipalities, which is a considerably large number. It represents, on one hand, the pervasion of an "automobile culture" in the Portuguese society, but it also reflects, on the other hand, a lack of offer in terms of public transport in peri-urban areas, both large and medium-sized cities. In absolute terms, more urbanised areas are the locations where automobile usage is more intense. This points to routine activities anchored in the use of the automobile; commuting as well as accessing diverse functions, as consumption and leisure.

It seems that the massive increase in infrastructure and the promotion of the automobile as a preferential mode of transport, supported the urbanisation development of the great metropolis. Along the western shoreline, between the cities of Viana do Castelo (north of Porto) and Setubal (south of Lisbon), the commuting patterns create functional regions and areas around metropolitan spaces and medium-sized cities, in a network-oriented urban system yet strongly anchored around Lisbon and Porto. However, particularly in the hinterland, the commuting fluxes and the polarisation levels are considerably weaker, a consequence of population decline and the small size of the urban areas (Carmo and Marques 2018; Fernandes and Seixas 2018; Marques 2004).

Box 3

Portugal's overall road network increased in about 2 thousand kilometres in the last two decades (from 12.010 km in 1998 to 14.325 in 2021), but what is interesting to note is that most of it was caused by the construction of freeways. In 1991, there were only 409 km of freeways in the country. By 2001, this total was of 1.659 km, and by 2021 3.065 km, making it the fourth country in Europe with the largest freeway network (behind Spain, Germany and France) and the second, only behind Spain, with the largest network per inhabitant.

Mobility by individual motorized transport also significantly augmented in this period. By 2001 it represented 45.7% of all population that travelled, increasing to 62% in 2011 and 67% in 2021. Of these, the total of population that moved to other municipalities to work or study in a daily basis amounted to a little less than half; 23% in 1991, 27% in 2001, and 32% in 2021. However, this has represented an increase of 87% between 1991 and 2021 in the number of persons that make such a commute; from 940 thousand to 1.756 thousand persons.

These fluxes have mostly intensified in the two metropolitan areas of Lisbon and Porto. In Lisbon, by 1991, around 312 thousand people were entering the city in a daily basis, with this number increasing to 442 thousand in 2001 and 425 thousand in 2011. In Porto, for the same three periods, the numbers were 114 thousand, 179 thousand and 172 thousand. For both cities, the numbers decreased in 2021 (to 378 thousand in Lisbon and 160 thousand in Porto), but it should be taken into account that the 2021 Census was made in 2020, during the pandemic period, so it is uncertain whether this is representative of a tendency.

There is, however, an important difference. In Lisbon, the fluxes display a clearly polarized metropolis, completely oriented towards the central city (the Lisbon municipality), while for Porto, the metropolitan model is more polycentric and less dependent on the central municipality of Porto.

3.5 The Welfare State Supports Urbanisation by Pooling Opportunities

Increasing urbanisation has led to the concentration of facilities and services, creating advantages and offering new opportunities for urban areas. Cities and metropolitan areas have thus been the focus of welfare state intervention to regulate or counterbalance tensions or imbalances in the market economy. The Portuguese case is again atypical in this respect (Carmo et al. 2014; Domingues 2003), given that Portugal has never really developed a welfare state. Before 1974, during the dictatorship, the focus was on "university cities" (mainly Lisbon and Coimbra) and on building a network of schools and high schools, hospitals, courts and other public buildings located in some regional towns (district centres). With the April 74 Revolution, political instability, the opening up of the economy to international influences and successive oil shocks hampered the development of a genuine welfare state.

Accession to the European Union (through pre-accession funds and the Community Support Framework) has strengthened the health system, expanded the network of educational facilities, built social facilities and developed some housing policy initiatives. Cities have seen a surge in the construction of public facilities, namely universities and polytechnic schools, primary and secondary education networks, hospitals and health centres, libraries, museums, sports and leisure facilities, multipurpose centres and municipal buildings. Investments have been made according to sector-specific or local objectives, with insufficiently articulated and integrated spatial planning. This has resulted in some contexts being over-equipped and others under-equipped (Carmo and Marques 2018; Silva 2002).

More recently, with the economic and financial crisis of 2008, there has been increased pressure to liberalise and minimise the role of the state, a trend that has been countered by the impact of the COVID-19 pandemic and the recent trend towards a much-debated municipal decentralisation (provision of essential services in the areas of education, health and social services) (Falanga and Ferrão 2023).

This evolutionary narrative has revealed a differentiated territorial dynamic (Box 4). The concentration of people and facilities has reduced distances and made services closer and easier to access (Marques et al. 2020) (Fig. 3.5a). However, in the Portuguese health sector, population concentration has also led to increased congestion and lower quality of some services. Currently, metropolitan areas and major cities are facing severe constraints in providing services, particularly in the case of medical emergencies. As Portugal is one of the countries with the highest life expectancy, but where the elderly have the poorest health and quality of life, urban concentration has led to a huge demand from an ageing population in a very fragile state of health (Naughton et al. 2023; Santinha et al. 2023; Tavares et al. 2023).

In terms of higher education provision (Fig. 3.5b), the model focussed on a traditionally polarised supply around Lisbon, Porto and Coimbra, balanced by a decentralised supply in the main medium-sized cities of Portugal. The supply on the coast has helped to densify urban growth between Viana do Castelo and Setúbal, enhancing the processes of urban and metropolitan agglomeration (Carmo and Marques 2018;

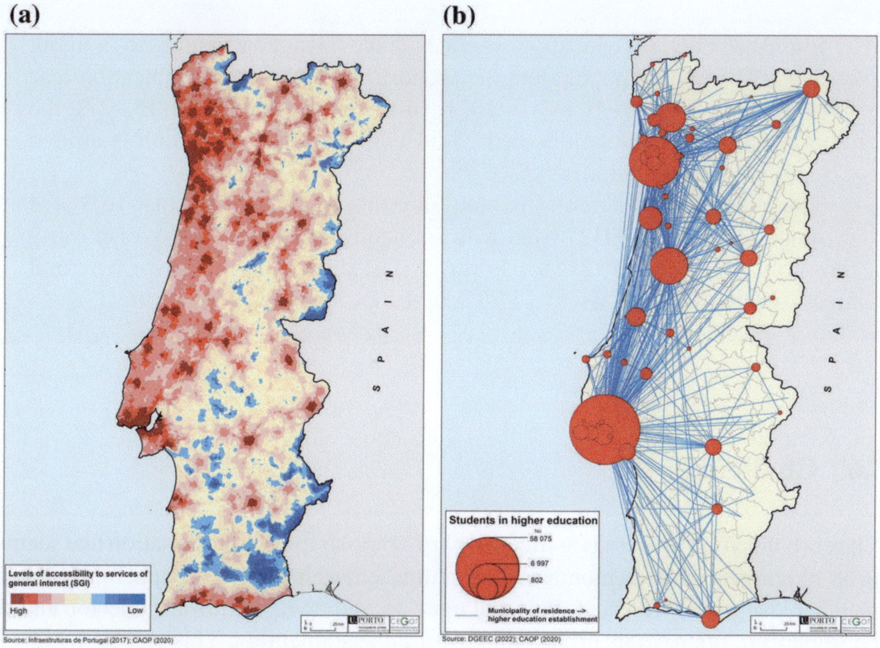

Fig. 3.5 **a** Levels of accessibility to SGI (2021). *Sources* Infraestruturas de Portugal (2017), DGEEC (2022). **b** Higher education students (2021). *Sources* Infraestruturas de Portugal (2017), DGEEC (2022)

Fernandes and Seixas 2018). The most prestigious universities, located mainly in the two metropolises, attract young students from different parts of the country. However, according to Fonseca (2023), a significant number of young people also attend higher education institutions in more peripheral areas, helping to strengthen the human capital and economic and social innovation capacity of these territories.

Box 4

The provision of education, health and social services in Portugal has improved significantly with European funding, contributing to major social change. However, service infrastructures have accompanied the processes of urban concentration and have also contributed to territorial segmentation.

Great progress has been made in the health sector: in 1991 there were 280 doctors per 100,000 inhabitants, while in 2021 this figure doubled to 562. Portugal is better placed than the rest of the European Union (EU-27: 405.8 doctors per 100 000 inhabitants). However, access to hospitals varies considerably from one area to another. In the Lisbon metropolitan area, for example, 2.4 per cent of the population lived more than 20 minutes away by car, while in the Baixo Alentejo region this figure rises to 54 per cent of the population.

Improvements in education services have also contributed to a more educated Portuguese population: the proportion with a tertiary education rose from 2.8% in 1991 to 19.8% in 2021. In the European Union (EU-27), the proportion of the population aged 25-34 with a tertiary education is around 42%, while in Portugal it is 44%.

The analysis of the risk of poverty in cities in 2022 shows that 16.3% of the population in the EU27 lived with an income below the at-risk-of-poverty threshold, compared to 12.8% in Portugal. The situation is much worse in rural areas, where the figure rises to 17.5% in the EU and 22.7% in Portugal

3.6 Challenges

A late urbanisation process is associated with a bipolar metropolitanisation that seems to be moving towards a monopolar structure, emerging from an apparently "dormant" or "forgotten" territory. Portugal must accelerate its territorial cohesion and at the same time improve its urban and territorial sustainability. This can be achieved by pooling wills, combining policies, articulating actions and the commitment to collaborative leadership.

How can Portugal move faster, how can it leapfrog stages? How can this delay become an opportunity for urban and territorial policies?

First, we must recognise that the "seeds" or "anchors" of change are located in metropolitan areas, regardless of the context. Metropolitan areas, especially the most dense and compact, are the structuring elements of the national urban system. National and international performance and the ability to collaborate at different scales will determine the country's trajectory. At the same time, by locating services, facilities, housing, employment and places to meet and socialise, medium-sized cities structure a diversity of territories with very different dynamics and challenges. The power of agglomeration can also be built by small urban centres that pool resources, alliances, wills and projects. Whatever the size of the agglomeration, it is essential to build relational capital or collaborative capacity at different scales (urban; sub-regional and regional; national or international) and across different sectors and spheres of action. Policies and measures should be promoted that combine the social, economic and environmental dimensions in a fair and sustainable way.

Today, the urban paradigm focuses on the need to create more equitable and sustainable spaces. The interplay between different models and challenges highlights the need to act in urban contexts with different challenges. Existing models must be improved, particularly in terms of living conditions, the quality of public spaces, the regeneration of the economic base, the strengthening of ecological structures and transport and mobility networks, where public space is able to induce a more communal urban life (shared and convivial) and where networks favour physical and relational proximity. In this way, urban contexts must promote diversity and inclusion of activities, uses, cultures and species. At the same time, it is essential to continue to reinvent the city, with and for people, to accommodate new and different lifestyles and to ensure quality of life and well-being for all. We need to plan with more ambitious visions for everyday life and bring about change by activating more experimental innovation. But we also need to be able to foresee and address externalities, inequalities and unfair and unsustainable practises.

Overcoming backwardness means "leapfrogging", taking advantage of the current moment of structural techno-economic paradigm change. In socio-economic terms, transitions in the broadest sense of the term (energy, environmental, digital, technological, in lifestyles and consumption, etc.) should be seen as an opportunity to make this "leapfrogging". We need to avoid the mistakes that have already been made in the policies that have been implemented in other European countries, without going down the path of the same mistakes that others have made. Urban densification, multifunctionality, the 15-min commuting, smooth mobility, alternative economies, social and economic innovation, inclusion, services of general interest, regeneration, the circular economy, the urban water cycle and energy efficiency are just some of the concepts underpinning policies closely linked to urban development and that needs to be explored. To do this, we need to contextualise, collaborate and experiment in order to "leapfrog" towards a more sustainable and inclusive urban and territorial development.

Acknowledgements The authors wish to thank Diogo Ribeiro and Catarina Maia for his contribution to database curation and cartography.

References

Ascher F (1995) Métapolis ou l'Avenir des Villes. Éditions Odile Jacobs, Paris

Ascher F (1999) Une Nouvelle Révolution Urbaine? In : Lacour C, Puissant S (eds) La Métropolisation. Croissance, Diversité, Fractures. Anthropos, pp 183–190

Ascher F (2001) Les Nouveaux Principes de l'Urbanisme. La Fin des Villes n'est pas à l'ordre du Jour. Éditions de l'Aube, Paris

Bairoch P (1985) The Jérico à Mexico. Villes et Économie dans l'Histoire. Gallimard, Paris

Balland P-A, Jara-Figueroa C, Petralia SG, Steijn MP, Rigby DL, Hidalgo CA (2020) Complex economic activities concentrate in large cities. Nat Hum Behav 4:248–254

Black WR (2010) Sustainable transportation: problems and solutions. Guilford Press

Camagni R (2001) The Economic role and spatial contradictions of global city-regions: the functional, cognitive, and evolutionary context. In: Scott AJ (ed) Global city-regions: trends, theory, policy. Oxford University Press, Oxford, pp 96–118

Carmo A, Ferrão J, Malheiros J (2014) Geografias do Estado social: reorganização territorial, habitação e urbanismo. In: Carmo RM, Barata A (eds) Estado social: de todos para todos. Tinta da China, Lisboa, pp 167–200

Carmo F, Marques TS (eds.) (2018). Programa Nacional da Política de Ordenamento do Território (PNPOT). Diagnóstico. Proposta técnica de alteração. 20/07/2018. Direção-Geral do Território, Lisboa

Carvalho L, Vale M (2018) Biotech by bricolage? Agency, institutional relatedness and new path development in peripheral regions. Camb J Reg Econ Soc 11(2):275–295. https://doi.org/10.1093/cjres/rsy009

Domingues Á (1993) Serviços às Empresas – Concentração Metropolitana e Desconcentração Periférica (o contraponto entre a Área Metropolitana do Porto e as áreas periféricas de industrialização difusa do Noroeste Atlântico de Portugal Continental). Dissertação apresentada à Faculdade de Letras da Universidade do Porto para Doutoramento em Geografia Humana, Porto

Domingues Á (2003) A cidade-providência. Revista aa Faculdade de Letras: Sociologia 13:171–183

Domingues Á, Marques TS (1987) Breve caracterização do «Vale do Médio Ave». Geografia: Revista da Faculdade de Letras da Universidade do Porto, 3

Drago A (2021) Habitação entre crises: partição das classes médias, políticas de habitação acessível e o impacto da pandemia em Portugal, Cadernos do Observatório Centro de Estudos Sociais. Universidade Coimbra, Coimbra

Falanga R, Ferrão J (2023) COVID-19 and the city: reframing our understanding of urbicide by learning from the pandemic. In: Carrión Meda F, Cepeda Pico P (eds) Urbicide: the death of the city. Springer, Cham, pp 107–124

Fernandes JAR (1998) A cidade 25 anos depois: o comércio e a reestruturação urbana do Porto. Cadernos De Geografia 17:203–207

Fernandes, J. A. R., & Seixas, J. (2018). Cities and urbanisation in democratic Portugal. *Méditerranée. Revue géographique des pays méditerranéens/Journal of Mediterranean geography* (130). https://doi.org/10.4000/mediterranee.10698

Ferrão J (1987) Indústria e território: breve história de uma união feliz. Revista Crítica De Ciências Sociais 22:55–68

Ferrão J, Jensen-Butler C (1986) Industrial development in portuguese regions during the 1970's. Tijdschrift Voor Economische En Sociale Geographie 77(2):132–148. https://doi.org/10.1111/j.1467-9663.1986.tb01305.x

Ferrão J, Marques TS (2003) National urban system: oveview. Direção-Geral do Ordenamento do Território e do Desenvolvimento Urbano, Ministério das Cidades, Ordenamento do Território e Ambiente, Lisboa

Ferrão J, Ribeiro JMF (Eds) (2014) Noroeste Global. Fundação Calouste Gulbenkian, Lisboa

Ferrão J, Falanga R, Liz C (2022) COVID 19—Preparar as cidades para riscos globais. In: Ribeiro L, Noronha J, Rodrigues, J, Oliveira R (eds) Metrópole e pandemia: presente e futuro. Rio de Janeiro: Letra Capital, pp 102–135

Florida R (2005) The world is spiky: globalization has changed the economic playing field, but hasn't leveled it. The Atlantic, 48–51 Oct 2005

Fonseca M (2001) A indústria têxtil algodoeira no Norte de Portugal. Tese de doutoramento em Geografia Humana, apresentada à Faculdade de Letras da Universidade do Porto, Porto

Fonseca M (2023) Innovation in the peripheries: counter-flows of students to second tier cities in Portugal. Geoforum 141:103732. https://doi.org/10.1016/j.geoforum.2023.103732

Fontes M (2005) Distant networking: the knowledge acquisition strategies of 'out-cluster' biotechnology firms. Eur Plan Stud 13(6):899–920

Francisco JM (2018) Design inclusivo na cidade: Um contributo ao nivel do equipamento urbano. Lisboa: Tese de Doutoramento em Arquitetura, apresentada à Faculdade de Arquitetura da Universidade de Lisboa

Friedmann J (2010) Place and place-making in cities: a global perspective. Plan Theory Pract 11(2):149–165. https://doi.org/10.1080/14649351003759573

Gama A (1987) Indústria e produção de um espaço peri-urbano. Revista Crítica De Ciências Sociais 22:33–54

Garreau J (1991) Edge city. Life on the new frontier. Anchor Books, New York

Gaspar J (1987) Portugal nos próximos 20 anos: A Ocupação e a Organização do Território - análise retrospectiva e tendências evolutivas. Fundação Calouste Gulbenkian, Lisboa

Gaspar J (1999) Economic restructuring and new urban form. Finisterra XXXIV(67–68), 131–152

Gaspar J, Brito Henriques E, Vale M (1998) Economic restructuring, social re-composition and recent urban changes in Portugal. GeoJournal 46:63–76

Jacobs J (1969) The economy of cities. Vintage Books, New York

Lacour C (1998) Formes Urbaines et Dynamique Métropolitaine des Emplois. In : Lacour C, Puissant S, Gaschet F (eds) Formes Urbaines et Dynamique Métropolitaine des Emplois. Pueble, 34° Colloque Annuel de l'ASRDLF

Louro A, Marques da Costa N, Marques da Costa E (2021) From livable communities to livable metropolis: challenges for urban mobility in Lisbon Metropolitan Area (Portugal). Int J Environ Res Public Health 18(7).

Lundvall B-A, Johnson B (1994) Learn Econ. J Ind Stud 1(2):23–42

Mangin D (2004) La Ville franchisee - Formes et structures de la ville contemporaine. La Villette Eds De, Paris

Marques TS (1988) Sistema produtivo industrial e território: um estudo da têxtil em Guimarães. Geografia: Revista da Faculdade de Letras da Universidade do Porto 4:55–103

Marques TS (2003) Dinâmicas territoriais e as relações urbano-rurais. Geografia: Revista da Faculdade de Letras da Universidade do Porto 19:507–521

Marques TS (2004) Portugal na Transição do Século: Retratos e Dinâmicas Territoriais. Edições Afrontamento, Porto

Marques TS, Maia C (2022a) Pensar en las ciudades en un contexto de cambio: de la concentración urbana a la gobernanza territorial. In: Fernandes JAR, Lois R, Sposito ME (eds) El mundo visto de las ciudades. Tirant Humanidades, Valencia, pp 765–780

Marques TS, Maia C (2022b) Retratos de la injusticia espacial en Oporto: de la persistencia temporal a los impactos de las crisis. In: Fernandes JAR, Lois R, Sposito ME (eds) El mundo visto de las ciudades. Tirant Humanidades, Valencia, pp 341–355

Marques TS, Queiroz JP (eds) (2022) AMP 2030. Estratégia e Prioridades de Ação. Área Metropolitana do Porto, Porto

Marques TS, Matos F, Maia C, Ribeiro D (2021a) Densidades populacionais e territórios em mutação. In: Cunha L, Santana P, Lourenço L, Santos N, Nossa P (eds) Geografia, Turismo e Território: Livro de homenagem a Fernanda Delgado Cravidão. Imprensa da Universidade de Coimbra, Coimbra, pp 285–308. https://doi.org/10.14195/978-989-26-2058-9

Marques TS, Santos H, Honório F, Ferreira M, Ribeiro D, Torres, M (2021b) O mosaico territorial do risco ao contágio e à mortalidade por Covid-19 em Portugal Continental. Finisterra 55(115):19–26. https://doi.org/10.18055/Finis20383

Marques TS, Santos H, Ribeiro D, Maia C, Ribeiro P (2016). Portugal, a Crise Económica e as duas Macrorregiões. In: Carballada A, Piñeiro A, Doval A, Pazos, Lois R (eds) Los escenarios económicos en transformación. La realidad territorial tras la crisis económica. VII Jornadas de Geografía Económica. Universidade de Santiago de Compostela, Santiago de Compostela, pp 175–186

Marques TS, Saraiva M, Matos F, Maia C, Ribeiro, D, Ferreira M, Amante A (2019) Assessing territorial vulnerabilities and spatial inequalities: the case of Portugal. In: AESOP 2019 Conference—Book of Papers. Veneza: Università Iuav di Venezia., pp 566–576

Marques TS, Saraiva M, Matos F, Maia C, Ribeiro D, Ferreira M, Van Heerden S (2022) Property investment and housing affordability in Lisbon and Porto. Publications Office of the European Union, Luxembourg

Marques TS, Saraiva M, Ribeiro D, Amante A, Silva D, Melo P (2020) Accessibility to services of general interest in polycentric urban system planning: the case of Portugal. Eur Plan Stud 28(6):1068–1094. https://doi.org/10.1080/09654313.2019.1658718

Marshall A (1890) Principles of economics. Macmillan, London

Matos F, Marques TS, Saraiva M, Maia C, Ribeiro D (2019) Housing in Portugal after the economic crisis: challenges for a new policy. In: Bargelli E, Heitkamp T, Simón-Moreno H, Vásquez Varela

C (eds) *Housing policy and tenure types in the 21st century a southern european perspective.* Pisa University Press, Pisa, pp 53–72

Monteiro J, Para M, Sousa N, Natividade-Jesus E, Ostorero C, Coutinho-Rodrigues J (2023). Filling in the spaces: compactifying cities towards accessibility and active transport. ISPRS Int J Geo-Inform 12(3)

Naughton L, Cunha F, Padeiro M, Santana P (2023) What the pandemic and its impact on the mobility and well-being of older people can teach us about age-friendly cities and communities. Soc Sci Med 338:116329. https://doi.org/10.1016/j.socscimed.2023.116329

Nechyba TJ, Walsh RP (2004) Urban Sprawl. J Econ Perspect 18(4):177–200. https://doi.org/10.1257/0895330042632681

Pacheco E (2001) Alteração das Acessibilidades e Dinâmicas Territoriais na Região Norte: Expectativas, Intervenções e Resultantes, Porto: Tese de Doutoramento em Geografia, apresentada à Faculdade de Letras da Universidade do Porto

Randolph GF, Storper M (2023) Is urbanisation in the Global South fundamentally different? Comparative global urban analysis for the 21st century. Urban Stud 60(1):3–25. https://doi.org/10.1177/00420980211067926

Reis J (2018) A Economia Portuguesa - Formas de Economia Política numa periferia persistente (1960–2017). Edições Almedina, Coimbra

Ribeiro JMF, Moura F, Chorincas J (eds) (2015) Uma Metrópole para o Atlântico. Fundação Calouste Gulbenkian, Lisboa

Ribeiro JMF, Moura F, Chorincas J, Marques TS (eds) (2016). Portugal no Centro. Fundação Calouste Gulbenkian, Lisboa

Salgueiro TB (1992) A cidade em Portugal: uma geografia urbana. Afrontamento, Porto

Santinha G, Soares C, Forte T (2023) Strategic planning as the core of active and healthy ageing governance: a case study. Sustainability 15(3):1959. https://www.mdpi.com/2071-1050/15/3/1959

Santos H (2022) Território, Conhecimento e Inovação Económica: O Espaço Relacional das Redes de I&D+I Dirigidas à Saúde Humana com Amarração em Portugal. Porto: Tese de doutoramento em Geografia Humana, apresentada à Faculdade de Letras da Universidade do Porto.

Scott A (1998) Regions and the world economy. The coming shape of global production, competition and political order. Oxford University Press, Oxford

Scott AJ, Storper M (2015) The Nature of cities: the scope and limits of urban theory. Int J Urban Regional Res 39(1):1–15. https://doi.org/10.1111/1468-2427.12134

Silva PA (2002) O modelo de welfare da Europa do Sul: reflexões sobre a utilidade do conceito. Sociologia, Problemas e Práticas 38:25–59

Tavares J, Santinha G, Rocha NP (2023) Unlocking the WHO age-friendly healthcare principles: Portugal quest and recommendations. Int J Environ Res Public Health 20(22):7039. https://doi.org/10.3390/ijerph20227039

Vale M (2005) Os lugares da Indústria. In: Medeiros CA (eds) *Geografia de Portugal - Sociedade, Paisagens e Cidade*). Círculo de Leitores, Lisboa, pp 375–392

Vale M, Caldeira J (2007) Proximity and knowledge governance in localized production systems: the footwear industry in the North Region of Portugal. Eur Plan Stud 15(4):531–548. https://doi.org/10.1080/09654310601134854

Vulevic A, Castanho RA, Naranjo Gómez JM, Loures L, Cabezas J, Fernández-Pozo L, Martín Gallardo J (2020) Accessibility dynamics and regional cross-border cooperation (CBC) perspectives in the Portuguese-Spanish borderland. Sustainability 12(5)

Weilenmann B, Seidl I, Schulz T (2017) The socio-economic determinants of urban sprawl between 1980 and 2010 in Switzerland. Landsc Urban Plan 157:468–482. https://doi.org/10.1016/j.landurbplan.2016.08.002

Chapter 4
The Road to Extended Urbanization in Spain

Oriol Nel·lo

Abstract In recent decades, the urbanization process in Spain has undergone very notable transformations. For historical reasons, Spain is a country in which the contemporary urbanization process suffered a certain delay compared to other European countries. Thus, the process of concentration of population and activity on the territory, typical of the implosive phase of the urbanization process, reached its peak in the mid-seventies of the twentieth century. Since then, urbanization has been characterized by the expansion of urban areas, the integration of space and the extension of built areas. This has led to the assertion of preeminence over the entire urban system of the main metropolitan areas: Madrid, Barcelona, Valencia, Bilbao, Seville, Malaga and Zaragoza. At the same time, it has caused notable problems of land artificialization, exacerbation of mobility, residential segregation and metropolitan governance.

Keywords Urbanization · Land artificialization · Segregation

4.1 Premise: Urban Dispersion and Urbanization Process

Alice Munro, when searching for the grave of an ancestor on the outskirts of Chicago swallowed up by the metropolis, noted that "human burial is one of the very few reasons that any land is undisturbed, nowadays, when all the land around is put to use" (Munro 2006). The process of dispersion of urbanization in Spain in recent decades has been so vigorous and accelerated that even this basic restriction pointed out by the novelist has come to be questioned. Thus, the expansion of Madrid, to give an example, has involved the urbanization of the surroundings of more than a dozen cemeteries located in old peripheral towns—such as Carabanchel, Vallecas, Vicálvaro, San Blas, Hortaleza or Barajas—without said proximity seems to have significantly affected the prices of the housing built there (Salido 2015). The fact is far from being anecdotal, on the contrary: in Spain, as in other southern European

O. Nel·lo (✉)
Department of Geography, Autonomous University of Barcelona, Bellaterra, Spain
e-mail: oriol.nello@uab.cat

R. C. Lois-González and J. A. Rio Fernandes (eds.), *Urban Change in the Iberian Peninsula*, The Urban Book Series, https://doi.org/10.1007/978-3-031-59679-7_4

countries, the extension of the artificialized land surface has reached unprecedented magnitudes in recent decades. This has involved the occupation for urban uses of areas far from pre-existing urban settlements and spaces that—due to their agrarian value, natural interest, environmental risks or symbolic character—had remained free of urbanization.

The extension of land artificialization is one of the most visible expressions of the urbanization process in Spain. A process that, in the last decades of the twentieth century and in the early stages of our century, has been characterized by the expansion of urban areas, the interdependence of the space that comprises them and the dispersion of urbanization. This development has had some consequences that are normally considered positive, such as greater accessibility to services, the spread of amenities associated with urban lifestyles, the growth in economic activity and the increasing attractiveness of urban areas. However, metropolitan development, often lacking guidance and governance, has also entailed highly negative effects, particularly such as the inefficient distribution of land uses and the increase in the environmental footprint; the fragmentation of natural habitats and the destruction of landscape values; the exacerbation of mobility and the increase in energy and resource consumption; as well as the worsening of social inequalities and residential segregation (Nel·lo 2016).

In this chapter, the main features of the urbanization process in Spain in the period 1975–2020 will be briefly explained. Thus, in the following pages, we will try to highlight that this process cannot be analyzed solely in morphological terms, that is, in terms of the artificialization of land and settlement density, but must be understood as the expression of a historical phase: that in which the urbanization process tends to integrate the entire regional areas and cancel the old opposition between city and countryside, between urban and rural space. The chapter consists, in addition to this introduction, of three sections: the first describes the phases that the process of contemporary urbanization in Spain has gone through. In the second, the magnitudes of the phenomenon of dispersion of urbanization are analyzed. Finally, the last section considers some of the main consequences—territorial, environmental and political—of the described evolution.

4.2 From Implosion to Urban Explosion: The Phases of the Urbanization Process (1959–2020)

The evolution of the urbanization process in Spain from the mid-twentieth century to the present corresponds well with the figures of urban "implosion" and "explosion" coined by Lefebvre (1970) and recently recovered by geographers such as Brenner (2013), i.e. the consecutive juxtaposition of an acute process of concentration of population and activity, followed by the extension of urban networks, the growing dispersion of urbanization and the integration of all regional and supra-regional spaces.

In fact, as we have explained on other occasions (Nel·lo 2004, 2016), in Spain the contemporary urbanization process had historically been characterized by its relatively late development compared to other European countries, as well as by the maintenance of strong regional inequalities. The convergence of these factors led to difficulties in forming an integrated urban system throughout the country (Vilà and Capel 1970; Bellido et al. 1987; Reher 1994; Terán 1999). However, these features experienced a very notable inflection starting in the late 1950s. Since then, the Spanish urban system has undergone accelerated transformations that have radically altered both the distribution of population and land uses. In this evolution, various phases can be distinguished that, as we said, respond well to the image of urban implosion and explosion (Nel·lo 2001, 2004, 2016).

Thus, the first phase of this process was characterized by the concentration of population and activities, with the consequent growth of the largest cities. This concentration had antecedents that date back at least to the nineteenth century and reached its most acute phase in the decade of the sixties and seventies of the 20th, subsequently to the approval of the Economic Stabilization Plan of 1959, which promoted economic modernization and the growth of industrial activity. This, in a country poorly equipped with infrastructure, took place in a highly concentrated manner, particularly in the urban areas of the Mediterranean coast, the Ebro valley, the Basque Country and Madrid, which attracted large contingents of population from of rural areas. As a result, large cities, densified and increasingly crowded, expanded, in most cases exceeding their municipal limits, and conurbating with neighboring towns. Thus, the seven main Spanish metropolitan areas—Madrid, Barcelona, Valencia, Bilbao, Seville, Málaga and Zaragoza—which in 1961 housed, together, 7,434,051 people (24.4% of the total population), reached in 1981 a population of 12,933,784 inhabitants (34.3%).

This concentrating dynamic experienced, however, a notable inflection starting in 1975, because of the oil crisis and the beginnings of the relocation of industrial production to third countries, with the consequent outsourcing of the economy, the brake on demographic growth and the decline of rural/urban migrations. The large urban areas, which, as we have seen, had grown by 3.5 million inhabitants between 1960 and 1981, increased by barely 0.5 million until 1996, to reach a population of 13,428,825. Now, this outstanding slowdown in growth in absolute terms was accompanied by a very notable redistribution of the population within metropolitan areas. Thus, in these three decades the central cities of each metropolitan area reduced their demographic growth very sharply and some of them—Madrid, Barcelona, Bilbao—even lost important contingents of population in absolute terms. These dynamics contrasts with the increase in the population of their metropolitan surroundings: between 1981 and 1996 the central cities lost close to half a million inhabitants while the rest of their metropolitan areas gained, as a whole, close to a million. Furthermore, from the morphological point of view, if in the previous phases urban growth had taken place in the form of an oil stain, without a solution of continuity, in this period low-density extensions proliferated, without physical connection with the pre-existing urban fabrics. Thus, the main Spanish cities turned from concentration, growth in continuity and high density to deconcentration, scattered growth and low density: the Spanish urban system was moving from urban implosion to urban explosion.

Starting in 1996, the metropolitan process entered Spain into what we have called its third phase (Nel·lo 2004). This was characterized by a new increase in the population of large metropolitan areas in absolute terms, up to 15,132,179 inhabitants. However, its weight on the country remained stable, around 33.8% of the population. Since the period of great internal rural migrations, typical of the 1959–1975 phase, was finished, the renewed demographic growth was the result of the existence of international migrations associated with the rapid growth of the economy, particularly in services and construction. In the metropolitan areas, the central cities not only erased the losses of the previous period, but also grew again. However, this growth was due, above all, to the arrival of contingents of foreign population, while the Spanish population continued to decrease: if in the year 2000 the Spanish population residing in the central cities of the seven main metropolitana areas was 7,130,304 inhabitants, in 2008 it had reduced to 6,820,291. On the other hand, between the same years the foreign population increased from 179,019 to 1,101,763 people (Nel·lo 2016). Thus, unlike the period 1975–1996, the third phase of the metropolitan process was characterized by the simultaneous presence of demographic growth in the central areas and in the metropolitan crowns, with the peculiarity that while the Spanish population decreased in the former and grew in the latter, the foreign population increased in both areas. From the morphological point of view, this third phase, which has been dubbed "the prodigious decade" of construction activity in Spain (Burriel 2008), meant the exacerbation of land artificialization.

This dynamic stopped abruptly in 2008, with the beginning of the global economic crisis. The growth of metropolitan areas, which in the decade 1996–2006 added 1.7 million inhabitants, was reduced to just 0.6 million between 2006 and 2015, the year in which they reached 15,739,227 inhabitants. Now, this new brake in absolute growth slowed down, but did not imply stopping the process of internal redistribution of population. Thus, while the central cities remained stagnant as a whole, the increase in population occurred in the metropolitan crowns, which in 2015 already surpassed the central cities in absolute population. In this period, the increase in social inequality resulting from the economic crisis and the austerity policies exacerbated the processes of residential segregation of social groups (Albertos and Sánchez 2014; Méndez et al. 2014; Nel·lo 2021). The urban explosion has thus been combined with an intensification of the separation of social groups over the urban space, with the consequences that will be mentioned later.

Barely overcoming the effects of the 2008 economic crisis, large Spanish cities have had to face the effects of a new historical shock: the health crisis caused by the Covid-19 pandemic (IGN 2021). The effects of lockdown and fear of contagion associated with high residential density were particularly evident in large Spanish cities. This raised expectations that there would be an intensification of population migration from urban areas to areas traditionally considered rural. However, although some movements in this direction took place during 2020 (González-Leonardo et al. 2022), they do not appear to have had lasting effects. Three years after the start of the pandemic, a certain equilibrium has been reestablished, characterized, above all, by the growing presence of the foreign population to urban centers, accompanied by migratory movements from these centers to the urban peripheries and towards some rural areas close to the main cities (Gil-Alonso et al. 2021; López-Gay 2017).

4.3 The Extension of Land Artificialization and the Generalization of Urban Lifestyles

The vicissitudes of the urbanization process that has been described, with the succession of phases of concentration and dispersion of population and activity, are in no way exclusive to the Spanish case, but common to many other spatial configurations in Europe and outside of Europe. Perhaps what most visibly distinguishes its evolution in Spain is the speed and virulence in the transformation of land uses (see Nel·lo et al. 2016 and Nel·lo 2016, which we follow in part in this section). Thus, over the last half century the successive phases of urbanization have implied simultaneously to the abandonment of very notable extensions of agricultural surfaces, and the advance of the artificialization of land.

The first phenomenon has led in many areas to a decline in surfaces devoted to crops and pastures, as well as an increase in forest. Paradoxically, among the regions where this process has been most prominent are some of the most intensely urbanized: in Catalonia, for example, the agricultural area was reduced between 1993 and 2005 from 1.1 million to 935,000 ha, while the forest area grew on more than 100,000 ha to exceed 64% of the total (Nel·lo and López 2016). Thus, in just over a decade and in the middle of a metropolitan dynamics characterized largely by the dispersion of urbanization, the forest area increased in a region as urbanized as Catalonia by an area equivalent to 10 times the surface of the city of Barcelona. The decline of crops and pastures, and the consequent advance of scrubland and forest, has led, where it has occurred, not only to a loss of landscape values, but also, in many cases, to a decrease in biodiversity.

On the other hand, the extension of the urbanized surface and infrastructure has translated, as was said, into an accelerated process of artificialization of land covers. Thus, according to data from the Corine Land Cover project, based on satellite images of the European countries, between 1987 and 2005 the surface of artificial land in the whole of Spain increased from 669,222 to 1,017,356 ha, with an increase of 54.9%. The average annual growth of the land thus transformed during this period is 19,340 ha (Prieto and Díaz 2011). In this way, the 348,134 ha artificialized in these two decades represent the transformation of an area that is practically equivalent to that of the island of Mallorca. The analysis of the data corresponding to the five-year period 2000–2005 is especially relevant. In this period, which corresponds to the peak of the real estate expansionary cycle, the average volume of artificialized land per year rose to 27,666 ha, that is, 75.8 ha per day. To have a comparison: St. Peter's Basilica in the Vatican would fit more than 30 times the surface area artificialized daily in Spain during the second five-year period of the real estate bubble. The processes of land squandering have been especially intense in the areas of the main urban areas, but they have also decisively affected the coastal strip, especially the Mediterranean and the Balearic and Canary archipelagos (Nogué 2005; Rullan 2011, 2022; Zornoza 2013; Checa and Nel·lo 2018).

Now, the artificialization of land is the most visible expression of the extent of urbanization, but by no means the only one. The trend towards the expansion of urban uses of space is more than confirmed by other indicators. Thus, for example, the analysis of the nighttime satellite image of the Earth allows us to verify that the surface of the land with brightness levels typical of urban uses tends to expand at an even greater rate than the direct occupation of the land. The land with urban brightness levels located within a radius of 50 km from the center of the Spanish provincial capitals has increased by 133% between 1992 and 2012, going from an area of 6992–16,249 km^2 (Nel·lo et al. 2016). This rhythm contrasts notably with the evolution of the increase in the extent of urban luminosity in European metropolitan areas, which was, in the same period, 54% (Nel·lo et al. 2017).

In Spain, therefore, the extension of urban brightness has been much faster than the artificialization of land in this period, thus showing how the diffusion of urban uses far exceeds the expansion of urban forms. Unsurprisingly, the greatest absolute growth in urban luminosity surface area has occurred around the largest urban areas. Thus, of the 12 provincial capital areas that experience the greatest expansion of urban brightness, 8 correspond to the Mediterranean axis (Girona, Barcelona, Tarragona, Castellón, Valencia, Alicante, Murcia and Málaga), 3 in the metropolitan area of the capital of the State (Madrid, Toledo and Guadalajara) and the last to Seville. To understand the evolution of the urbanization process, it is also particularly interesting to note that urban brightness confirms the tendency towards the structuring of large urban axes and constellations: the Mediterranean coast from the French border to Alicante, accompanied in the Balearic Islands by the great spot. urban of the Bay of Palma (Checa and Nel·lo 2018); the axis of the Ebro and the nebula of the Cantabrian and the Basque cities (what has been called Eukal Hiria) (Gárate 2010); the Costa del Sol and the Guadalquivir valley in Andalusia (Rodríguez and Sánchez Escolano 2010); the Atlantic urban region, which extends at least from A Coruña to Porto (Lois and Pino 2015); and finally, presiding over the center of the peninsula, the urban galaxy of Madrid, spread over more than half a dozen provinces and three autonomous communities (Solís 2011).

4.4 The Territorial, Environmental and Political Implications of Urban Dispersion

The dispersion of urbanization, succinctly described so far, is one of the main expressions of the contemporary urbanization process in Spain. The phenomenon has often been interpreted in exclusively morphological terms, considering the physical extension of buildings and urban fabrics. Its origins and evolution show that it is about much more than this. Dispersion is one of the main consequences, but not the only one of the contemporary urbanization processes. A process characterized by the growing integration and interdependence of the territory, the extension of urban lifestyles and the commodification of land and natural resources, from which decisive territorial, environmental and political implications are derived.

As far as territorial structures are concerned, in Spain, as in other European countries, the urbanization process is giving rise not only to the configuration and consolidation of metropolitan areas (Feria and Martínez 2016), but to the conversion of urban areas into true regional spaces. These integrate both artificialized land and extensive territories destined for agricultural uses and operational spaces, that is, areas that support those extractive, tourist, recreational, energy production or even symbolic activities, essential for the urbanization process (Brenner 2013). We are witnessing the configuration in Spain of socio-spatial formations that in other contexts have been called regional urbanization (Soja 2016) or metropolitan archipelago (Indovina 2009). The dispersion of urbanization and the integration of space are in no way incompatible with the abandonment of some settlements, the decline of cultivated areas and the advance of the forest area. As has been explained, the depopulation of extensive areas of the peninsular interior, the phenomenon that has come to be called "*la España vacía*" (Molino 2016; Recaño 2017), has been one of the most notable consequences of the concentration phase of the urbanization process. In the current stage in which the extension of urban areas makes the regional space more integrated and interdependent, the settlements on the peripheries of the main urban centers and some rural areas receive the influx of population and activities from them, while the rest of the regional territory, despite being integrated and closely interdependent on urban dynamics, is mainly used for second homes or constitutes "operational spaces" that provide energy and resources.

The effects of the dispersion of urbanization and the accelerated transformation of land uses constitute one of the main environmental problems that Spanish society must face. Part of the alarms raised by these transformations may be due to nostalgia for an urban model that ceased to exist decades ago, if it never came to exist (Amenós 2015). But the truth is that the environmental, functional consequences and economic aspects of urban dispersion are not at all banal (Camagni et al. 2002). Firstly, the artificialization of land decisively alters natural habitats, makes it difficult to recharge aquifers and contributes to fragmenting spaces of natural interest, thus putting their biodiversity at risk (Naredo and Frías 2007). On the other hand, the dispersion of settlements is directly related to the increase in mobility needs and energy consumption in urban areas (López 2017; Pazos 2022). Likewise, the cost of urbanization and the provision of public services tends to be notably higher in dispersed urban settlements than in dense areas. It must also be taken into account that, in increasingly extensive urban areas, the combination of administrative fragmentation with the intensification of residential segregation favors the maintenance and reproduction of social inequalities (Nel·lo 2021). Finally, dispersion bequeaths to present and future generations a set of urban spaces whose equipment and maintenance costs are in many cases unaffordable for those who inhabit them, so that they will tax the public coffers for many years (Burriel 2014). In any case, urban dispersion is a very difficult phenomenon to reverse, representing a considerable challenge for spatial planning and policies (Indovina 2007; Muñoz 2011).

Let's conclude. In Spain, the urbanization process has brought notable benefits in terms of improving living conditions and access to services by the population. Now, to combine these benefits with the environmental, functional, and social challenges associated with the urbanization process itself, three requirements should be met: the existence of adequate administrative structures, efficient instruments for land use planning and political will to use both for the well-being of most of the population. However, the Spanish situation in these three areas presents some notable shortcomings.

As far as administrative structures are concerned, it should be noted that metropolitan government experiences in Spain are scarce and have had very limited successes to date (Sorribes and Romero 2006; Nel·lo 2016; Hildenbrand 2016; Tomàs 2023). The progressive convergence of urban areas with regional spaces—or even their overcoming as in the case of Madrid (Solís 2011)—could facilitate the more direct involvement of the autonomous communities that govern them in urban management. This would require, however, the existence of a culture of federal loyalty between local, regional, and state governments that—due to historical circumstances that it would be too long to detail here—is not present in most cases (Romero 2009).

Despite some advances that have occurred in recent decades, in general terms, territorial planning at a supralocal scale does not have a long and successful track record either (Benavent 2012; Feria 2011; Hildebrand and Rodríguez 2010; Farinós and Olcina 2022). In this way, the urbanization process—no matter how vigorous it may have been in Spain—lacked, in most cases, adequate guidance regarding the protection of open spaces, infrastructure planning and development of the settlements. The lack (and/or ineffectiveness) of instruments capable of guiding urban planning has been one of the factors that have facilitated the accelerated and disorderly occupation of land.

Finally, the government of spatial transformations in pursuit of collective well-being requires a very notable political will. Over the last decades, notable progress has been achieved in some fields, particularly regarding the provision of services in urban areas and the protection of areas of natural interest. The actions of public administrations in these fields have been supported largely by citizen movements (Jiménez Sánchez 2005; Nel·lo 2015). However, on many other occasions the dynamics of land squandering, with the mentioned environmental and social corollaries, have been left to the discretion and impulse of the real estate market. Public powers have not always been able to (or wanted to) tame this impulse in order to subject it to the general interest. On the contrary, on many occasions, when managing land uses, they have contributed (by action or omission) to private interest prevailing over environmental or social considerations.

The urbanization process in Spain is, as we have seen, in a stage characterized by the integration of the regional spaces and widespread urbanization. This situation entails on the one hand, problems of environmental sustainability and social cohesion and, on the other, great potential for collective well-being. Whether society can overcome the former and use the latter will depend on the capacity and will of citizens and public powers.

References

Albertos JM, Sánchez JL (eds) (2014) Geografía de la crisis económica en España. Publicacions de la Universitat de València, València, 770 p

Amenós J (2015) El mito legal de la ciudad compacta. Universitat Autònoma de Barcelona, Bellaterra (167 p)

Benabent M (2012) Treinta años de ordenación del territorio en el Estado de las autonomías. In: Castañer M (ed) (2012) El planejament territorial a Catalunya a inici del segle XXI. Una nova interpretació i projecció del país. Societat Catalana d'Ordenació del Territori, Barcelona, pp 140–165

Brenner N (ed) (2013) Implosions/explosions: towards a study of planetary urbanization. Jovis Verlag, Berlin

Burriel E (2008) La "década prodigiosa" del urbanismo español (1997–2006). Scripta Nova. Revista electrónica de Geografia y Ciencias Sociales XII, 270(64)

Burriel E (2014) El estallido de la burbuja Inmobiliaria y sus efectos sobre el territorio. In: Albertos JM, Sánchez JM (eds) Geografía de la crisis económica en España. Publicacions de la Universitat de València, Valencia, pp 101–140

Camagni R, Gibelli MC, Rigamonti P (2002) I costi collettivi della città dispersa. Alinea, Milano, 169 p

Checa J, Nel·lo O (2018) Urban intensities. The urbanization of the Iberian Mediterranean coast in the light of nighttime satellite images of the earth. Urban Sci 2:115

de Terán F (1999) Historia del urbanismo en España. Siglos XIX y XX, Madrid, Cátedra, 397 p

del Molino S (2016) La España vacía. Viaje por un país que nunca fue, Madrid, Turner

Farinós J, Olcina J (eds) (2022) Ordenación del territorio y medio ambiente. Triant lo Blanch, València

Feria JM, Martínez L (2016) La definición y delimitación del sistema metropolitano espanyol: permanencias y cambios entre 2001 y 2011. Ciudad y Territorio. Estudios Territoriales, XLVIII(187):9–24

Feria JM (2011) La ordenación del territorio en las áreas metropolitanas españolas. In: Almonte JMJ (coord) Ordenación del Territorio y Urbanismo: conflictos y oportunidades. Universidad Internacional de Andalucía, Sevilla, pp 127–158

Gárate MM (2010) La cornisa cantábrica. ¿convergencia de modelos en el largo plazo? Historia Contemporánea 42:303–345

García Bellido A et al (1987) Resumen histórico del urbanismo en España. Instituto de Estudios de Administración Local, Madrid, 289 p

Gil-Alonso F, López-Villanueva C, Bayona-i-Carrasco J, Pujadas I (2021) Towards an even more spatially diversified city? New metropolitan population trends in the post-economic crisis period. Urban Sci 5(2):41 (1–19)

González-Leonardo M, López-Gay A, Recaño J, Rowe F (2022) Cambios de residencia en tiempos de COVID-19: un poco de oxígeno para el despoblamiento rural. Perspectives Demogràfiques 26:1–4

Hildenbrand A (2016) Gobernanza y planificación territorial en las áreas metropolitanas: análisis comparado de las experiencias recientes en Alemania y de su interés para la práctica en España. Tesis doctoral, Universidad de Sevilla, Sevilla

Indovina F (2007) La ciudad de baja intensidad. Lógicas, gestión y contención. Diputació de Barcelona, Barcelona, 540 p

Indovina F (2009) Dalla città diffusa al arcipelago metropolitano. Franco Angeli, Milán, 288 p

Instituto Geográfico Nacional (2021) La pandemia Covid-19 en España. Primera ola: de los primeros casos a junio 2020, Madrid, Instituto Geográfico Nacional, 159
Jiménez Sánchez M (2005) El impacto político de los movimientos sociales. Un estudio de la protesta ambiental en España, Centro de Investigaciones Sociológicas, Madrid
Lefebvre H (1970) La révolution urbaine. Gallimard, Paris
Lois R, Pino D (coords) (2015) A Galicia urbana. Edicións Xerais, Vigo, 776 p
López J (2017) Mobilitat quotidiana, consum de carburants i forma urbana a la regió metropolitana de Barcelona. Documents d'Anàlisi Geogràfica, 63/2
López-Gay A (2017) Hacia un patrón territorial complejo de la movilidad residencial. El caso de la Región Metropolitana de Barcelona. Papers 102(4):793–823
Méndez R, Abad LD, Echaves C (2014) Atlas de la crisis. Impactos socioeconómicos y territorios vulnerables en España. Tirant lo Blanch, Valencia
Muñoz F (2011) Estrategies vers la ciudad de baixa densitat: de la contenció a la gestió. Diputació de Barcelona, Barcelona, 515 p
Munro A (2006) The view from the castle rock. Vintage, London
Naredo JM, Frías J (2007) El metabolismo económico de la conurbación madrileña (1984–2001). In: Francesco Indovina, La ciudad de baja intensidad. Lógicas, gestión y contención. Diputació de Barcelona, Barcelona, pp 131–184
Nel·lo (2004) ¿Cambio de siglo, cambio de ciclo? Las grandes ciudades españolas en el umbral del siglo XXI. Ciudad y Territorio. Estudios Territoriales XXXVI(141–142):523–542
Nel·lo O, López J (2016) El procés d'urbanització. In: Giner S, Homs O (eds) La societat catalana. Enciclopèdia Catalana, Barcelona
Nel·lo O, López J, Martín J (2016) La luz de la ciudad. El proceso de urbanización en España a partir de las imágenes nocturnas de la Tierra. Universitat Autònoma de Barcelona, Bellaterra
Nel·lo O, López J, Martín J, Checa J (2017) Energy and urban form. The growth of European cities on the basis of night-time brightness. Land Use Policy 61:103–112
Nel·lo O (2001) Ciutat de ciutats. Reflexions sobre el procés d'urbanització a Catalunya, Barcelona, Empúries, 239 p
Nel·lo O (2015) La ciudad en movimiento. Crisis social y respuesta ciudadana. Díaz & Pons, Madrid, 2005 p
Nel·lo O (2016) El proceso de urbanización: motor y expresión de las transformaciones sociales y territorials. In: Romero J (coord) Geografía humana de España. Tirant lo Blanch, Valencia
Nel·lo O (ed) (2021) Efecto barrio. Las consecuencias sociales de la segregación espacial en las grandes ciudades ibéricas. Tirant lo Blanch, València
Nogué J (2005) La transformació del territori i del paisatge de la Costa Brava (1956–2003). Situació actual i propostes d'actuació. In: Debat Costa Brava. Congrés: Un futur sostenible. Col·legi Oficial d'Arquitectes de Catalunya, Girona, pp 88–121
Pazos M (2022) Perpetuum Mobile. Movilidad en tiempos pandémicos. Tirant lo Blanch, València
Prieto F, Campillo M, Díaz JM (2011) Tendencias Recientes de la evolución del territorio en España (1987 a 2005). Ciudad y Territorio. Estudios Territoriales XLIII(168)
Recaño J (2017) La sostenibilidad demográfica de la España vacía. Perspectives Demogràfiques 7
Reher DS (1994) Ciudades, procesos de urbanización y sistemas urbanos en la Península Ibérica. In: Guardia M et al (eds) Atlas histórico de las ciudades europeas (Vol. 1, La península Ibérica). Centre de Cultura Contemporània, Barcelona, pp 1–29
Rodríguez F, Escolano LMS (2010) La planificación territorial y el sistema urbano de Andalucía. notas para una reflexión desde la Geografia. Cuadernos Geográficos 47/2
Rodríguez F (ed) (2010) El estado de la ordenación del territorio en España. Cuadernos Geográficos 47/2
Romero J (2009) Geopolítica y gobierno del territorio en España. Tirant lo Blanch, Valencia
Rullan O (2011) La regulación del crecimiento urbanístico en el litoral mediterráneo español. Ciudad y Territorio. Estudios Territoriales XLIII(168):279–297
Rullan O (2022) Gegantisme turístic i nanisme urbanístic a les Illes Balears. Treballs de la Societat Catalana de Geografia 94

Salido J (2015) La vivienda no teme a los cementerios. El Mundo 30(10):2015
Soja E (2016) Regional urbanization and the end of the metropolis era. In: Nel·lo O, Mele R (eds) Cities in the 21st century. Routledge, New York, pp 41–56
Solis E (2011) Del área metropolitana hacia la región urbana policéntrica madrileña: cambio de escala, estructura y articulación territorial. Tesis doctoral, Universidad Complutense de Madrid, Madrid
Sorribes J, Romero J (2006) El fracaso de las experiencias de gobierno metropolitano en España. En: Romero J, Farinós J (coord) La gobernanza territorial en España. Claroscuros del proceso a partir del estudio de casos, Universitat de València, Valencia
Tomàs M (ed) (2023) Metrópolis sin gobierno. La anomalía española en Europa, València, Tirant lo Blanch
Vilà J, Capel H (1970) Campo y ciudad en la geografía española. Salvat, Barcelona
Zornoza C (2013) Crecimiento urbanístico en la zona costera de la Comunidad Valenciana (1987–2009). Análisis y persepectivas de futuro, Universitat de València, Valencia

Part II
Inequality

Chapter 5
The Contemporary Geographies of Urban Inequality: Insights from Portugal and the Lisbon Metropolitan Area

Jorge Malheiros, Diogo Gaspar Silva, and Leandro Basílio Júnior

Abstract This chapter explores the problematics of contemporary urban inequality in Portugal, assuming a spatial reading where segregation plays a didactic role. The Lisbon Metropolitan Area (LMA) constitutes the empirical case that is framed in a brief analysis of the recent evolution of inequalities, underlining income in relation with other dimensions. The issue of inequality has become a focal point in academic and political discourse. Stiglitz or Piketty have highlighted the widening income gap and its societal repercussions, showing how neoliberalism has amplified inequality by reducing regulation, increased financialization and promoted the concentration of wealth. Within this context, urban areas emerge as the most unequal spaces, with housing expulsions and spatial segregation being the major expressions of this inequality. In Portugal, income inequality remains high, despite the reduction observed along the twenty-first century. Urban regions, particularly Lisbon and Porto, display the highest inequality levels. From a geographical perspective, urban inequality in the LMA reflect socioeconomic changes, particularly in labour and housing markets, that experienced liberalization, internationalization and a huge increase in prices. Though we find a trend towards spatial desegregation, city divisions are not reducing, with top earners and EU migrants concentrating in central Lisbon and its affluent and lower-income groups and non-EU migrants pushed to suburbia with lower accessibility and limited urban resources.

Keywords Regional income inequalities · Urban inequalities · Spatial inequalities · Socio-urban segregation · Portugal · Lisbon Metropolitan Area

J. Malheiros (✉) · D. G. Silva
CEG-IGOT, Universidade de Lisboa, Lisbon, Portugal
e-mail: jmalheiros@edu.ulisboa.pt

D. G. Silva
e-mail: diogosilva4@edu.ulisboa.pt

L. B. Júnior
Universidade Federal do Rio Grande do Norte, Natal, Brazil
e-mail: leandrojr7@hotmail.com

R. C. Lois-González and J. A. Rio Fernandes (eds.), *Urban Change in the Iberian Peninsula*, The Urban Book Series, https://doi.org/10.1007/978-3-031-59679-7_5

5.1 Around Inequality Issues: Income, Society, Space and the Urban Scale

In the twenty-first century, especially in the aftermath of the 2007/2008 financial and economic crisis, the issue of inequality gained prominence both on the academic and political agendas. Reference authors such as Wilkinson & Pickett (2010), Stiglitz (2012) or Piketty (2014 and 2020), just to mention a few, drew attention to the increasing concentration of income among the extreme top social groups and the growing hardship faced by middle classes leading to social distress and political dissatisfaction. Over the last few decades, this process has emerged as an integral feature in many Western countries.

Whilst its fundamental origins are multi-causal, there has been a general consensus within the literature that the rising inequality seems to be a symptom of the lubrification of neo-liberal political-institutional and ideological stocks (Peck et al. 2013). A number of critical reviews and empirical studies have widely debated the ways in which neo-liberal processes have reshaped policies, institutions and spaces and, of course, fueled inequality issues. The easing of property access, the concentration and reproduction of patrimony (Piketty 2020: 43–45), the opportunities generated by the expanding and the less regulated digital sector are often mentioned as causes for this process. But the appropriation of natural resources and profitable former public companies by national and international monopolies, the changing labour market relations now marked by precarity and big difference in the pay role as well as the weakening of traditional trade unions have also played a relevant role. Unsurprisingly, the longer-term consequences of these practices and agendas have been easy to gauge. For instance, in France, between 1983 and 2015, the average income of the healthiest 1% has increased 100% and the average income of the healthiest 0,1%, 150%, whereas the average increase of the remaining population increased only 25% (Garbinti and Goupille-Lebret 2017, in Piketty, 2020: 124–125). In the United States, between 1980 and 2010, the wages of the bottom 10% have shown a growth of only 15%, against an increase of 150% of the wages of the top 1% and 300% of the 0.1%. Around 2009, an US household belonging to the wealthiest 1% had 225 times the wealth of an average American household, a value that almost duplicates the register of 1983 (Stiglitz, 2012: 8).

But income and wage distributions are just one way—perhaps the most classical—of measuring inequality. However, the differences in access to social services, such as education and health, are also an expression of inequality and have differential consequences for the opportunities of the various social groups, particularly for lower classes, racialized migrants and national ethnic groups.

Whilst access to education has become nearly universal almost everywhere, in countries like France or Portugal, we have witnessed an increasing gap between public and private schools. For example, the latter often exhibit a lower turnover of teaching staff, more learning resources and a selective admission process, which favors upper middle and upper class families. This produces better results in the selection exams for these students widening their opportunities and future success that is amplified by the higher levels of family cultural and social capitals. The gap between private and public schools cannot be dissociated from a de facto reduction of public expenses in education that took place in several countries. For instance, in France, only between 2007 and 2017, the public expenses per capita in higher education reduced 10% (Piketty 2020: 124). In Portugal, between 2009 and 2023, though the absolute amount of the public budget dedicated to higher education remained stable, both the value spent *per student* and the percentage in the national GDP declined (Aguiar-Conraria et al. 2022: 4). Concerning the health care system in Portugal and elsewhere, we have also assisted to a progressive privatization and commodification. This "restructuring" is making the access more difficult and the responses of the public services harder, exposing the ones with poorer resources to longer waiting queues at the urgencies and longer periods to obtain a surgery or other medium/major intervention. This originates a wider gap in the life expectancies of richer and poorer and a lower health quality of the later (D'Eramo 2023).

In addition to the financial-monetary and the social expressions of inequality, housing, both in terms of access and quality, is another sector where inequality is growing. This has been particularly obvious in the aftermath of the 2007/2008 world financial and economic crisis when rents and prices sky-rocketed in several cities around the world. Because it is not possible to separate housing from land, one of the dimensions of housing inequality corresponds to location, an issue that involves the access to urban resources and the quality of the place. It is not the same to live in a peripheral neighbourhood disconnected from the remaining urban fabric, with poor transport connections and limited economic resources, and in a central space, well connected with a myriad of facilities and job opportunities. This limits the opportunities of the residents, both materially and symbolically, because poor peripheries, slum areas or social housing neighbourhoods become segregated spaces that carry the stigma of *no go places* where only unskilled workers, unemployed and poor migrants, often seen as excluded, are capable to live.

Actually, socio-spatial segregation can be seen as the spatial expression of exclusion and inequality, particularly of urban inequality, and is being affected by the recent changes in the mechanisms underlying the most recent stage of financialized capitalism (Malheiros and Madeira 2023). With the reduction of interest rates and instability in stock markets, urban real estate emerged, as in other historical periods, as a safe arena for global capital investment, seeking new forms of profitability based on the transfer of capital from the primary accumulation circuit (the productive sphere in a strict sense) to the secondary circuit (production of built environment) (Salgueiro 1994; Mendes 2008). In this new phase, where housing market liberalization is intense, new real estate products marked by volatility emerge (such as urban tourist accommodation or shorter-term lease agreements that target, for instance, digital nomads), buyers and investors internationalize and deepen the finance-housing nexus within a framework that refers to global capital circuits (Mendes 2017). In this context, city centres, desirable due to the rent gap, become preferred investment spaces, whereas in previous phases it was the suburbs (and suburban expansion) that fuelled accumulation processes based on land appreciation. And with this, real estate speculation and housing price escalation are fuelled, initially in the city centres of metropolises and subsequently in other areas, while inequalities are accentuated, assuming new expressions. Attention must also be paid to the processes of territorial expulsion of less privileged social groups—which translate and accentuate urban inequality (and socio-spatial segregation)—as a consequence of financialization processes. For instance, the incapacity to cope with rising rents and the increasing interests of mortgages are forcing low, middle and even middle-high classes to move to less central spaces. In some cases, evictions occur, a process that should be interpreted as a form of territorial expulsion that has intensified in many cities worldwide. Perhaps that's why, beyond socio-spatial segregation, residential expulsion and the inability to access certain city neighbourhoods (and the resources here present—facilities, commerce, services) by the poor and even the middle classes, are today one of the most visible territorial expressions of inequality. While new forms of segregation include micro elements marked by unequal contact (e.g. contiguity between buildings or residential blocks occupied by different social groups) resulting from contemporary urban processes, evictions, as territorial expulsions, seem to constitute the more striking form of loss and relegation, meaning a forced new life in the shadow areas of the city.

In metropolises such as Lisbon and Porto, the processes of physical rehabilitation and regeneration of the centres that lead to direct and indirect displacement of working classes and labour migrants (because they no longer have access to these residential areas) correspond to the advances of gentrification due to the arrival of upper-middle and upper classes, that are transforming the social patterns of metropolises through a kind of filtering up process, leaving the centre to the wealthy and transient populations, such as tourists and digital nomads (Brollo and Celata 2023). Peripheralization of the working and middle classes take place, often with the emergence of segregated spatial pockets where poor, migrant, and racialized ethnic groups cluster. These processes, largely commanded by finance, give new relevance to the relationship between contemporary capitalism and spatial (in)justice (Madeira and Vale 2015; Estevens 2017; Silva and Carmo 2023), which has urban segregation as one of its most relevant and evident dimensions. Because the transition to cities marked by a gentrified centre, where the working classes have no place, is slow and subject to constraints (for example, due to the role of public action in promoting social mixing or the resilience of some neighbourhoods), traditional processes of segregation—based on large homogeneous neighbourhoods—may actually decline and be replaced by fragmentation (Salgueiro 1999 and Caldeira 2000) or powerful forms of micro-segregation. This leads different groups to occupy small contiguous social spaces (blocks, buildings), but interaction between them remains very low or non-existent, since they do not share the same public space or shopping areas, or don't do it at the same time (Fernandes and Chamusca 2020), reinforcing social distance and the loss of sociabilities and social cohesion. This point to the loss of one of the essential and distinctive characteristics of urban space: the ability to facilitate interaction between people with a myriad of different characteristics (Malheiros et al. 2012). Overall, it points to the need to reassert the notion of centrality that goes beyond the logic of the easy access to a certain point well served by the transport network and "implies the availability of manifold possibilities and access to social resources" (Schmid 2012: 57).

Having into consideration this panorama, this chapter explores the problematics of contemporary urban inequality in Portugal, assuming a spatial reading where segregation processes play a didactic role. Although the Lisbon Metropolitan Area (LMA) constitutes the empirical case, a previous section about the recent evolution and the features of inequalities in Portugal and the main Portuguese Urban regions provide the set of contextual elements that are necessary to understand the specific situation.

5.2 Inequalities in Portugal: Key Features and Recent Evolution at Various Geographical Scales

In their seminal work about the impact of inequality on health and social problems in advanced societies, Wilkinson and Pickett (2010) show that Portugal, along with the UK and US, display the highest levels of income inequality and also the highest values of health and social problems in most of the indicators analyzed. Using another index (the percentage of gross income possessed by the poorest 40%[1]), between 2005 and 2022, Portugal has shown a positive evolution (from 17.9 to 21%). However, this was not sufficient to remove it from the group of countries with the less fair distributions, and the post-2010 trend is not very encouraging. In fact, only the Baltic States, Bulgaria, Romania and in lesser extent Greece and Spain display unfairer distributions.

Having into consideration the analytical framework briefly outlined in Sect. 5.1 (structural higher expression of inequalities in urban spaces and its aggravation in the present context of financialization that exacerbates the exploitation of urban space and its extreme commodification) and the aforementioned indicators that set the context and underline the position of Portugal as a country marked by persistent inequality levels, this section discusses three questions: Are Portugal and the Portuguese regions increasing or reducing their inequality levels, measured in terms of income? Are the main urban regions, and in particular the two Metropolitan Areas of Lisbon and Porto (LMA and PMA), more unequal than the rest of the country? Has inequality increased in the twenty-first century and particularly in the aftermath of the 2008 crisis in these Metropolitan Areas?

Figure 5.1 shows the relationship between the amount of the total income held by the 20% richest and by the 20% poorest in Portugal. Over the last three decades, we have seen a steady decline in income distribution inequality, which has reduced from 7.4 to 5.6 between 1994 and 2022. Likewise, the Gini coefficient applied to income and population, discloses the same trend, registering a decrease from 38.1% in 2005 to 34% in 2015 and 33% in 2021. This has contributed to a positive convergence with

[1] Source of data - Eurostat | European Union Household Panel (ECHP); European Statistics on Income and Living Conditions (EU-SILC).

the EuroZone value (Fig. 5.2), a process that also results from a negative global trend observed in this monetary macro-zone, where the Gini Coefficient—and therefore inequality—slightly increase in this period.

Despite the positive trend in the reducing income inequalities, which translated into an apparent improvement in the well-being of Portuguese households, particularly until the outset of the 2007/2008 financial crisis, a more detailed analysis of the process justifies some caveats (Rodrigues et al. 2012). First, and from an international perspective, inequality levels in Portugal remain high when compared to other mostly advanced, European countries to a persistent inequality that pairs with significant levels of poverty (Tavares and Carmo 2023; Rosa 2015). Second, the narrowing of the inequality gap can be primarily attributed to the advancements observed among households situated at the lowermost rungs of the social ladder, namely those in the first decile of income. On the one hand, this shows that social measures aiming at softening poverty and the scarcity of resources (e.g. increases both in social basic income and minimum wages and pensions) have elevated the income levels of the most economically disadvantaged individuals (Rodrigues et al. 2012). On the other hand, it indicates a relative decline for the middle classes, as their income tends to converge with lower income levels. Thirdly, while income distribution shows some signs towards equality, the same cannot be said for salaries. This becomes obvious as employees in the top decile experience the highest salary increases (Rodrigues et al. 2012). Fourth, economic and financial shocks, which seem to be happening more frequently in the contemporary stage of capitalist development, have wider income inequality implications, particularly on the lower social groups. Finally, the post 2007/2008 crisis period and in particular the most recent years (with the exception of the 2013–2019 period) point to a widening of the income gap (Tavares and Carmo 2023). If the COVID19 clearly had its role, more substantial elements such

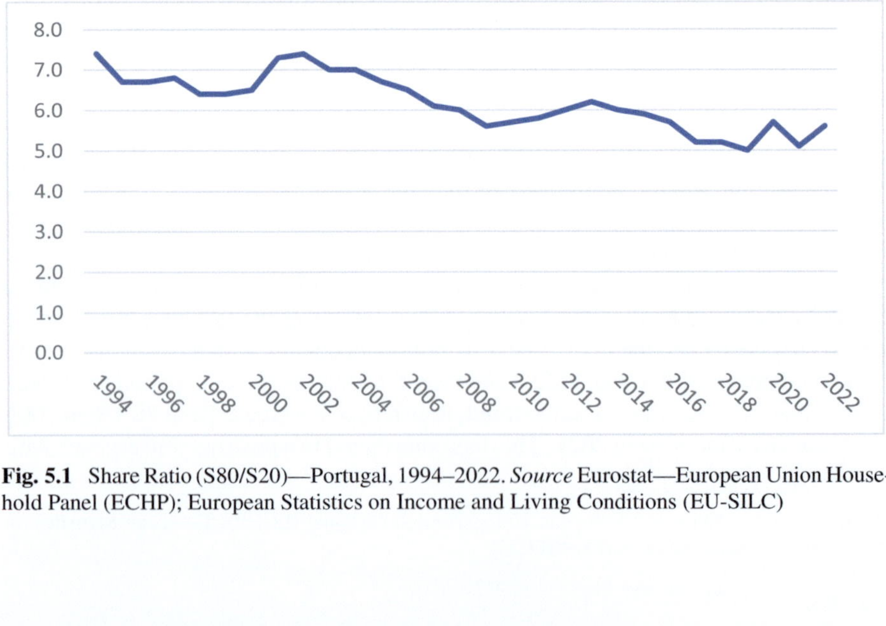

Fig. 5.1 Share Ratio (S80/S20)—Portugal, 1994–2022. *Source* Eurostat—European Union Household Panel (ECHP); European Statistics on Income and Living Conditions (EU-SILC)

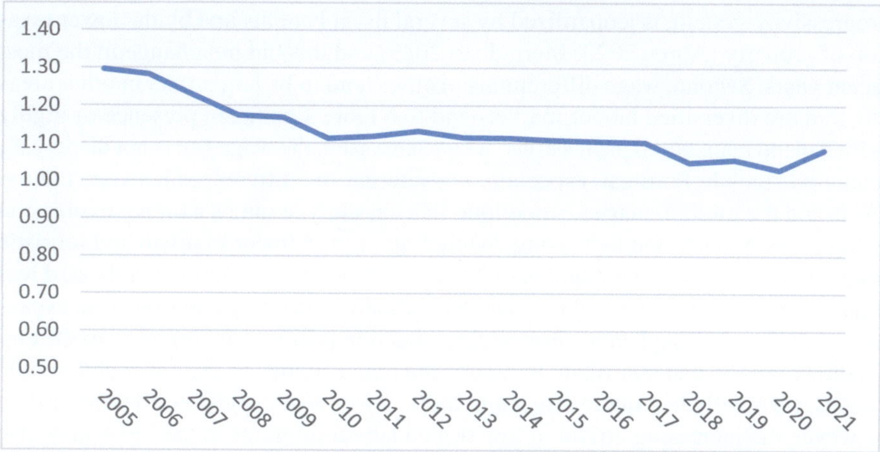

Fig. 5.2 Ratio between the Gini Indexes in Portugal and EuroZone (2005–2021). *Source* Eurostat—European Union Household Panel (ECHP); European Statistics on Income and Living Conditions (EU-SILC)

as the intensification in the use of the housing market as an investment target and a capital reserve with very powerful consequences over the fast increase in real estate prices, the maintenance of labour precarity that goes together with relative low wages even for many of the highly-skilled and the recent inflationary context, are certainly combining to block the progression of the low, middle and even middle-high groups, a phenomenon that pushes inequality.

An analysis by regions of the recent evolution of inequality based in the share decile[2] confirms the ideas advanced in the initial section, with some contextual specificities. The observation of Tables 5.1 and 5.2 shows that the two metropolitan areas of Portugal, and particularly the two head-cities, Porto and Lisbon, display the higher levels of inequality, with emphasis on the later. In structural terms, the combination of higher levels of urban population diversity with the capitalist strategies that historically have privileged the spatial concentration of capital as a key investment and rotation strategy largely explain the results found. Concerning recent evolution, every Portuguese region (with the exception of Madeira) displays a reduction in income inequality between 2015 and 2021, but the most recent years (post-2018) are marked by a reversion of this trend in the most urbanized and touristic regions (only the PMA displays a smooth decline). As mentioned before, the COVID19 pandemic direct and indirect effects (reduction of the activity of precarious workers; strong constraints to circulation) that had a higher impact in urbanized spaces characterized by mobility and interaction have certainly played a role here, but deeper reasons should also be convoked. First, the Portuguese fiscal system penalizes more labour than capital, and even the redistributive effect of household income taxes associated with the existing

[2] Relationship between the amount of the total income held by the 10% richest and the amount of total income held by the 10% poorest.

progressivity system, is jeopardized by several fiscal benefits and by the lower taxation of property (Abreu 2023; Mergulhão 2023) and this did not change in the most recent years. Second, wage differentials in cities tend to be larger than in other areas due to more diversified labour markets and to a more significant presence of highly skilled in top managerial professions. As we have seen, the wage gap is not displaying reduction signs in Portugal, especially considering the differences between the top levels and the middle salaries, something that certainly assumes a higher expression in big cities. Finally, the increasing channeling of investment to urban and touristic spaces (Lisbon, Porto, the Algarve or Madeira) benefitting from the overheated real estate market associated to the urban rehabilitation processes and the fast expansion of the touristic and short-term accommodation business are bringing to the city wealthier groups that contribute to reduce the opportunities of the lower and middle classes, eventually leading to direct and indirect forms of expulsion (Mendes 2017). However, the increasing arrival of low skilled labour migrants in the LMA may also be contributing to the widening of inequality gap.

Despite this situation, it must be said that the income levels in the most urbanized regions of the country are higher than in other areas, something that could supposedly contribute to alleviate the needs of the poor and the lower classes, even in a context of inequality. However, as a recent analysis of poverty and inequality in Portugal states (Tavares and Carmo 2023), the LMA (and the Azores) shows the highest poverty levels of the country and displays a worrisome evolution in the immediate post-pandemic period (after 2021). Having this into consideration, inequality levels won't probably reduce here, unless the processes of population transition and gentrification continue its expansion and assume such a proportion in the city of Lisbon that we reach a homogenization by the top. In this case, inequality levels are expected to

Table 5.1 Inequality in declared gross income after IRS deduction applied to fiscal households (P90/P10)—Portuguese Regions (NUTS 2), 2015–2021

	2015	2016	2017	2018	2019	2020	2021
Portugal	7.8	7.8	7.5	7.1	6.9	7.1	6.9
Norte	7.7	7.5	7.2	6.9	6.6	6.8	6.5
Porto Metropolitan Area	**8.0**	**7.8**	**7.4**	**7.1**	**6.9**	**7.0**	**6.9**
Porto (City)	**10.5**	**10.4**	**10.1**	**10.1**	**9.7**	**10.2**	**10.1**
Centro	6.7	6.7	6.6	6.3	6.1	6.2	6.0
Lisbon Metropolitan Area	**8.8**	**8.7**	**8.7**	**8.0**	**7.8**	**8.3**	**8.2**
Lisboa (city)	**12.4**	**12.9**	**13.7**	**10.5**	**10.3**	**11.3**	**11.1**
Alentejo	6.4	6.5	6.2	6.2	6.1	6.2	5.9
Algarve	8.1	8.1	7.8	7.2	7.0	8.8	7.9
Autonomous Region of the Azores	8.3	8.5	8.3	8.0	7.6	7.4	7.1
Autonomous Region of Madeira	7.7	7.8	7.7	7.4	7.2	7.6	7.7

Source INE, Income Statistics at local level based on information produced by the Finance Ministry—Tax and Customs Authority

Table 5.2 Inequality in declared gross income after IRS deduction applied to fiscal households—ratio of P90/P10 of each Region to the value observed in Portugal (2015–2021)

	2015	2016	2017	2018	2019	2020	2021
Norte	0.99	0.96	0.96	0.97	0.96	0.96	0.94
Porto Metropolitan Area	**1.03**	**1.00**	**0.99**	**1.00**	**1.00**	**0.99**	**1.00**
Porto (City)	**1.35**	**1.33**	**1.35**	**1.42**	**1.41**	**1.44**	**1.46**
Centro	0.86	0.86	0.88	0.89	0.88	0.87	0.87
Lisbon Metropolitan Area	**1.13**	**1.12**	**1.16**	**1.13**	**1.13**	**1.17**	**1.19**
Lisboa (city)	**1.59**	**1.65**	**1.83**	**1.48**	**1.49**	**1.59**	**1.61**
Alentejo	0.82	0.83	0.83	0.87	0.88	0.87	0.86
Algarve	1.04	1.04	1.04	1.01	1.01	1.24	1.14
Autonomous Region of the Azores	1.06	1.09	1.11	1.13	1.10	1.04	1.03
Autonomous Region of Madeira	0.99	1.00	1.03	1.04	1.04	1.07	1.12

gain a higher expression in the urban peripheries of Lisbon. An insight of the spatial consequences of this ongoing process will be given in the next section.

5.3 Rethinking the Geographies of Urban Inequality: Insights from the Lisbon Metropolitan Area in the Twenty-First Century

The geographical expressions of inequality in the LMA cannot be dissociated from housing and labour markets which are, themselves, expressions of the changes taking place in the twenty-first century capitalism. Concerning the housing market, the LMA experienced an intense process of liberalization and internationalization that goes hand in hand with the social and economic valorization of the city center (that benefits from the added-value associated to the rent gap) and the expansion of urban tourism. As a result, investment in housing, often as a financial asset, experienced a significant increase leading to a fast and intense growth in prices. According to the 2021 Census data, the average monthly charges associated with owning a home in the LMA reached almost 400 euros, a value almost 12% higher than in the neighbouring region of Oeste and close to 20% higher than in the Region of Cávado, in the north west of Portugal where the dynamic city of Braga is located. If we focus on rent tiers, the percentage of dwellings paying more than 650 euros per month reaches approximately 16% in the LMA, a percentage that decreases to less than half in the case of the PMA. Finally, if our analysis targets housing transactions, in 2022 the median values registered in the LMA were 30% higher than the ones registered in the PMA, 70% higher than the values observed in the Region of Cávado and 41% higher when compared to the median of the global transactions that took place in Portugal.[3]

[3] Data from INE, *Statistics of Housing Prices at the Local Level.*

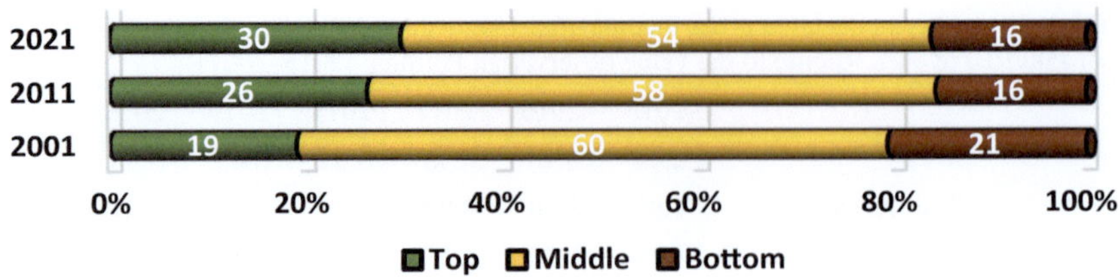

Fig. 5.3 Socio-professional composition in the functional urban area of Lisbon.[4] *Source* INE, Census 2001, 2011 and 2021

But the most striking numbers take place at the city level, with the median of the transactions in Lisbon reaching a value by square meter slightly below 4 400 euros in 2022, the highest of the country, that was 1.7 times higher than the median value observed in Porto, 3.3 times higher than the value registered in Braga and almost threefold the global value for Portugal (1484 euros per square meter).

Having this scenario into consideration, one would expect an increasing separation between a city progressively dominated by the presence of the wealthier and an increasing peripheralization of the low and middle classes. In addition, at the neighbourhood level, the hypothesis of an increase in segregation levels is on the table. However, because rehabilitation and gentrification processes are stepwise and often take place at a micro level, at least in their initial stages, and also because some measures protect tenants with more than 65 years, the process promotes a more fragmented metropolis, where micro spaces occupied by the rich juxtapose to spaces occupied by lower classes.

Considering the changes in the labour market composition along the first 20 years of the twenty-first century (Fig. 5.3), we may talk of an asymmetrical polarization because the reduction of the middle groups (clerical workers, sellers, manufacturing workers) occurs at the expenses of the growth in the top groups (managers and professionals) and also of an apparent trend towards the stabilization of the bottom groups (unskilled workers and operators/assemblage workers). Although this situation points to a skilling process that incorporates a growing number of professionals it also shows the shrinking of the middle categories and the maintenance of a stable number of the bottom categories, particularly in the last inter-census period (2011–2021), when the percentage of the specific category unskilled workers even experienced a slight growth in the FUA of Lisbon (from 12 to 13%). If we have into consideration what was mentioned about the widening of the wages gap, namely the increases registered in the top categories and its increasing distance in relation to the salaries of the middle groups (Ferreira et al. 2021), we add an extra element to the inequality process.

[4] Though our analysis uses as territorial references the Lisbon Metropolitan Area (LMA), the city (municipality) of Lisbon and its subunits, mostly the *freguesias* (administrative parishes), in this case, due to a simplification procedure in the data calculation process, we took as a reference the Functional Urban Area of Lisbon, that is slightly larger than the LMA, but completely overlaps this. Therefore, we assume that the interpretation made is didactical and correct, not introducing any substantial changes to what took place at the LMA, *strictu* sensu.

Having into consideration the short analysis of housing and socioprofessional composition that disclose progressive trends towards inequality in the LMA, to verify how this combines with the geographical location of the various groups in the metropolis, the segregation process and, consequently, the evolution of their right to the city is our final goal.

A look to Fig. 5.4 shows that the global distribution pattern of the top and bottom social groups in the LMA did not experience very significant changes between 2001 and 2021, something that confirms the resilience of the physical and social forms of the urban space. However, a thorough examination displays some signs of a more quartered city to use the term of Marcuse (1989). The top social groups cluster in the city of Lisbon, with a more visible over-representation in the west and central areas, expanding along the coastal strip that connects Lisbon and Cascais. This location is not a surprise, once this the area comprehends most of the high class neighbourhoods of the metropolis; however, the most significant element corresponds to the strengthening of the presence of the top socio-professional groups in this area and in some contiguous ones, as the over-representation in all but one freguesias' of the municipalities of Oeiras and Cascais shows, in 2021. This means that the residential space of the city of Lisbon (and also of the neighbouring municipality of Oeiras) is being appropriated by the top groups.

Looking to the "other side", that is the residential location of the bottom socio-professional groups, the traditional levels of peripheralization are much higher and

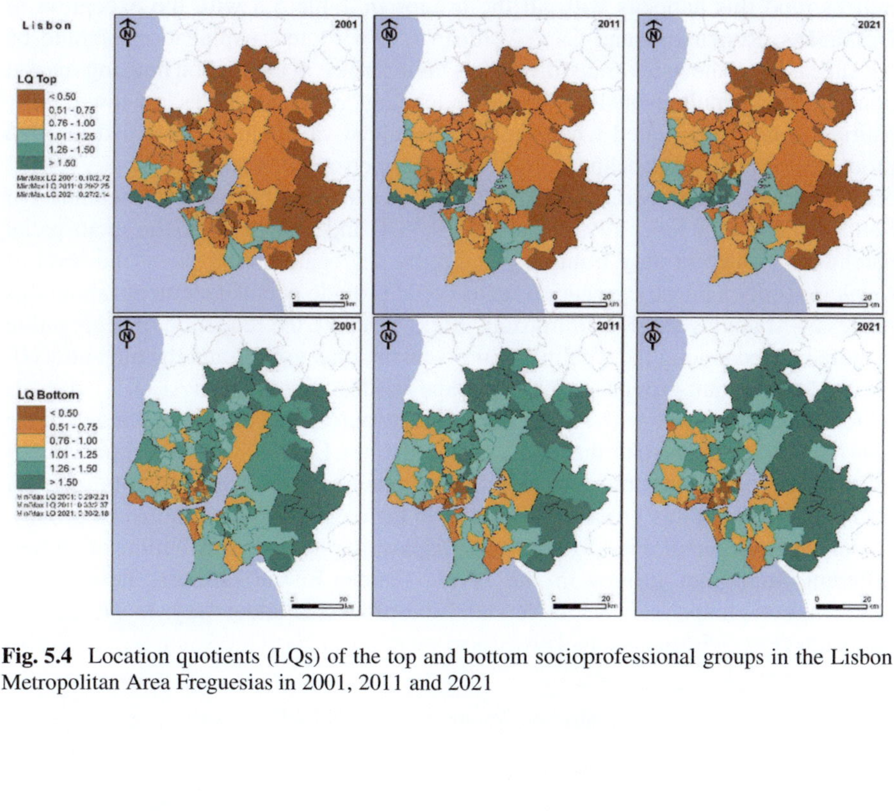

Fig. 5.4 Location quotients (LQs) of the top and bottom socioprofessional groups in the Lisbon Metropolitan Area Freguesias in 2001, 2011 and 2021

spread all over the LMA. Nevertheless, this peripheralization is steadily increasing, as the evolution of the under-representation spaces inside Lisbon and its immediate western suburb show. On the opposite position, we find the most peripheral areas of North and East of LMA, often the transition to the peri-urban, where the bottom social groups are increasing their over-representation.

From the overall pattern, we detect a trend towards a more divided city with a centre strongly appropriated by the wealthier and the peripheralization of the low and middle classes. Though, it should be said that this pattern cannot be read as a simple spatial continuum of each of the groups because several "intrusions" can be found.

It is important to note that the second decade of the twenty-first century has been marked by a significant increase in the number of foreign residents in the LMA (growth rate of 25% between 2011 and 2021) and especially in the city of Lisbon (growth rate of almost 60% in the same period), mostly fed by the arrivals of Western European EU citizens, Brazilians and South Asians (the group that displays the highest increase). This diverse geographical composition of the foreigners' arrivals, that also displays diverse trends in terms of labour market insertion (e.g. Europeans more represented in the top levels; South Asians at the bottom and also intermediate levels), shows different location patterns. Curiously, the segregation indexes reveal a global trend of decline (with the exception of the French), that is stronger among South Asian groups, though these remain as the most segregated in 2021 (Table 5.3). In addition to a technical issue that corresponds to the "effect of larger numbers" that normally contributes to scattered distributions and lower segregation indices (and this happens with all the groups in Table 5.3 with the exception of Ukrainians, Romanians and Cape Verdeans), two content explanations should be advanced. First, the quick transformations taking place in the Lisbon housing market don't happen simultaneously everywhere leaving "expectant holes" of low-quality waiting to be rehabilitated where immediate added-value can be obtained through processes of relatively cheaper renting (for the metropolitan and Lisbon city levels) combined with overcrowding. This justifies, for instance, the over-representation of South Asians in the inner city of Lisbon (Fig. 5.4) and contributes to the sociospatial patchwork, which is only limitedly caught by segregation analysis at the level of *freguesia* (only a microsegregation focus could provide detailed elements about this process—Maloutas and Karadimitriou, 2022). Second, the reduction of segregation is more limited among the wealthier European national groups and the old and well-established migrant groups such as the Cape Verdeans and other PALOP[5] nationals which points that those with higher economic resources or a better knowledge of the housing market and also of the urban structure of the LMA have wider choice possibilities and try to resist more to extreme peripheralization. Eventually, homophily also plays a role in this process, especially in the case of longer established groups.

An observation of Fig. 5.5a–c, that display the spatial distribution of 3 very different immigrant groups in the LMA, Guinea-Bissau citizens, Italians and South Asians, helps to better understand the processes under way. In the case of Guineans (Fig. 5.5a), a dispersal process to suburban areas of the South Bank and also

[5] PALOP stands for African Country with Portuguese as its Official Language.

Table 5.3 Segregation Indexes for the main foreign groups in the *freguesias* of the Lisbon Metropolitan Area, 2011 and 2021

	2011	2021	Difference
German	43.9	40.8	– 3.1
Spanish	36.8	35.0	– 1.8
French	34.1	42.5	8.4
Italian	**43.7**	**39.8**	**– 3.9**
British (UK)	48.4	44.5	–3.9
Romanian	34.9	29.6	–5.3
Ucranian	23.8	21.5	– 2.3
Angolan	34.1	28.9	– 5.2
Capeverdian	40.9	40.1	– 0.8
Guinean (Guinea-Bissau)	**46.8**	**45.0**	**– 1.8**
Sãotomense	50.4	47.0	– 3.4
Brazilian	19.8	14.9	– 4.9
Chinese	32.1	25.0	– 7.1
Bangladeshi	82.6	58.0	– 24.6
Nepalese	84.7	49.6	– 35.1
Indian	**61.9**	**43.3**	**– 18.6**

Source of the origin data: INE, Census 2011 and 2021

North East, along the Tagus River, justify the reduction observed in the segregation index but also show the reduction of their housing opportunities in the most accessible and valuable areas of the LMA: it is meaningful that their over-representation areas (LQ > 1) reduce inside the city of Lisbon, in the *freguesias* of the municipalities of the west strip, namely Cascais, and even in the north and northwest immediate peripheries of Lisbon. On the contrary, the Italians (Fig. 5.5b), one of the foreign groups that registered a very high growth in the past decade, are more concentrated in the city of Lisbon and in the west strip *freguesias*, somehow reproducing the geographical pattern of the top socioprofessional groups. Finally, South Asians (Fig. 5.5c), the group of the three that grew the most in the LMA, reveals a residential dispersal trend between 2011 and 2021, characterized by three dominant types of location: (a) the traditional presence in inner city spaces not yet taken by renewal and touristification that seems to be reducing (at least in relative terms), as the slight decline in areas with LQs over 3.0 points to; (b) a relatively well-established presence in the immediate suburban ring northeast and north of the city, where signs of a reduction of the concentration levels also seem to emerge; c) a growth in the cases of over-representation in various peri-urban *freguesias*, both in the northern and southern limits of the LMA, what may be related with the important presence of South Asians in agriculture activities but also showing limited access to the housing market: they may remain in the city centre while upwards transition is not complete but their housing opportunities, either temporary or permanent, seem to be more and more located in the less accessible and more peripheral metropolitan spaces.

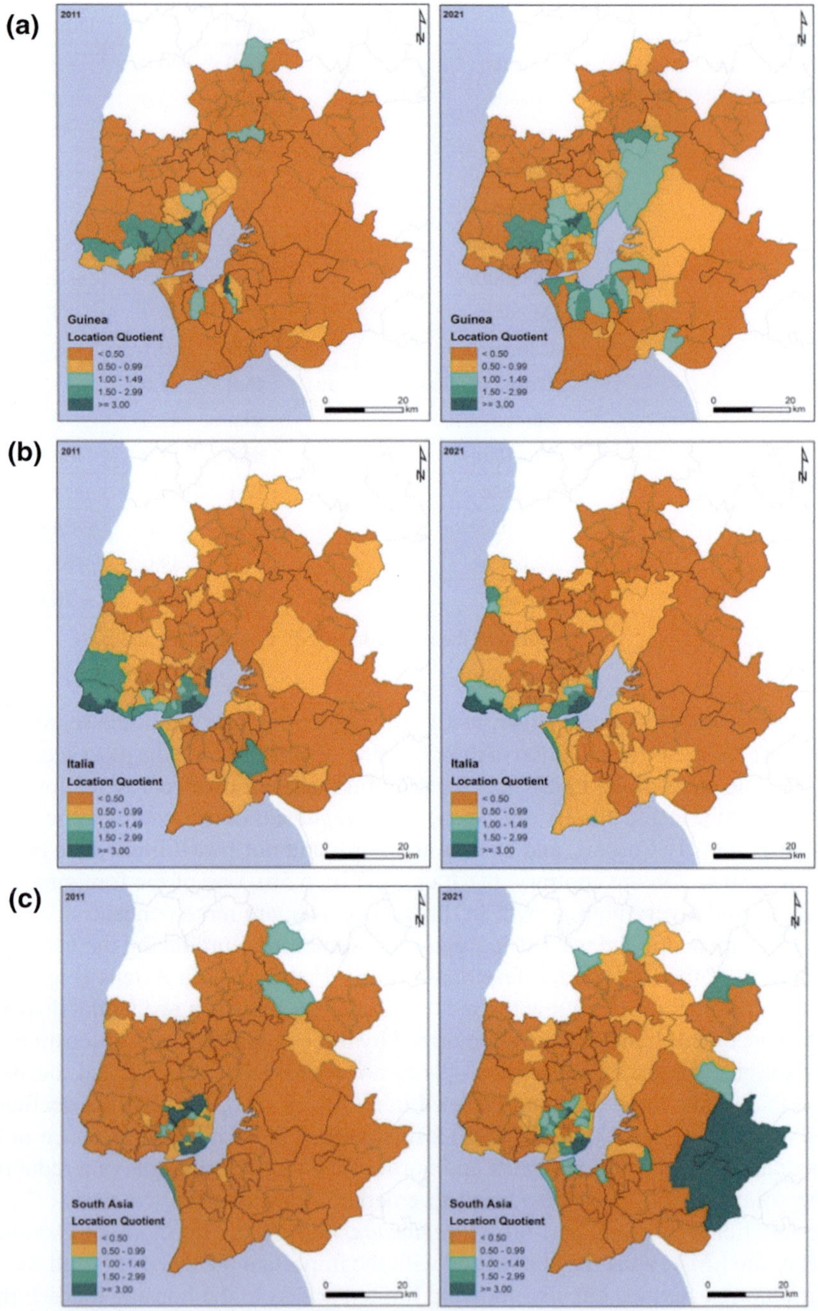

Fig. 5.5 Location quotients of three foreign groups in the LMA *freguesias* in 2011 and 2021. **a** Citizens from Guinea-Bissau, **b** Citizens from Italy, **c** Citizens from South Asia

5.4 Concluding Remarks

Interest in inequality lies less in its intrinsic characteristics and more in the social and health disadvantages associated with it. As Wilkinson and Pickett (2012) have demonstrated, higher levels of economic inequality lead to higher social and health distress, as urban geographical inequality, characterized by forms of segregation and limitations in the access of some groups to certain parts of the town and their resources, lead to spatial injustice putting into question social and territorial cohesion. In fact, the two forms of inequality are related and urban spaces, where people and capital cluster, tend to display higher levels of economic inequality. Both the neoliberal stage of capitalist development and its most recent evolution seem to contribute to the aggravation of inequality, namely in urban areas where socioprofessional polarization is more intense, real estate investment supported by the liberalization and internationalization of the housing markets overheats prices, and commodification of social services and urban space reaches unprecedented levels.

An analysis of the Portuguese case shows, contrary to other countries, that economic inequality, measured by the income distribution, experienced a slight reduction between the 1990s and the present. However, this was not enough to cause a significant convergence with the more developed EU countries. Furthermore, if the reduction of the inequality level is mainly due to the positive evolution of the income of the poorest, the middle classes have seen their incomes getting closer of the basis and more distant from the top.

The geographical analysis confirmed the metropolitan areas of Porto and particularly of Lisbon as the most unequal spaces, with the economic gap having their highest expression in its central municipalities. If economic inequality has displayed an overall trend towards decline in these metropolitan spaces, after 2018 a reversal of this tendency seems to be emerging, as the widening of the inequality gap in the regions that are more urbanized or display higher levels of tourist attraction shows. There, namely in the case of the LMA that was subject of a more detailed analysis, the wage differentials are higher and there is evidence of asymmetrical polarization with an increase in the professional/managerial groups, the relative maintenance of the low skilled and the squeezing of the middle. On the one hand, real estate investment and increasing housing prices, particularly in the most central and historical areas, attract wealthier skilled foreigners coming from developed countries. On the other hand, low skilled foreigners employed in undifferentiated jobs in the service and construction sectors and also in the new platformized services of distribution are finding residence in some transtion spaces that still wait to be gentrified or touristified.

Concerning the geographical distribution of socioprofessional and national groups in the LMA, the statistics show a trend towards desegregation at *freguesia* level for the first two decades of the twenty-first century. Though this points to a supposedly more even access to the metropolitan space, this result must be interpreted with some care. First, the slow and relatively fragmented process of rehabilitation and gentrification leaves "unattended spaces" and generates areas where highly skilled dominated blocks are contiguous to working class blocks. This does not mean that the members of the two groups effectively interact, but that eventual separation can only be measured at the microlevel, below the level of the *freguesia*. Second, the mapping of the various immigrant and socioprofessional groups points to the slow strengthening of concentration of the top social categories and EU migrants in the city of Lisbon, particularly in its central and western areas, and also in the western coastal strip that ends in Cascais. This corresponds to the areas that concentrate the neighbourhoods with higher housing prices and are associated with a better image and the highest accessibility to social and economic resources. On the contrary, lower social groups and non-EU immigrants (with the exception of North-Americans) tend to be more scattered in the metropolitan peripheries, a process that contributes to the decline observed in segregation indices. However, their presence in the city of Lisbon and in the better off western coastal areas of Oeiras-Cascais strip is slowly declining, something that points to a reduction of the capacity to have access to these residential spaces, potentially constraining their right to the city. New groups of South Asian migrants that are still over-represented in some inner-city spaces where rehabilitation and gentrification did not yet occur and the potential rent gap has not materialized evidence the only exception to this pattern. However, if centrality and accessibility are much higher in this case, the living conditions experienced in old buildings where overcrowding is the norm does not point to a fair access to urban well-being.

References

Abreu A (2023) Portugal's Inequality regime: many contradictions, multiple pressures. Revista Crítica De Ciências Sociais 130:127–156

Aguiar-Conraria L, Cerdeira ML, Sarrico C (2022) Financiamento Público do Ensino Superior em Portugal. Universidade do Minho e ISCTE—IUL, Lisbon and Braga

Brollo B, Celata F (2023) Temporary populations and sociospatial polarisation in the short-term city. Urban Stud 60(10): 1815–1832

Caldeira T (2000) City of walls: crime, segregation, and citizenship in São Paulo. University of California Press, Berkeley

D'Eramo M (2023) The death gap in New Left Review online, 18 Jan 2023. (https://newleftreview.org/sidecar/posts/the-death-gap

Estevens A (2017) A Cidade Neoliberal. Conflito e Arte em Lisboa e em Barcelona. Le Monde Diplomatique (edição portuguesa)/Deriva Editores, Lisboa

Fernandes J, Chamusca P (2020) Cosmopolitização do centro de cidade, transformação da oferta e uso do tempo, in Cachinho, Salgueiro and Guimarães (org) Comércio. Consumo & Governança Urbana, Lisboa, CEG 2021:31–50

Ferreira P, Lopes MC, Tavares LP (2021) O Salário Médio em Portugal. Retrato atual e evolução recente. Fundação Calouste Gulbenkian, Lisboa

Madeira PM, Vale M (2015) Desigualdade e Espaço no Capitalismo Contemporâneo: uma questão de (in)justiça territorial? Geousp – Espaço e Tempo (Online) 19(2):196–211

Malheiros J, Carvalho R, Mendes L (2012) Etnicização residencial e nobilitação urbana marginal: processo de ajustamento ou prática emancipatória num bairro do centro histórico de Lisboa?. Sociologia, Revista da Faculdade de Letras da Universidade do Porto, Thematic Number: Imigração, Diversidade e Convivência Cultural, pp 97–128

Malheiros J, Madeira PM (2023) Segregação Socioespacial Urbana. In: Santos O, Silva K, Malheiros J (eds) Geografia Urbana. Revisitando Conceitos e Temas. Editora Universitária da Universidade Federal Rural de Pernambuco, Recife, pp 47–62

Maloutas T, Karadimitriou N (eds) (2022) Vertical Cities—micro-segregation, social mix and urban housing markets. cheltenham, Edward Elgar
Marcuse P (1989) Dual city: a muddy metaphor for a quartered city. Int J Urban Reg Res 13(4):697–708
Mendes L (2008) Urbanização clandestina e fragmentação socio-espacial urbana contemporânea: o bairro da Cova da Moura na periferia de Lisboa. Geografia 2:57–82
Mendes L (2017) Gentrificação turística em Lisboa: neoliberalismo, financeirização e urbanismo austeritário em tempos de pós-crise capitalista 2008–2009. Cadernos Metrópole 19(39), 479–512
Mergulhão A (2023) Estudo – *Causa Pública —a Fiscalidade em Portugal*. Causa Pública, Lisbon, Nov 2023
Peck J, Theodore N, Brenner N (2013) Neoliberal urbanism redux? Int J Urban Reg Res 37(3):1091–1099
Piketty, T. (2014) O Capital no Século XXI. Temas e Debates/Círculo de Leitores, Lisboa
Piketty T (2020) Pelo Socialismo! Temas e Debates/Círculo de Leitores, Lisboa.
Rodrigues CF, Figueiras R, Junqueira V (2012) Desigualdade Económica em Portugal. Fundação Francisco Manuel dos Santos, Lisbon
Rosa E (2015) Os Números da Desigualdade em Portugal. Lua de Papel, Alfragide.
Salgueiro TB (1994) Novos produtos imobiliários e reestruturação urbana. Finisterra – Revista Portuguesa de Geografia 29(57):79–101
Salgueiro TB (1999) Ainda em torno da fragmentação do espaço urbano. Inforgeo, Espaços Urbanos 14:65–76
Schmid C (2012) Henri Lefebvre, The right to the city, and the new metropolitan mainstream. In: Brenner N, Marcuse P, Mayer M (eds) Cities for people, not for profit. Critical urban theory and the right to the city. Routledge, London, pp 42–62
Silva K, Carmo A (2023) Justiça Espacial e capital espacial. In Santos O, Silva K, Malheiros J (eds) Geografia Urbana. Revisitando Conceitos e Temas. Editora Universitária da Universidade Federal Rural de Pernambuco, Recife, pp 79–100
Stiglitz JE (2012) The price of inequality. W.W. Norton & Company Inc., New York
Tavares I, Carmo RM (2023) Análise da Pobreza, privação e Desigualdades em Portugal: Tendências Recentes num País Persistentemente Desigual. Observatório das Desigualdades, ISCTE-IUL, Lisbon, December 2023
Wilkinson R, Pickett K (2010) O Espírito da Igualdade. Editorial Presença, Lisbon

Chapter 6
The Construction of Residential Exclusion in Madrid and Its Metropolitan Area

Isabel González-García, Cristina Fernández-Ramírez, and Raquel Rodríguez-Alonso

Abstract Compared to other methodologies for measuring inequality in cities, urban vulnerability has two advantages: it allows its definition to be approached from multiple dimensions, such as housing. In addition, it anticipates the risks of certain urban areas suffering processes of segregation and exclusion. Taking the city of Madrid as a case study and the catalogue of vulnerable neighbourhoods as a reference, the article explains how the real estate market socially and economically orders the city, and how and where important pockets of vulnerability associated with the difficulty of access to housing have been forming.

Keywords Urban vulnerability · Urban inequality · Socio-spatial segregation · Residential exclusion · Deprived neighbourhood · Real state effort

---

This article is the result of research linked to the National Plan project: Strategies for the eco-social transition of large Spanish urban areas in a scenario of climate crisis and scarcity of resources (PID2021-126190OB-I00).

I. González-García (✉) · C. Fernández-Ramírez · R. Rodríguez-Alonso
Urban Planning, Department of Urban and Regional Planning, School of Architecture of Madrid, Universidad Politécnica de Madrid, Madrid, Spain
e-mail: isabel.gonzalez@upm.es

C. Fernández-Ramírez
e-mail: cristina.fernandez@upm.es

R. Rodríguez-Alonso
e-mail: raquel.rodriguez@upm.es

R. C. Lois-González and J. A. Rio Fernandes (eds.), *Urban Change in the Iberian Peninsula*, The Urban Book Series, https://doi.org/10.1007/978-3-031-59679-7_6

6.1 Urban Inequality and Socio-Spatial Segregation: A Long History in the Urban Area of Madrid

The trend towards inequality is a permanent feature throughout the history of cities and is inseparable from them. In fact, inequality has come to be considered one of the greatest challenges of contemporary urban areas (Sevilla-Buitrago 2013). These inequalities are increasingly associated with a growing social segregation that, combined with functional fragmentation—exclusionary zoning, hypertrophy of mobility infrastructures—produces an increasingly segmented and simplified urban space (Graham and Marvin 2002).

The concept of socio-spatial segregation has appeared in urban studies literature to explain "the growing division in society between the haves and have-nots; the socially included and excluded, and a reduction in the size of groups in between" (Hamnett 1998, p. 15). Inequality is neither new nor the result of globalization, although it has been strengthened significantly in recent decades. The philosopher Jean-Jacques Rousseau argued in the mid-eighteenth century that social and political inequality was the result of private property and the abuses of those who appropriate for themselves the wealth of the world and the private benefits derived from this appropriation (Rousseau ed. 2002).

The factors that increase segregation in cities are diverse and may vary according to the nature and intensity of social processes. These, in turn, depend on a wide range of factors. Historically, this phenomenon has been mainly related to three aspects: cultural elements (racial, ethnic, religious), status (income level, power relations) and that related to the division of urban functions (zoning according to urban use) (Marcuse 2022). These aspects overlap and reinforce each other. The most recurrent examples are often related to processes of segregation of disadvantaged groups such as the destruction of shantytowns undertaken by the Brazilian government in the 1960s during its military dictatorship, which led the less well-off inhabitants to seek refuge in other segregated areas (Perlman 2019); South Africa's apartheid laws (1984–1991), during which no African had the right to acquire a plot of land in the whole of South African territory, or was forbidden to hold a legal permit to reside in a city or to take his wife and children into his home (Strauss 2019); or more recently Madrid's shantytowns (1950–1960) caused by marginal urbanization during the Franco era (Burbano Trimiño 2020).

The official discourse of public administrations has taken up this challenge by proposing the need to implement policies aimed at alleviating these effects. For instance, policies for consolidated areas—like the Integrated Urban Regeneration model—insist on the need to promote the social and functional diversity of neighbourhoods and their articulation with the existing city from an integral approach (de Vivienda 2010). However, from other, more critical perspectives, these proposals are based on real estate revaluation strategies that take advantage of urban improvement and requalification interventions, triggering forms of displacement and social exclusion (Castrillo Romón 2014). In other words, spatial transformations that promote the incorporation of higher income groups not only fail to improve the diversity of the area, but also encourage gentrification and the expulsion of the most vulnerable or less integrated population (increases in housing rents, penetration of retail, leisure and, other activities alien to the area's original population, etc.).

All these processes are characterized by non-voluntary exclusion and segregation. However, the segregation of the most advantaged groups is characterized as a voluntary and sought-after process (García 2014). In fact, urban segregation can be said to be both a cause and an effect of inequality (Nel-Lo 2018). Urban segregation transfers social inequalities to the physical space and, in turn, the characteristics of urban areas can accentuate processes of inequality. Ultimately, these dynamics can be understood as mutually reinforcing.

In the Spanish case, this cleansing of urban space has been taking shape since the 1960s with "Francoist developmentalism" through a sophisticated process of residential segregation supported by a hegemonic real estate model and the active support of public authorities. As Roch (2008) suggested:

This process of "cleansing" is so powerful, and so "necessary", that it has not hesitated to undertake the transformation of traditional spaces […] It is completed with a spectacular deployment of another mosaic of similar characteristics but newly designed, supported by new and expensive transport infrastructures, faster and with greater capacity, which lend it an urbanized framework … for the insertion of powerful apparatuses of private centrality, which replace the traditional spaces of collective centrality. This is a space composed of loose pieces that serve as an anti-city since its strict real estate nature suppresses any trace of the traditional city and "liberates" it from all its conditioning factors as well as from any of its potentialities. The model of exclusion is thus established: a dismembered and exclusive body from which any traces of the urban have disappeared. Probably the most refined form of dispossession.

In the last decade of the twentieth century and until the 2008 recession, the segregation process was accelerated, supported by large-scale real estate operations. The subsequent crisis and the recent pandemic crisis have briefly slowed down the process. However, they have incorporated new mechanisms of social exclusion that often continue to rely on dynamics of urban transformation: financialization of housing, new small-scale urban developments on both vacant and built-up areas, changes in uses, new facilities and service spaces "adapted" to emerging lifestyles (commercial and leisure elitization tailored to specific groups, "hipsterization", "smart uses" linked to ICTs, etc.), the expansion of e-commerce, etc. These phenomena are producing new rounds of residential displacement and exclusion even in peripheral districts and fringe areas in large metropolitan areas in Spain. The mechanisms of segregation have intensified in this process, and it seems necessary to develop new tools to identify them and understand how they work.

For decades, in the Madrid area, a growing process of urban segregation has been taking place through a sophisticated mechanism of "social space cleansing" using access to housing and the real estate market. The effects in a city like Madrid, and by extension in its metropolitan area, are the consolidation of a system of "class" spaces. This process has been in the making since the previous century. Consequently, different parts of the metropolitan area have been configured around simplified real estate with a high degree of homogenization and specialization in the different dimensions (social, urban, functional…) (González-García 2013). The predominance of these models, highly dependent on the features of real estate approaches in different historical periods, accentuates ongoing segregation dynamics of both downward exclusion processes (vulnerable, disadvantaged neighbourhoods, the "surplus city", etc.) and upward self-exclusion processes (exclusive areas, "high standing" private housing developments and "alpha" commercial areas).

In the case of the city of Madrid, social segregation has a clear spatial manifestation between the rich north-west and the poor south-east (Fig. 6.1). This border, which the General Plan of 1985, in its preliminary studies, drew for the first time, has been maintained since then under the name of the "poverty line". In fact, with minor variations, it is a persistent frontier that marks two realities which have deepened the differences in recent decades.

This is a historical division that was originally linked to the environmental characteristics of the territory. It runs parallel to the Sierra del Guadarrma (from north-west to south-west along the edge of the Community of Madrid) generating a transversal gradient of wealth from the granite mountainous areas (high environmental quality, abundance of clean fresh water, livestock) to the gypsum plains of the south-east (sewage discharges, soil unsuitable for construction and agriculture).

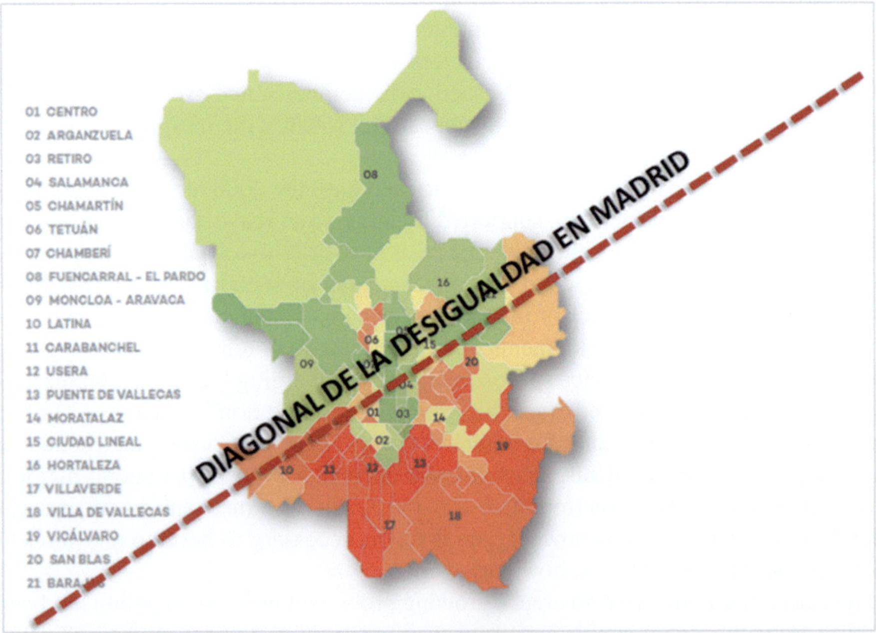

Fig. 6.1 Historical diagonal of inequality in Madrid that divides the rich north-western districts from the poor south-eastern districts. *Source* Decentralization and Territorial Action, Madrid City Council (2018)

This division will remain in the territory and will be reflected in the city of Madrid with the complicity of urban planning that will consolidate this trend. Thus, the General Plan of Madrid of 1946 (known as the Bidagor Plan, named after its promoter) consolidated the model, with a residential north-west for the middle and upper classes of the city compared to a south and south-east relegated to a service space to house industry, worker accommodation, water purification facilities, waste dumps, etc.

This pattern has been maintained and even reinforced since then and is shown repeatedly in all analyses related to inequality and segregation.

6.2 Urban Vulnerability, a Multidimensional Process of Risk of Exclusion and Disadvantage.

According to the dictionary, *vulnerable* is an adjective that defines something or someone that can be damaged, hurt or receive physical or moral injury. It therefore refers to two aspects: on the one hand, to a certain susceptibility to be affected by some circumstance (related, therefore, to the endogenous characteristics of the subject or object); and, on the other hand, to the intensity of the affection and its capacity to produce damage or injury (related to circumstances exogenous to the subject or object). Applied to city and urban areas, as a social space, there seems to be a consensus in understanding urban vulnerability as "the potential for the population of a specific urban space to be affected by some adverse circumstance, so that the concept alludes not so much to the existence of a critical situation that is currently verified as to the existence of certain conditions of risk, fragility, disadvantage that would make it possible to enter into that critical situation" (Hernandez Aja 2007).

In a further elaboration of this concept, Hernández Aja states that urban vulnerability refers to "that process of unease in cities produced by the combination of multiple dimensions of disadvantage, in which any hope of upward social mobility, of overcoming their social condition of exclusion or close to it, is seen as extremely difficult to achieve" (Hernandez Aja 2007).

This is also interpreted by the United Nations when it considers vulnerability "as a state of high exposure to certain risks and uncertainties, combined with a diminished capacity to protect or defend oneself from them and to cope with their negative consequences" (UN 2003).

In the specialized literature, there also seems to be a certain consensus that urban vulnerability can be approached from at least two perspectives (De Santiago 2010): a more objective or measurable perspective based on certain parameters related to the conditions of social disadvantage that express the structural disadvantages of a population to develop vital projects; and a second more subjective perspective, related to a psychosocial situation that derives from the perception that citizens have of the territorial area where they develop their lives and their own social conditions.

In this sense, vulnerability could be understood as a relative, contextual concept, which must be framed in a specific territory.

In line with all this, there are numerous proposals to assess, measure and identify populations or individuals who, as a consequence of the combination of a series of conditions, are immersed in a multidimensional process of disadvantage and vulnerability.

In general, these proposals are formulated from two different approaches: starting from the individuals and groups affected (person-based) or starting from the territorial spheres where these conditions of vulnerability are concentrated and fed back into each other (place-based). In the field of both urban studies and urban planning and policies, the second approach predominates, with numerous proposals having been developed for the delimitation of vulnerable areas and/or neighbourhoods in Europe. Among these numerous initiatives, it is important to highlight the vulnerability observatories (Córdoba Hernández et al. 2021) due to their importance and the possibilities for the design of urban policies that counteract these disadvantage processes.

With regard to the conditions that must affect a territory or individual in order to be considered vulnerable, although this is a multidimensional combination of aspects that must be framed in a specific territory, as has already been stated, most methodologies and studies on the subject show that the lack of access to a basic level of material conditions is one of the determining conditions of vulnerability. This means that the concept of vulnerability is directly related to income level, access to the labour market, educational level and access to adequate housing. From this point of view, vulnerability can be understood as a stepping stone to poverty and social exclusion.

Since 1996, the Spanish National Authority has promoted a continuous study on urban vulnerability in Spain, which has resulted in the Observatory of Urban Vulnerability[1] since 2009. Since its inception, our Research Group (Grupo de Investigación en Arquitectura, Urbanismo y sostenibilidad_GIAU + S,[2] de la Universidad Politécnica de Madrid) has been in charge of feeding the observatory through an agreement with the Ministry. The concept of urban vulnerability, which has been used, has been built around two dimensions. The first of these is based on the conditions of structural disadvantages of a population to develop vital projects in contexts of security and trust. The second dimension considers the psychosocial conditions that affect citizens' perception of the territory in which they live. For this reason, there are four perspectives (Hernández Aja et al. 2014):

- *Socio-demographic vulnerability* refers to demographic phenomena that influence the increase in social vulnerability in terms of population characteristics and dynamics. Selected indicators would cover key areas exploring demographic dynamics, migration, ageing, family structure, gender, youth population, health, security and social networks in urban contexts. Other possible indicators would

[1] https://www.mitma.es/arquitectura-vivienda-Y-suelo/urbanismo-Y-politica-de-suelo/observatorio-de-la-vulnerabilidad-urbana.

[2] http://vps181.cesvima.upm.es/

be related to household structure, where the gender perspective should be at the centre of the characterization of the population, together with the child and youth population. A third set of indicators will be related to residents' quality of life and the presence of physical and psychological illnesses.

- *Socioeconomic vulnerability* addresses the capacity of households to ensure material and emotional well-being, linked to unemployment, job insecurity, poverty and educational levels. The main indicators would be related to income and working conditions. Those related to youth (representation in the labour market, emancipation rate or housing tenure regime in relation to young people) and social welfare services (persons, families or households receiving social assistance in terms of pensions, subsidies, housing, etc.) should also be included. Indicators on poverty exclusion (with respect to the pre-set poverty line), on the digital divide and on the network of the public as well as private and third sector resources should also be included.
- *Residential vulnerability* refers to housing as a physical support. Thus, a degraded housing stock can lead to the appearance of conditions of vulnerability. The indicators to be considered are those related to the surface area, the functional and constructive characteristics of the dwellings, age, state of conservation, property value, tenure (ownership by individuals or legal entities, dwellings owned by public administrations, rental, other regimes), situations of a financial or mortgage nature (outstanding payments, evictions…) or questions related to the level of use and occupation of the residential stock (empty dwellings…). In recent analyses, as a consequence of real estate bubbles, an indicator related to the economic effort that households make to buy or rent housing in relation to their disposable income has gained strength.
- *Subjective vulnerability* is related to residents' perception of their urban and social environment. This dimension reflects urban malaise as well as influences the image of the neighbourhood in a wider urban context. The analysis of this subjective vulnerability is crucial to promoting public policies that address the real needs of communities, prioritizing territorial balance and citizen participation. Some of the possible indicators would be related to noise, pollution, mobility and accessibility to services, the quality of public space-green areas or subjective safety.

Residential vulnerability can therefore be considered as one of the dimensions of vulnerability which we will analyse in depth later on, but first it is worth taking a brief look at how urban vulnerability has been measured and delimited in Spain from these perspectives.

6.3 Measuring Urban Vulnerability in Spanish Cities: The Catalogues of Vulnerable Neighbourhoods (1991–2011): Case Study—Madrid

The various Catalogues of Vulnerable Neighbourhoods in Spain (1991, 2001 and 2011) (Hernandez Aja et al 2018) were developed between 1996 and 2017 through collaboration agreements between the Government of Spain, through the Ministry responsible at the time for housing and urban policies, and the Polytechnic University of Madrid (Juan de Herrera Institute/giau+s). Technical assistance for data processing was provided by the company Analística for the first 1991 Catalogue, and by Argea Consultores for those of 2001 and 2011. The statistical data used were provided by the National Statistics Institute (INE).

The aim of the Catalogues is to locate neighbourhoods in which the conditions of residents are significantly worse than those of the Spanish population as a whole. The Catalogues constitute a reflection on urban vulnerability from an urban planning perspective, focusing on the identification and analysis of specific areas that are in a worse situation than the rest of the Spanish urban fabric. These areas may constitute pieces which, due to their size, autonomy and homogeneity, may be the object of intervention plans and/or programmes to improve their position.

The Catalogues also provide an overview of the urban vulnerability of Spanish cities over two decades, allowing an analysis of both the imbalances existing in a city, region or the country as a whole, and their evolution over the period.

The first two catalogues, as they included field work, made it possible to get to know the cities analysed and to establish contact with the municipal technical services. They constituted a training laboratory for town planners at the Madrid School of Architecture, as they required intensive work carried out over several years by a large group of students, recently qualified architects and professors. Some of the team members have participated in the elaboration of more than one of the Catalogues.

The methodology of the first two Catalogues, 1991 and 2001, is the same, with variations and adjustments in the definition of some of the indicators, depending on the information available in each Census. In the 2001 Catalogue, the contents were extended by incorporating more detailed information for each of the delimited areas, as well as specific reports for each Autonomous Community and city.

In 2013 work began on the 2011 Catalogue. The methodology had to be revised in depth with respect to that used for the 2001 data due to changes in the preparation and availability of information for the 2011 Census. While the two previous censuses, 1991 and 2001, were based on universal fieldwork, the 2011 Census was based on a sample, which prevented the availability of some of the variables at the census section scale. Furthermore, in the latest edition of the catalogue, fieldwork was dispensed with and all the research was carried out in the office, with the full incorporation of Geographic Information Systems (GIS) and database management, which in previous works were still at an embryonic stage.

The three Catalogues were produced for cities with more than 50,000 inhabitants and provincial capitals.

The delimitations of Vulnerable Neighbourhoods are carried out by means of a double process of statistical and spatial urban analysis, following the methodology below.

- Definition of the Basic Indicators of Urban Vulnerability (IBVU).
- Obtaining the variables disaggregated by the census section that allows the IBVU to be calculated.
- Establishment of the Reference Values (RV) for the IBVU that allows an area to be considered vulnerable.
- Delimitation of Vulnerable Statistical Areas (VSA), by calculating the core indicators for census tract groupings exceeding any of the reference values.
- Delimitation of Vulnerable Neighbourhoods (VN). Spatial areas have been delimited with a sufficient degree of homogeneity and spatial continuity to be able to intervene with the tools of urban planning. To this end, the VSAs were reviewed by means of desk- and fieldwork. Their delimitation was verified and contrasted by the municipal technical services. The aim was to encompass urban fabrics without significant spatial discontinuities, so that the end result would be a Deprived Neighbourhood, on which socio-urban projects could be developed.

The three indicators selected as Basic Indicators of Urban Vulnerability (IBVU) were level of education (IEST) and unemployment (IPAR) (both related to socioeconomic conditions) and building deficiencies (IVIV) (referring to the quality of housing).

The dimensions of vulnerability analysed have been maintained in the three Catalogues, for simplicity, ease of interpretation and communication, and to allow temporal comparison. The definition of the indicators is based on the proposal made in 1996 by the OECD in which two indicators were proposed: unemployment rate and income level. The first could be reproduced as data were available, but, for the second, statistical information was not available in Spain, so it was decided to use two proxy variables that showed a high correlation with income: the level of education and housing deficiencies.

To determine the vulnerability of an area, Reference Values (VRs) were established at the national level for each of the indicators and years. Thus, the areas detected as vulnerable are those that present worse indicators than the state (Table 6.1).

Table 6.1 Criteria for vulnerability levels of basic vulnerability indicators

Vulnerability	IEST y IPAR	IVIV
0	< 1	< 1
A	$\geq$ 1 and < VR. Reference value	$\geq$ 1 and < VR. Reference value
B	$\geq$ VR. Reference value and < 2.5	$\geq$ VR. Reference value and < 4
C	≥ 2.5	≥ 4 y < 8
D		≥ 8

The multidimensional nature of vulnerability is captured with the use of three indicators, two of them referring to the socioeconomic conditions of the population and the other to those of residential support. In each delimited area, one or more of these dimensions of vulnerability may occur.

The intensity levels are evaluated through a classification system that allows grading the various situations, from the mildest, in risk positions before real consequences may occur for its inhabitants, to the most acute, in which the population is already in a situation of social exclusion.

Once the levels for each basic indicator are established, the classification is carried out in two phases using double-entry matrices. In the first, the area is classified based on its level of socioeconomic vulnerability, through the intersection of the levels of the two indicators referring to these dimensions, educational level and unemployment (Table 6.2).

This sociodemographic classification allows us to characterize the populations that live in each of the delimited neighbourhoods based on their capacity to face a possible crisis, through their relationship with employment and training, associated with economic and relational integration. The Catalogue seeks to expand these dimensions by incorporating the quality of housing since the characteristics of housing can increase the vulnerability of some populations compared to others which are better housed. To do this, the level of socioeconomic vulnerability of the population is crossed with the level of residential vulnerability, thus obtaining the global classification of the degree of vulnerability of the area.

The global vulnerability classification is defined, therefore, as the intersection of socioeconomic vulnerability, understood as structural vulnerability, with residential vulnerability, understood as contextual vulnerability, so that the latter can contribute to increasing the vulnerability of the population depending on the quality of the housing in which they live. The system contemplates four degrees of vulnerability: Mild (VL), Medium (V), Severe (VS) and Critical (VC) (Table 6.3).

In terms of the results obtained (Table 6.4), the scope of the Catalogues was 116 cities in 1991, 139 in 2001 and 147 in 2011. There were 370 vulnerable neighbourhoods in 81 cities in 1991, 6216 vulnerable neighbourhoods in 103 cities in 2001, and 918 vulnerable neighbourhoods and statistical areas in 132 cities in 2011. While

Table 6.2 Classification of vulnerable neighbourhoods according to their degree of socioeconomic vulnerability

Unemployment Ipar			0	A	B	C
			0-MN	MN-VR	VR-2.,5MN	>2.5M
Educational level IEST	0	0-MN	NV	NV	VL	VS
	A	MN-VR	NV	NV	V	VS
	B	VR-2,.5MN	VL	V	V	VS
	C	>2,.5MN	VS	VS	VS	VC

MN national average; *VR* reference value; Degrees of vulnerability: *VC* critical; *VS* severe; *V* medium; *VL* mild

Table 6.3 Global classification of vulnerable neighbourhoods according to their degree of vulnerability

		Residential Vulnerability_IVIV-				
		0	A	B	C	D
		0-MN	MN-VR	VR-4MN	4MN-8MN	>8MN
Socioeconomic vulnerability	NV. Non vulnerable	NV	NV	VL	VL	V
(Educational Level Unemployment)	VL. Mild vulnerability	VL	VL	VL	V	VS
	V. Medium vulnerability	V	V	V	VS	VC
	VS. Severe vulnerability	VS	VS	VS	VC	VC
	VC. Critical vulnerability	VC	VC	VC	VC	VC

MN national average; *VR* reference value; Degrees of vulnerability: *VC* critical; *VS* severe; *V* medium; *VL* mild

the Catalogues increased their scope of study on all three dates analysed, the population living in vulnerable areas and their spatial concentration increased further. The population analysed increased by just over 22% between 1991 and 2011 and the population in vulnerable areas by over 136%. The number of vulnerable neighbourhoods increased even more significantly, from 370 in 1991 to 918 in 2011. In 2011, almost 6.7 million people lived in vulnerable neighbourhoods, 27% of the population analysed, compared to 2.8 million and 14% in 1991.

The territorial distribution of vulnerability follows the known patterns of socioeconomic inequality in the geography of Spain. The northern half of the country concentrates the lowest vulnerability indices, housing less than 10% of its urban population in neighbourhoods and delimited areas (this is the case of Asturias, Euskadi, Navarra, La Rioja, Asturias and Aragon), while the highest indices are found in the autonomous city of Ceuta and the Canary Islands. Andalusia improved between 1991 and 2011, although it still had one of the highest rates in Spain in 2011. Cataluña, Comunidad de Madrid and Comunitat Valenciana were the regions where vulnerability worsened the most, and, of these, only Madrid was below the national average in 2011.

The results for the Madrid area are shown in the map below (Fig. 6.2). In 2011, the average values of three of the basic indicators of vulnerability in the city of Madrid show a reality that, in general, presents better conditions than the Spanish average.

The location of vulnerable neighbourhoods follows the historical logic of sociospatial segregation in Madrid. In 2001, some neighbourhoods appeared in central areas of the city linked to the poor state of residential buildings that have subsequently disappeared, thanks to rehabilitation processes. The new vulnerable areas are located in the south-west–east arc, between the A5 and A2 motorways. The vulnerable neighbourhoods that remain in the so-called *Almendra Central* (central

Table 6.4 Study area and vulnerable neighbourhoods and statistical areas detected in 1991, 2001 and 2011

	1991	2001	2011	Increase		
				1991–2001 (%)	2001–2011 (%)	1991–2011 (%)
Total population	38,872,268	40,847,371	46,815,916	5.1	14.6	20.4
Population analysed	20,049,737	21,550,443	24,534,189	7.5	13.8	22.4
Population in BBVV	2,832,080	4,441,041	6,697,400	56.8	50.8	136.5
No. cities analysed	116	139	147	19.8%	5.8%	26.7%
No. cities with BBVV	81	103	132	27.2	28.2	63.0
No. BBVV	370	621	918	67.8	47.8	148.1

BBVV Deprived Neighbourhoods

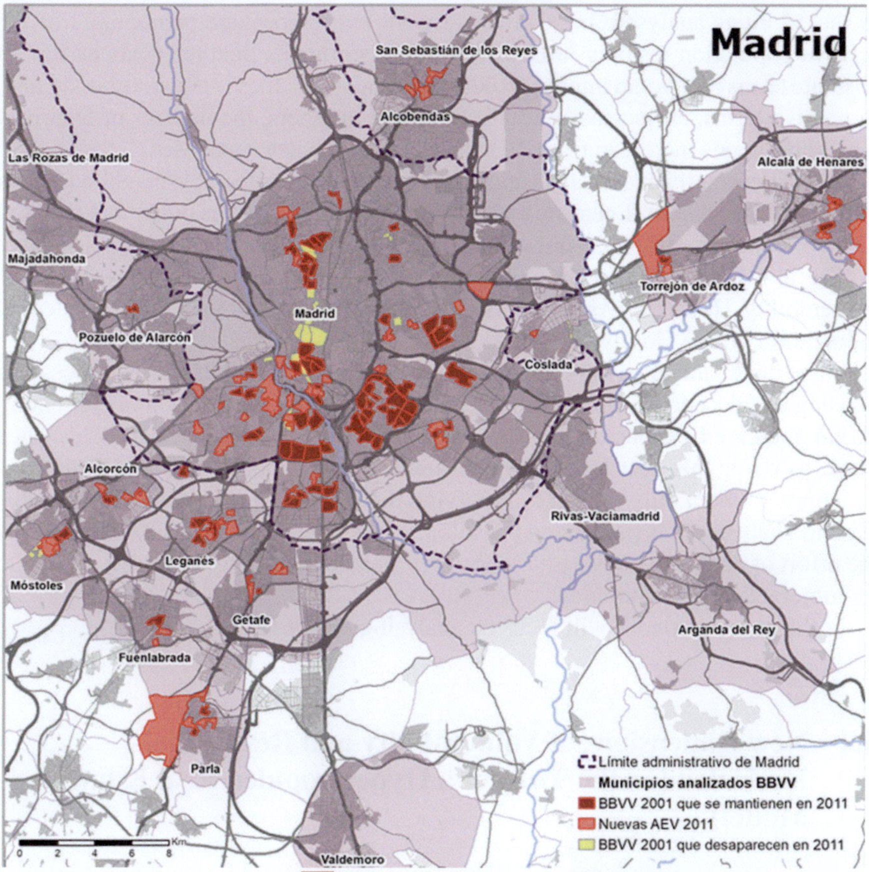

Deprived neighbourhoods identified in 2001 and maintainedin2011 | New deprived areas identified in2011 | Deprived neighbourhoods identified in 2001 that disappearedin2011

Fig. 6.2 Delimited deprived neighbourhoods in the Madrid area 1991–2001–2011

area of the city, inside the M-30 ring road) correspond to areas historically related to the working classes and low-income neighbourhoods.

In addition to the catalogue of vulnerable neighbourhoods, a detailed analysis of social and residential vulnerability has been carried out in 13 Spanish cities, including Madrid.

To analyse residential vulnerability, two indicators were selected: housing accessibility (referring to collective housing properties without a lift) and the state of conservation of the building.

In 2011, 24.87% of Madrid's primary dwellings were in buildings with three or more floors without a lift, compared to 14.08% in Spain and 20.27% on average in the 13 cities analysed. In 2011, there were two percentage points more of the relative population residing in Deprived Neighbourhoods than ten years earlier. In absolute terms this would imply 115,000 new residents in these neighbourhoods, and, in relative terms, an increase of 19%. In global numbers, the number of Deprived Neighbourhoods increased from 79 to 91 between the dates analysed.

Work is currently underway to update the catalogue of vulnerable neighbourhoods with data from the 2021 census, which is not a universal census either, but is based on a combination of sample and administrative records.

The aim is to maintain the possibility of analysing the temporal evolution of vulnerability. However, work is being done on new indicators that reflect residential vulnerability more closely.

Since the financial crisis of 2008, the historical problem of difficulty in accessing housing and decent accommodation, whether through purchasing a home or renting it, has increased exponentially. The economic effort that many households must make to cover this cost has brought to light a new dimension of residential and economic vulnerability related to satisfying the need for accommodation. For this reason, analysing this process and developing a vulnerability indicator that reflects the effort of access to housing has been seen as essential for updating the catalogue of vulnerable neighbourhoods.

In the next section we will focus on analysing this process.

6.4 New Dimensions of Vulnerability and Residential Exclusion [or the Twist of a Hypertrophied Real Estate Model]

Much has been written about the hypertrophy of the Spanish property market and the dismantling of the already meagre material economy. The new notion, which gained strength with the birth of the twenty-first century, is based on the model of the global city and immaterial economies. In the Spanish case, it was based on the deployment of a real estate machine that is capable, on the one hand, of moving large amounts of capital and, on the other, of generating employment for the social strata coming from the factory economy.

This "new ideology of always" (Roch 1993) has not only generated a notion of great social roots but has also managed to modify the socio-residential physiognomy. The result is a sort of social mosaic with a tendency towards homogeneity, where lower incomes are displaced towards the most distant and least desirable tesserae of this mosaic. The capacity to choose is subordinated to the income-location relationship, while the social metabolism is oriented towards extractivism and social and territorial polarization (Naredo 2022).

For all these reasons, the analysis of the functioning of the real estate market is presented as one of the most interesting approaches to the problem of differential access to housing. This implies the conversion of the product, housing, from a basic necessity into a shelter product for income (from family income to, of course, high investment) with help from both institutions and the financial market (Rodríguez and Espinoza-Pino 2017).

In this context, the term "real estate city" was coined (Roch 2008: 117) to define the conversion of a space that should be the representation of the social space, with its conflicts—as a complex collective construction with its rules of composition, more or less appropriate—to conversion into a financial product, an exchange good. This process delves into pre-existing socio-spatial segregation and urban vulnerability.

The fact that cities have privileged spaces and others that carry the weight of a certain stigmatization could be understood as linked to more or less objective conditions, such as centrality, the better or worse quality of parks built or the proximity to public spaces, interest, quality natural areas or great facilities. For these reasons, we must add public interventions that generate and enhance new centralities and spaces of desire (Castrillo et al. 2014). This has become a kind of increasingly refined spatial classification. The crisis of the economic-financial system and the increase in unemployment showed, in a stark way, the subordination of the income of the inhabitants of Madrid to the payment of housing. This triggered a wave of evictions and a highly vulnerable social situation. As a consequence of all this, currently, not only has the gap not been reduced, but, in light of the evolution of real estate values, we find ourselves facing a more segregated Madrid. Socio-spatial mobility is curtailed and urban spaces, consolidated as aspirational economic representations, appear increasingly stereotyped.

What has happened to housing prices? (Fig. 6.3). Through the offers published on real estate portals over the last twenty years, it is possible to analyse the evolution of second-hand housing prices. This exercise offers us an image of the valuation, seen as natural, of the real estate stock, and ends up telling us (as if it were a self-fulfilling prophecy) what the social valuation, converted into price, of the city's spaces is. In a first analysis of the time series, we see how the century begins with an upward sequence of prices supported by the vision of housing as the greatest refuge for family income and supported by the ease of credit that triggered, in 2008, the first great crisis of the recently inaugurated century. Despite this, the twenty-one districts of the city have a very stable location in the series, with a maximum unit value (€/m^2) of the city represented by the district of Salamanca in most years and a unit value minimum in the Villaverde district. The first coincides with the place where historically the highest income population of the city has been located, while Villaverde is one of the districts on the southern periphery of the city, where industry and workers' accommodation have also historically been located.

The series begins with a certain constancy in the distance between the smallest value and the largest value. Thus, the second duplicates the first. Around 2004, the Villaverde district, accompanied by all those that make up the south-south-east of the city, managed to reduce its distance with those of greater value, with growth of between 20 and 30% annually. This, far from signifying the closing of the historical north–south fracture, reflects the titanic effort of the southern districts, boosted by the ease of access to credit, and the increase in the dependence of income on mortgage payments in these districts. In the following years, 2005 and 2006—in which the

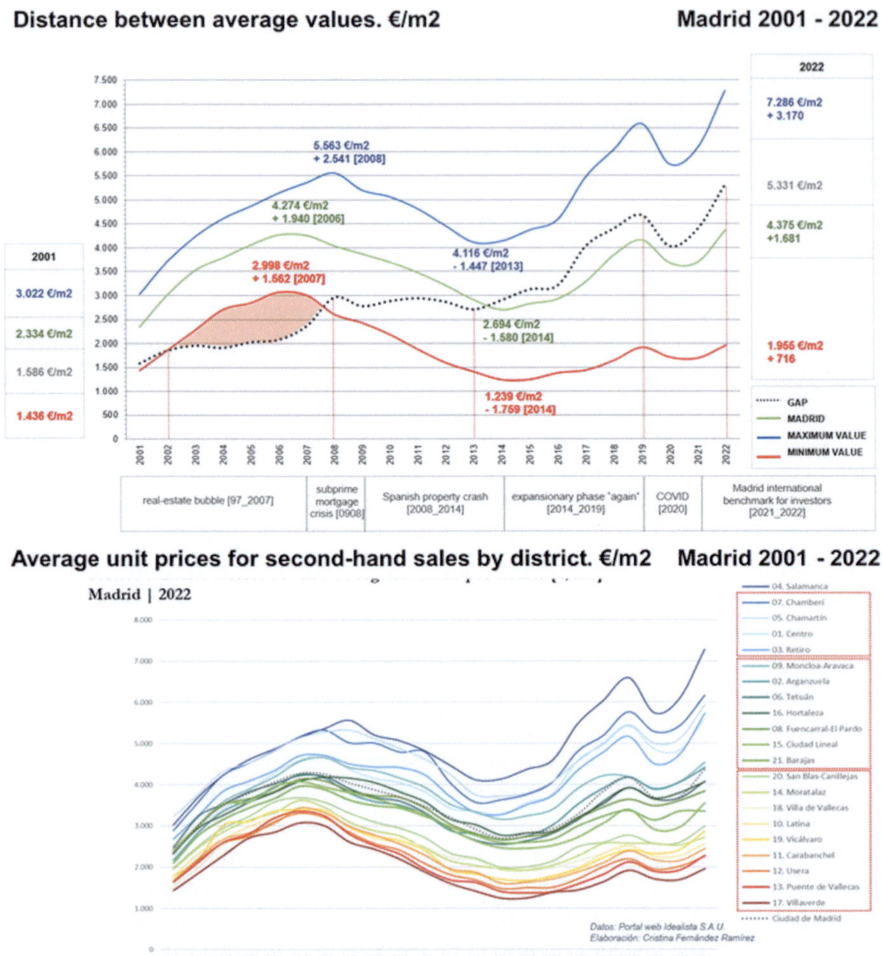

Fig. 6.3 Distance between average unit prices (€/m^2) and evolution of average unit prices (€/m^2) of second-hand housing in Madrid by districts during 2001–2022. *Source* Cristina Fernández, based on data from the web portal Idealista real estate SAU

highest price was reached in the Villaverde district (€3000/m^2)—an even growth of around 6% was maintained until 2007. From that moment on, first in a gentle way and then with rates that exceeded 10%, there was a sharp drop in prices in the southern districts (with rates that reached 15% between 2014 and 2015). Although this drop was generalized throughout the city, it had lower percentages in the rest of the districts. For example, in the district of Salamanca, decreases above 10% were never achieved and with a time lag of two years.

Starting in 2017, the price distance between districts resumed the same path taken between 2007 and 2009, with a return to increases above 10% in the southern districts and above 15% in the most central ones, reaching 20% in the case of the Centro district.

It is notable that in the years prior to the bursting of the bubble, districts such as Villaverde had already reached the maximum point of their valuation, establishing a ceiling that could hardly be exceeded by the income that accessed these homes (€3000/m^2, which corresponds to a cost of €180,000 for a home of just 60 m^2 built, 50 m^2 usable).

Analysing the complete series, we verify that a sort of classification into blocks (clusters) has occurred that is clearly shown in recent years. The average unit prices (€/m^2) show that in 2001 there was a hierarchical structure where the most prestigious districts took first place, and those that were more peripheral, poorly built or poorly communicated, were located at the base of the pyramid. This, which perfectly describes what we had called the Fordist city, becomes, at the end of the process, the image of a city that has deepened the dynamics of socio-spatial segregation. The possibilities of socio-spatial mobility have been reduced through the generation of blocks of value, blocks of income.

We can see defined blocks already in 2009, with a higher group formed by the districts of Salamanca, Chamartín and Chamberí, with a jump of 12% until Retiro. The latter, together with Moncloa, Centro and Hortaleza, once again establish distances with respect to a third block, where the districts of Fuencarral, Tetuán, Ciudad Lineal, Arganzuela and Barajas are located. The fourth and last block is made up of San Blas, Moratalaz, Villa de Vallecas, Carabanchel, Vicálvaro Latina, Usera, Puente de Vallecas and Villaverde.

From the recovery of the sharp falls that had occurred around 2008, between the years 2012–2013, a path of much more pronounced growth began in the districts with the highest income, which led to a definitive separation between the upper block and the rest, until a significant drop occurred in 2020 in the context of the COVID-19 pandemic, whose recovery to previous values occurs just a couple of years later.

Recent studies (Musterd et al. 2017) show that socio-residential restructuring processes do not occur at the same speed as changes in the social structure. Increases in social inequality are followed by an increase in residential segregation with lags of up to ten years. The speed of the process will be closely related to the structure of the tenure regime present in the area. Ownership without pending payments, inheritances and other ways of accessing housing outside the market generates greater resistance to the social structure of a space. Added to this is the capacity for the intervention of public administrations to curb the excesses of the market or the global financial context and their need to seek refuge spaces for their income.

6.5 Segregation and Real Estate Effort

The *real estate effort indicator* aims to measure the financial effort that households make to access a home in a specific location, in this case, in a census section of the more than 4,000 in the Community of Madrid. This access effort is a very illustrative indicator since it relates the income of the resident population in a territorial context to the price of housing in the same place. That is, it shows us the ability or inability of residents to remain in the space they inhabit, with all the consequences that this can generate in terms of difficulties in taking root in a space or losing previous networks.

The real estate effort indicator for rental access is of great value to detect the spaces in the city where the population may see its continuity compromised, due to the tension generated by the voracity of the real estate market. The rent gap expected by the owners could cause the expulsion of their current residents who are looking for tenants with a higher income level to continue with the upward trend.

Access to housing under ownership has been the majority for decades in Spain. According to data from the population and housing censuses (INE), in 1950 the percentage of home ownership did not reach half of the homes (49%), rising each decade until reaching its peak in 2001 with 82%. From that moment on, it began to decrease, reaching 78.9% in 2011 and 75.8% in Spain and 63.8% in the Community of Madrid in 2021. In the specific case of the Community of Madrid, we find an average percentage of rental tenure of 17.8% of the homes, which increases to almost 18.5% in the metropolitan area (14.5% in the metropolitan south, 15.4% in the metropolitan east, 17.7% in the metropolitan north, 16.3% in the metropolitan west and more than 20% in the municipality of Madrid). The increase in the percentage of rentals and the lack of official territorialized data on sales prices make the indicator increasingly consistent.

The prepared indicator uses the median monthly rent of the property, considering the typology of collective housing as the most representative of the Madrid metropolitan area, with respect to the median income per consumption unit. We understand that this income parameter is the one that best adjusts to household income when considering the composition by age and situation and its contribution to the economies that make up the household.

To set the reference values for the real estate effort, specific bibliography and official sources have been analysed. Thus, regulatory entities such as the Bank of Spain consider a household's debt level greater than 30% of total net income as risk, and can increase it to 35% if they do not have other acquired debts. Other official indicators, such as those collected in the Living Conditions Survey, use 40% of disposable income as a limit reference, considering, as expressed in their methodology, tax, financial or maintenance expenses. Therefore, in our case, it would be reasonable to consider 30% or 35%, given that only the price of the property is used, without including associated expenses.

In this analysis (Fig. 6.4), an acceptable real estate effort has been considered below 30% of the income, conditionally acceptable up to 35%, high up to 40%, at risk up to 45% and critical if it exceeds this value.

The area that brings together a greater percentage of census sections with a high rental real estate effort (ranging 40–45%) or very high (more than 45%) appears in the municipalities of the metropolitan south (14.5%), in the districts of the southern periphery of the municipality of Madrid (17.2%), the eastern districts (10.6%) and in the Central Almendra (13.1%). With a more careful look, we can observe the tension in the districts of the southern periphery of the city, especially those bordering the

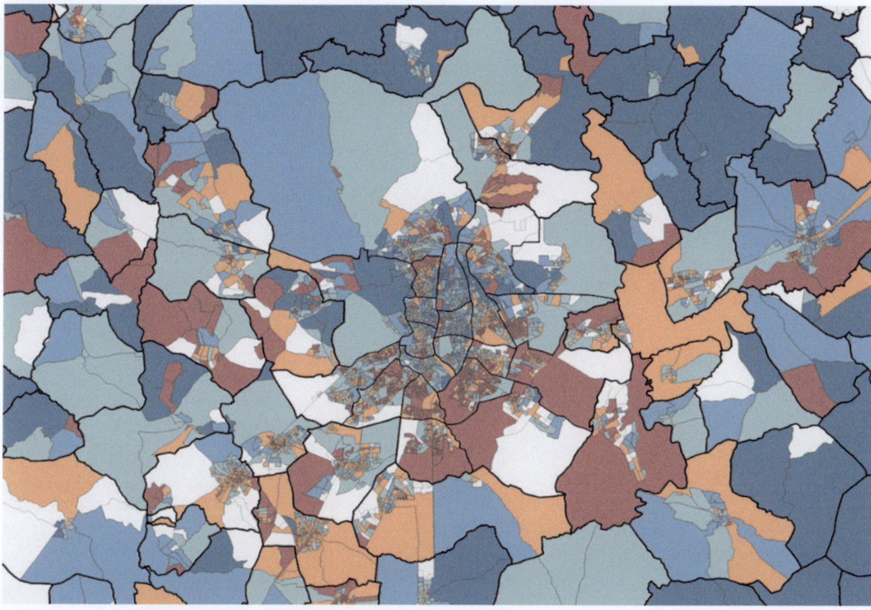

Fig. 6.4 Real estate effort on a rental basis. 2021. Comunidad de Madrid. *Source* Own elaboration based on data from the Atlas of Household Income Distribution and the State System of Housing Rental Indices

central area, where we find a greater concentration of census sections that exceed 45% of the effort to pay rent.

In view of the data, it seems clear that the real estate effort of rental access is relevant data, but knowing the lack of homogeneity in the distribution of the rental percentage in the Madrid region, it is necessary to incorporate a new indicator that allows us to measure the impact of this tenure regime on the territory. This would allow us to know the relevance of renting as a tenure regime and, therefore, its exclusive capacity. Therefore, we will talk about.

- High risk, when the real estate effort exceeds 35% in census sections with a rental tenure regime higher than the Madrid average.
- Increasing risk, when the real estate effort exceeds 35% in census sections with a rental tenure regime lower than the Madrid average.
- Limited risk, when the real estate effort is less than 35% in census sections whose rental tenure regime is higher than the Madrid average.
- Low risk, when the real estate effort is less than 35% in census sections whose rental tenure regime is lower than the Madrid average.

The final result (Fig. 6.5) is that the high risk (coincidence of a real estate effort greater than 35% and a percentage of rental housing greater than 18.7%) colonizes an important part of the south-eastern arc of the region and a large part of the Almendra Central. The central districts of Centro, Chamberí and Tetuán stand out in a notable

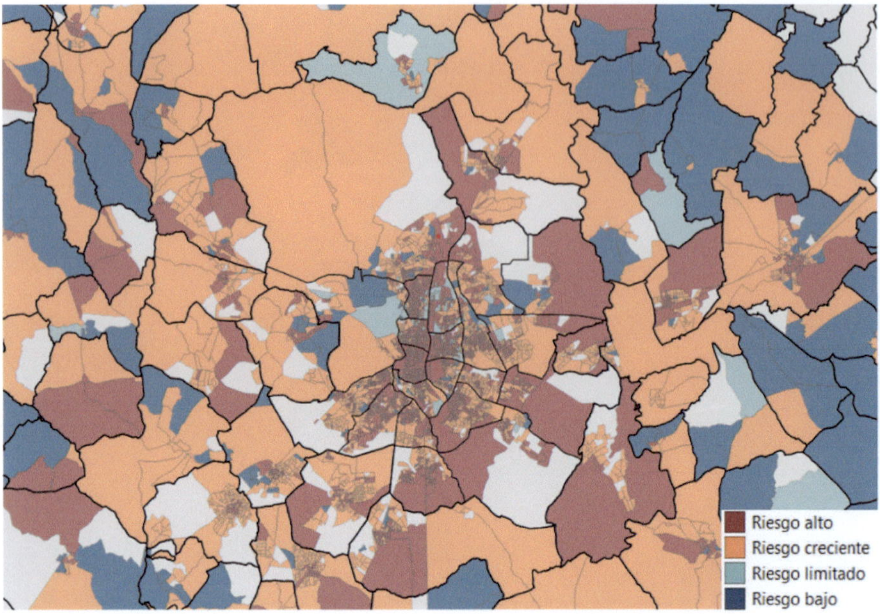

Fig. 6.5 Real estate effort on a rental basis, 2021. Comunidad de Madrid. *Source* Own elaboration based on data from the Household Income Distribution Atlas, State Housing Rental Indices System and 2021 Population and Housing Census

way. The ability to radiate both from the Central district and from structuring roads such as Paseo de la Castellana to the north or Alcalá Street to the east, dispersing its influence to the San Blas-Canillejas district, can be observed.

Among the municipalities of the southern arc, the situation of the first ring (Getafe, Leganés and Alcorcón) stands out, with a high risk in the oldest centres and closest to the commuter train network and increasing in new developments. This process is reproduced in the most remote municipalities such as Fuenlabrada or Parla and with less intensity in Móstoles, Pinto or Valdemoro.

The Henares corridor, which in the previous analyses did not seem to have intense effort rates, in this case, after incorporating the presence of the rental tenure regime, shows high-risk situations. These areas are directly related to the commuter train route. The municipality of Alcalá de Henares, the metropolis centre of the corridor, appears as a space with great real estate risk when it comes to rentals.

Finally, it should be noted that in the territorialization of the indicator, spaces with high incomes appear associated with a critical risk, which we can attribute to one extent or another to the relationship between high rental prices, the difficulties of statistically describing high incomes (despite that the data is derived from tax records) and the incorporation of globalized incomes in the market of these spaces, which is also generating processes of substitution of the resident population.

6.6 Some Concluding Thoughts

As this article has shown, urban vulnerability makes it possible to measure inequality in cities using multiple interrelated variables that seek to address the complexity of the problem. Moreover, by the methodological approach itself, it is intended to go beyond the mere description of a process at a specific moment in time.

One of the objectives is to anticipate the risks to which certain urban areas and the population living there are subjected, which can serve as a tool for public policies to anticipate processes of segregation and exclusion. Housing in its many facets has always been one of the basic pillars of integration in today's society. Therefore, it becomes a fundamental indicator for defining urban vulnerability. The way in which residential vulnerability is measured has changed over the 40 years covered in the article.

Today, dwellings lacking some basic services such as water are very rare, so they can be considered outdated for an analysis of this type. However, aspects such as insecurity of tenure or difficulty of access to housing have become everyday problems that affect a significant percentage of the population.

In the current context of the reconfiguration of the real estate system as a capitalist actor (in terms of agents and ways of proceeding), of changes in the forms of access to housing (from ownership to renting), of profound and growing social inequality in cities, the analysis of vulnerability could be a fundamental tool in the service of public policies. It would make it possible to slow down or remedy undesired processes and even eliminate the negative effects of certain interventions in the city.

References

Burbano Trimiño FA (2020) La urbanización marginal durante el franquismo: el chabolismo madrileño (1950–1960). HISPANIA NOVA. Primera Revista de Historia Contemporánea on-Line En Castellano. Segunda Época 301. https://doi.org/10.20318/hn.2020.510

Castrillo M, Matesanz Parellada A, Sánchez Fuentes D ,y Sevilla Buitrago A (2014) ¿Regeneración urbana? Deconstrucción y reconstrucción de un concepto incuestionado" Papeles de relaciones ecosociales y cambio global Núm 126:129–139

Castrillo Romón MA (2014) Urbanistas "en transición"? El enfoque antropológico del mejoramiento urbano y sus efectos sobre la concepción urbanística. *Ciudad y territorio: Estudios territoriales*, ISSN 1133-4762, N° 179, (Ejemplar dedicado a: Por la rehabilitación, la regeneración y la renovación urbanas), pp 81–97

Córdoba Hernández R, González García I, Guerrero Periñán G (2021) Observatorios Europeos de Vulnerabilidad y Pobreza urbana. Características y transferibilidad. Ciudades 24:207–224. https://doi.org/10.24197/ciudades.24.2021.207-224

De Santiago, E: El Observatorio de la Vulnerabilidad en España: una herramienta para la regeneración urbana integrada. In X Congreso Nacional del Medio Ambiente, 2010. http://www.conama10.vsf.es/web/generico.php?idpaginas=&lang=es&menu=87&id=79&op=view&inicio=&idactividad=79&pestana=326&abierto=1

de Vivienda M (2010) Declaración de Toledo

García Carballo A (2014) Urbanizaciones de lujo y segregación residencial de las clases altas en Somosaguas, Pozuelo de Alarcón (Madrid). Ería: Revista cuatrimestral de geografía 94:125–144, ISSN: 0211-0563, ISSN-e: 2660-7018

González-García I (2013) Una aproximación a la definición de la variedad urbana desde la complejidad: aplicación al análisis urbanístico de tres barrios de Madrid. Tesis Doctoral. https://oa.upm.es/20230/

Graham S, Marvin S (2002) Splintering urbanism: networked infrastructures, technological mobilities and the urban condition. Routledge, London

Hamnett C (1998) Social polarisation, economic restructuring and welfare state regimes. Urban Stud 33(8):1407–1430

Hernández Aja A, Alguacil Gómez J, Camacho Gutiérrez J (2014) La vulnerabilidad urbana en España. Identificación y evolución de los barrios vulnerables. Empiria Rev De Metodología De Cienc Sociales 27:73–94

Hernández Aja A, Rodríguez Alonso R, Rodríguez Suárez I, Gómez Giménez JM, González García I, Córdoba Hernández R, Jaramillo Cáceres S (2018) Barrios vulnerables de las grandes ciudades españolas. 1991/2001/2011. Instituto Juan de Herrera. IJH

Hernández-Aja A (2007) Áreas vulnerables en el centro de Madrid. Cuad De Inv Urbanística 53:5–97

Marcuse H (2022) La lucha contra el liberalismo en la concepción totalitaria del Estado. Constelaciones. Rev. De Teoría Crítica, (13). Recuperado a partir de https://constelaciones-rtc.net/article/view/4587

Musterd S, Marcińczak S, Van Ham M, Tammaru T (2017) Socioeconomic segregation in European capital cities. Increasing separation between poor and rich. Urban Geogr 38(7):1062–1083

Naredo JM (2022) La crítica agotada. ISBN: 9788432320422. Siglo XXI

Nel-lo O (2018) La segregación urbana en Cataluña. In: Barrios y Crisis: Crisis económica, segregación urbana e innovación social en Cataluña. Tirant Humanidades, pp 71–110

Perlman J (2019) Ciudades sin tugurios, ciudades sin alma. Repensando los conceptos y las consecuencias de la marginalidad en las favelas de Río de Janeiro. Andamios 16(39):207–233

Roch Peña F (1993). "Nuevas ideologías de siempre". Ciudades: Revista del Instituto Universitario de Urbanística de la Universidad de Valladolid, ISSN 1133-6579, ISSN-e 2445-3943, Nº. 1, 1993 (Ejemplar dedicado a: La ideología urbanística, 25 años después), pp 115–130

Roch F (2008) La deriva patológica del espacio social en el modelo inmobiliario neoliberal madrileña. Scr Nova nº 12:270

Rodriguez Alonso R, Espinoza P (2017) De la especulación al derecho a la vivienda. Más allá de las contradicciones del modelo inmobiliario español. Ed. Traficantes de Sueños. Diciembre de 2017. ISBN: 978-84-947196-7-7

Rousseau JJ (ed. 2002) Discours sur l'origine et les fondements de l'inégalité parmi les hommes (1754) (Collection). J.-M. Tremblay. https://doi.org/10.1522/cla.roj.dis3

Sevilla-Buitrago A (2013) Debating contemporary urban conflicts: A survey of selected scholars. Cities 31:454–468

Strauss M (2019) A historical exposition of spatial injustice and segregated urban settlement in South Africa. Fundamina 25(2). https://doi.org/10.17159/2411-7870/2019/v25n2a6

United Nations (2003) Report on the world social situation, 2003. Soc Vulnerability Sources Challenges

Chapter 7
Gender-(in)sensitive Cities? Progress and Setbacks in Portugal

Margarida Queirós and Joana Pestana Lages

Abstract This chapter examines Portugal's journey in fostering gender-sensitive cities and territories, and spotlighting progress and obstacles from the last two decades. It aims to call for comprehensive strategies that acknowledge gender dynamics and foster cities where women and other discriminated groups feel safe, empowered, and fully recognized and integrated.

Keywords City planning · Municipal equality plans · Gender lenses · Gender-aware planning · (in)sensitive cities

7.1 Introduction

The feminist critique of urban theory and planning that has been unleashed since the 1970s has shown how urban environments have been created to predominantly suit the needs of men and the heteronormative family. *Our cities are patriarchy written in stone, brick, glass and concrete* (Kern 2020: 13, quoting Jane Darke). This kind of account would lead the architect and urban planner Dolores Hayden to question what a "nonsexist city" might look like and how cities may differ if they were designed equally by and for men and women (Hayden 1980). Although fundamental, the question she raises remains little explored. And likewise, there are other identities that question the heteronormative structure of the cities. A meaningful example as a starting point for thinking about the experiences in urban spaces of a transgender woman is that written by Petra Doan in 2010. The author uses autoethnography, combining cultural analysis and interpretation with narratives of the self, resulting in a dense description of life experiences in the city that reveal disturbing emotions and revelations about the location and position of someone who doesn't fit the binary

---

M. Queirós (✉)
Instituto de Geografia e Ordenamento do Território, Centro de Estudos Geográficos, Universidade de Lisboa, Lisbon, Portugal
e-mail: margaridav@edu.ulisboa.pt

J. P. Lages
DINÂMIA'CET, Iscte—Instituto Universitario de Lisboa, Lisbon, Portugal

R. C. Lois-González and J. A. Rio Fernandes (eds.), *Urban Change in the Iberian Peninsula*, The Urban Book Series, https://doi.org/10.1007/978-3-031-59679-7_7

standard and finds themselves on the periphery of the power network. Recognizing the danger of generalization inherent in the process of reflexive research, Doan (2010) feels that her subject position as a visibly transgender woman provides a unique perspective and opportunity for a detailed description of the ways in which the "tyranny of gender" shapes perceptions of gendered spaces in the city. One of the main challenges lies in urbanism based on the categorization of men/women associated with the attachment to dichotomous boundaries and rules that fix the structures that maintain the fundamental inequalities of the heteronormative system. The gender fluidity that the author adopts implies a refusal to remain in one gender or another—isn't the city home to voices silenced by subalternity related to race, ethnicity, social status, and sexuality?

What accounts can help us understand how women and other individuals who do not fit the "norm" experience city life, and what role does urban planning play in supporting their meaningful lives? How do everyday spatial practices contribute to clarifying the mediation of space in gender relations and vice versa, and how can this knowledge help to build inclusive and caring cities?

Despite the progress that feminism has brought to the theory and practice of city planning, the integration of gender lenses, which observe and recognize its fluidity and performativity in professional practice, remains limited. Experiences of the "tyranny" of heteronormative spaces vary between places, contexts, and political regimes, with the recognition of intersectionality, factors that, in combination, reveal the complexity of gender experiences in tension with, for example, race, ethnicity, class, age, sexuality, and space. In feminist geography, architecture, and urbanism, there must be space to struggle against this violence, and to recognize women and people like Doan.

Considering Le Corbusier's "Modulator", a system of proportions that restores harmony to space in relation to the male human body, we can conclude that the main scale, measure, and source of beauty of all things in the city exist in male bodies (Muxí 2018). This is how modernity and the patriarchal system (that the Charter of Athens replicates) have highlighted the values and experiences of men, who have established what is good, rational, and memorable in the city. This explains why at least half of humanity has been excluded from the possibility of demonstrating and living other values, experiences, and knowledge.

If the questions we ask are the outcome of different knowledge, then the answers to the challenges we face will be equally different. It is therefore appropriate to reflect on the need to build a new, inclusive, and diverse discourse, where other visions, values, and concepts have a place and are observed in urban theory and planning. One of the ways to respond to this omission lies in the study of everyday spaces. Everyday life translates revealing spatiality and temporality for feminist analysis, from which urban planners must learn—or activity from which they must be an integral part (Queirós and da Costa 2020; UN Habitat 2012).

Although women and oppressed and subalternized groups are producers of the city, they are not on an equal footing when compared to men's standards. Claiming "everyday caring spaces" presents itself as an indicator where specific elements (objective and subjective) convey familiarity and security for women, children,

LGBTQIA+, certain groups, religious or ethnic, etc. The perception of ethnic, racial, and gender associations to space complicates how rights or a sense of belonging are mediated in tension in the city. As Mollet et al. suggest (2018: 190), *everyday spaces are imbued with gendered meaning*. The authors demonstrate that activities such as walking in the street, sitting in a public garden, going shopping or a football game, and enjoying reading a book at home are moments and spaces where our identities as gendered individuals are produced.

Being aware of persistent gender inequalities in accessing all the familiar everyday spaces (related to the structural organization of society or gender systems), in the next section we explore the disparities and challenges within urban living that neglect the diverse needs and experiences of genders, centered in the Portuguese urban context, making an exploratory review of works and authors. Then, we present short but critical thoughts on the Lisbon Metropolitan Area's Municipal Equality Plans fostering gender-sensitive places in Portugal, and, finally, spotlighting progress, and persistent obstacles from the last two decades.

7.2 A Trifold Exploration of Gender-(in)sensitive Cities

Making a trifold approach on key elements of "gender-(in)sensitive cities", as proposed in this chapter, this section sheds light on the disparities within urban spaces, exposing the shortcomings in providing inclusive and safe spaces. Three axes are presented, giving a broad view on some features of urban life: from mobility to homelessness, the selected cases illustrate the status quo of spatial issues affecting women and people with non-hegemonic identities.

The research was led by three principles: (i) a gender-sensitive city is feminist, not patriarchal; (ii) gender-sensitive city must be autonomous, not dependent; (iii) a gender-sensitive city does not invisibilize, rather recognizes the life and experience of all individuals. Following these principles, a more detailed view on each one is offered, focused on urban spaces through a lens of equity.

7.2.1 Feminist, not Patriarchal

Is the fallacy of the "neutral" still operating?

Unlike other European countries, Portugal has not yet adopted a robust gender mainstreaming approach in urban planning. Countries such as Austria, Spain, and the United Kingdom have implemented such strategies and plans for more than two decades. For instance, Vienna's Gender Mainstreaming approach toward social housing was initiated in the 1990s. In Spain, since the 2000s, Inés Sánchez de Madariaga's (Madariaga, 2004) concept of "infrastructure for daily life" has prioritized the value of reproductive work alongside productive work, with urban spaces designed to support this appreciation. Similarly, the United Kingdom has developed a toolkit called "Gender Equality and Plan Making", created by urbanist Clara Greed and others in 2003 (Greed et al., 2003). These authors and strategies also warned of the urgency of deconstructing the dominant history of urban planning, contesting the universality of the transmission of knowledge based on a feminist epistemology, rescuing from oblivion women who were denied, or silenced, voices in the making of cities (Lages 2020; Muxí 2018).

The issue of equality is primarily protected by the Portuguese Constitution, in its article 13 (Equality Principle), which prohibits discriminatory treatment between individuals,[1] but until today it has not been convincingly taken up with full expression in the territory in plans and political guidelines at the urban level. A first step toward this goal was taken with an exploratory assessment to diagnose gender inequalities with a spatial perspective (Gaspar et al. 2009a) and a guide to promote gender mainstreaming with a territorial and environmental viewpoint (Gaspar et al. 2009b). Back then the authors showed that in Portugal, there was a lack of policies geared toward promoting gender in spatial planning, proposing a set of indicators to foster gender mainstreaming. It was also considered that only a multidisciplinary and intersectoral approach, in which the participation of women is fundamental, numerous times absent, could overcome this absence. Later, a "Guide to Municipalities" was published to support city councils toward gender mainstreaming through strategic planning (Queirós et al. 2010). As strategic plans for combating inequalities and instruments for the shared resolution of problems in territories, the municipal plans for gender equality emphasize the responsibility of municipal strategic planning to think "relationally" about space as a product of interventions and a powerful instrument to improve social relations and quality of life, and restore identities. Choosing a territory that houses all people in the best way that the contemporary world allows implies reflecting about space and place in terms of its qualities (belonging, access, and use, to name a few).

Afterwards Monteiro and Ferreira (2016) point out that, within the Portuguese context, urbanism and urban planning do not incorporate the diversity of all people living in urban settings, relying instead on neutral urbanism. They also drew attention to the lack of awareness, either by political decision-makers and urban planners, on issues relating to usually discriminated groups (women, people with disabilities, older people, or those belonging to ethnic minorities). The authors concluded that there are no training or capacity-building measures in place to overcome these shortcomings.

Distant from an academic viewpoint, "Integrating gender equality into the New Urban Agenda" was published by the NGO PpDM (Portuguese Platform for Women's Rights), as part of a project called "Feminismos no Centro" (Feminisms at the Centre).

---

[1] The innovative Constitution of the Portuguese Republic (CRP), enacted on April 2, 1976, prohibits discriminatory treatment between individuals. The capacity to integrate each person and guarantee equal citizenship rights is a fundamental role of the State. Whether more regulatory or interventionist, the State must ensure relations between citizens are free from prejudice of a relational, political, ideological, or territorial nature, thus guaranteeing that no form of discrimination exists.

Responding to "What is the New Urban Agenda?" and its relation to other international agendas promoting women's human rights, the work pointed to *pathways for action on gender mainstreaming in urban planning and the design and construction of inclusive and safe cities for women* (PpDM 2017: 2).

Promoting awareness campaigns, educational programs, and cultural initiatives that challenge stereotypes and foster change, as those just pointed, are essential strategies. Nevertheless, if urban planners do not have access to any training and guidance to question the "standard person", the patriarchal society, and the power imbalance we live in, gender inequalities will endure. Still there is no dedicated course in any university curricula, either in urban planning, geography, or architecture in Portugal, specifically dedicated to gender topics. For instance, there is no questioning about how gender bias influences height and position of buildings and tunnels (Kern 2020) and *the inverse, that once built, our cities continue to shape and influence social relations, power, inequality and so on* (Kern 2020: 14), meaning that their form influences the choices for individuals and groups.

In Portugal, most urban planning policies ignore the fact that women and men use space differently due to reconciling home life with work or, for cultural reasons, presenting different needs and concerns (Queirós et al. 2010: 55). For example, in the municipality of Vila Nova de Gaia within the Porto Metropolitan Area (PMA), women's entry into the job market is hindered by the lack of public transport within the municipality itself. Before 2024, the 14 transport companies operating in the PMA only had Porto as their destination. Thus, V. N. Gaia was faced with the lack of public transport serving only travel within the municipality, which affected particularly the mobility of women within the municipality of Gaia, often leading them to reject job offers, as they did not have access to accessible and adequate public transport (Queirós et al. 2010: 59).

Learning that urban space is not neutral and was not designed neutral is perhaps a good step worth taking under the Portuguese urban planning context. For that, valuing the experience of women, and therefore valuing the experience of everyday life, can lead to meaning of urban planning sensitive to gender: *We define urban planning from a gender perspective as a discipline that conceives the city and the territory as a space that facilitates and accompanies the realisation of everyday tasks and the sustainability of life, which is equitable for all people and communities* (Paricio and Regional 2019: 7).

Gender-based violence: overcoming the public–private dichotomy

Violence against women and other non-hegemonic identities is the most vicious side of sexist societies. Gender-based violence in urban settings, viewed through feminist lens, exposes systemic inequalities embedded in city dynamics. Cities, while bustling with opportunities, often serve as hotbeds for various forms of violence against women—street harassment, sexual assault, and intimate partner violence (Gordon and Riger 1991; Pain 2001). A feminist perspective unveils how urban spaces perpetuate power imbalances, restricting women's freedom, opportunities, and safety.

Kern (2020) states that through the socialization process, women are given instructions related to a spectra of unknown danger: never talk to or accept gifts from a stranger, never say you're home alone. The male "predator" has plagued women since childhood. The messages that girls and women receive about their bodies, clothes, hair, makeup, weight, and correct behavior (how they should sit, talk, walk, etc.) carry a narrative of caution and restraint. However, Kern (2020) reveals that there is enough data to show that women are much more vulnerable and experience violence at the hands of men (or partners) they know and in private spaces, such as the home and the workplace.

So, one urban dimension of domestic violence is related to home itself. Women experiencing domestic abuse and gender-based violence often face additional difficulties in securing safe and stable housing. Leaving an abusive environment is complex and can be dangerous, as some of the most frightening statistics on the number of murdered women unfortunately show that in 2022 there were 22 femicides and the Annual Internal Security Report (RASI) shows more than 30,000 reports on domestic abuse, 72.4% of the victims being women. In 2022, 22 women (and zero men) in Portugal were victims of voluntary manslaughter in the context of domestic violence.[2]

According to the Portuguese Commission for Citizenship and Gender Equality (CIG) website, only 172 out of 278[3] mainland municipalities are signatories to the protocol aimed at setting up a pool of affordable rental housing, to provide accommodation for victims of domestic abuse. This means that, in 106 Portuguese municipalities, in 2023, there is still no response that guarantees the autonomy and empowerment of victims of domestic abuse, through a housing response in the municipality where they live, so they must leave their home and their networks (social, labor, family, etc.). Even though legislation has contemplated measures since 2015 such as the removal and surveillance of the accused or aggressor, in most cases it is still the victim who leaves the family home to protect their life.

This situation is aggravated by the current housing crisis that makes it more difficult for victims of domestic abuse to leave the home where the aggression took place, and consequently break the cycle of violence. NGOs and other institutions that run shelters have also reported difficulties in finding housing provisions, which leads to women staying in support structures for longer. This situation does not allow institutions to open new vacancies, thus reducing their effective response capacity. This example scrutinizes the inefficiency of policies which fail to prioritize women's security and needs. Nevertheless, the last two decades have been marked by major advancements. The most significant is certainly the legal change that made domestic abuse a public criminal offense, in 2000. However, it wasn't until 2007 that it was established as an autonomous offense under the Portuguese Penal Code.

To dissect the public–private dichotomy regarding gender, we must also consider the extreme situation where it really disappears: the experience of homeless women. The most recent official figures point to 10,773 homeless people in Portugal, which

---

[2] Statistics available at https://www.cig.gov.pt/area-portal-da-violencia/portal-violencia-domestica/indicadores-estatisticos/#title9.

[3] Azores and Madeira as autonomous regions are not covered.

represents an exponential growth of 78% in four years.[4] Women continue to be less represented than men, but there has also been an increase. The "National Strategy for the Integration of People Experiencing Homelessness" reports 20% of women in 2020, and 22% in 2022. Women experiencing homelessness often have a different profile to men in the same situation: they don't usually sleep rough, they have fewer problems with drug addiction, alcoholism, and mental illness, and many have a job. Although everyone faces social stigma, homeless women may be more invisible due to survival strategies that lead them to be less present (visible) on the streets, which can make it difficult to get help and support.

Regarding available shelters and accommodation, women can face additional barriers to access, often due to the lack of specific spaces for them or fear of being in mixed environments that can be unsafe. It is therefore essential to create a more articulated and robust response that takes women's issues into account, expanding both the number of accommodations available immediately and the type of response provided. Housing First projects targeted to chronically homeless people have proved to be suitable for promoting more inclusive integration. In Lisbon, the "Mouraria Housing First" was implemented in 2013. The results were so positive that it has extended to the entire city of Lisbon, with the name of "É Uma Casa", "Lisboa Housing First" (This is a Home). However, given that some of these programs are mainly based on renting dwellings on the open market, the sharp rise in market prices has made it difficult to bring in new units in central areas that are well-served by urban infrastructure, as recommended by the program.

Based on women's experiences, there have been recent developments on the topic: the foundation of the SOMOS Association in 2022 by women who have experienced homelessness, Lages' taxonomy on gender and housing precarity (Pestana Lages and Jorge 2023), and the first doctoral thesis in Portugal on women experiencing homelessness (Nobre 2021). They both emphasize the urgent need to listen to and give voice to women, their experiences, and journeys, countering the prevailing reductive and stereotyped views of people living in this condition. This leads us to the next principle, centered on autonomy.

7.2.2 *Autonomous, not Dependent*

Mobility, movement

In the twenty-first century, most urban transport policies are still assumed to be neutral and universal, without a gender perspective. Nonetheless, urban mobility is intrinsically linked to gender dynamics, shaping the experiences and opportunities of individuals within cities and often fails to accommodate the diverse needs and realities of genders. Women and men have substantially different profiles when it comes to using transport and their time. Women's travel patterns and their use of

---

[4] Data available at https://www.enipssa.pt/enipssa, the web platform for the National Strategy for the Integration of People Experiencing Homelessness 2017–2023 (ENIPSSA 2017–2023).

transport are more complex than those of men, meaning multi-stop trips considering the gender division of labor within the family and community (Queirós and da Costa 2012).

As a result of the daily combination of household chores, care, and work, women have a more complex travel chain than the traditional (male) home-work pattern, using walking and public transport more often, for example (Queirós 2022). Data collected from a sample of the population living in the Lisbon Metropolitan Area in 2016 (http://genmob.ceg.ulisboa.pt/) revealed that the percentage of weekly car journeys is higher for men (81%) than for women (66%), while women use public transport more (30% women and 16% men) and active modes more often.

This means that city transportation planners still pay little attention to gender differences in their design and evaluation, particularly regarding the purpose and frequency of journeys, mode, distances traveled, stopping points, and associated times. And when they do, these policies assume that the majority of women are in the domestic field and men in the public sphere, and that the territory through which they circulate is a stage structured on the basis of hierarchies (Queirós 2022; Queirós and da Costa 2020).

Madariaga (2013) revealed that in urban areas, women in full-time employment had fewer opportunities than men in full-time employment, implying that women suffer from a 64% reduction in their potential opportunities, and gender roles were identified as the main reason for this inequity. In Portugal, mobility patterns are closely associated with the definition of employment catchment areas: where the dynamics between home and work is greater, it happens that men travel out of the municipality of residence more often than women, and this movement is associated with the use of individual transport (Queirós et al. 2010; Queirós and da Costa 2020). In metropolitan areas, however, the mobility differences for reasons of employment between men and women are fewer outside the municipality. In these areas, women use public transport more often (the higher difference between men and women in public transport use occurs in the region of Greater Lisbon, a functional area).

As gender-disaggregated data on transportation becomes available, evidence supporting general trends in gendered mobility patterns and needs is likely to be increasingly supplemented by regional differences. It is this awareness and careful documentation of regional peculiarities that becomes so essential for the development of local, gender-aware transport planning strategies. The specific balance among these will vary from region to region, and policymakers and transport operators need to gather information locally in line with best gender-balancing practice (Queirós and da Costa 2020: 86). But the analysis of spatial mobility really needs to go beyond diagnoses of gender differences and consider the mechanisms underlying the production of gender inequalities.

Challenging mobility: who has an active mode?

A feminist city ponders sustainable and active mobility, such as walking and cycling. Souza (2019) studied women's walkability and developed comparative research—carried out in Brasilia and Lisbon which explores infrastructure and public space features (comfort, safety, etc.) that influence and limit women's ability to walk. The analysis focuses attention on the insecurity felt by women in both cities, alerting to the urgency of urban planning to attend to the creation of safe and attractive environments for walking—she devised the "walkability index of fear" (Souza 2019). The motivation for this study is linked to the perpetuation of masculine patterns in planning and governing cities, with transport being based on "standard" norms and measures that in theory provide a better quality of life for the entire population, but which are not applied in practice. The author concludes that women, in both cities, are oppressed and limited in their choices and freedom of movement by foot. Another important finding of the research was that women also suffer walking limitations due to sociocultural issues, a structuring machismo, and a system of affirmation of patriarchal privileges.

Findings from Lisbon attested to the index result for the city: "mild fear" was found in women when they referred to their fear of walking at night. Also, women referred to the discomfort they suffered due to the inefficient infrastructure of the pavements, lighting, and the lack of public policies to deal with the irregular parking of vehicles on the pavements. The culture of the Portuguese, which is also sexist and male chauvinist, influences the development of the commuting strategies they adopt daily in the city (for example, planning their journeys by avoiding certain routes and timetables).

Cycling constitutes a good observation topic for the last two decades: climate change is asking for transformations regarding the way we move, and cycling is healthy, cheaper, and is growing in numbers. Pascoal (2023) lists the factors that lead women not to use bicycles in the Portuguese context: historical and cultural factors combined with a lack of cycling infrastructure and a sense of insecurity experienced by women are major obstacles. In addition, there is a lack of sex-disaggregated data and studies on cycling patterns and trends considering women, leading to a scarce visibility of this issue.

According to Eurostat data, Portugal became the largest producer of bicycles in the EU in 2022, despite being one of the European countries where this means of transport is least used. Data shows that only 1% of Lisbon's population cycles regularly. Félix (2019) points out the factors that encourage and dissuade the transition to cycling, particularly by women. In the case of the city of Lisbon, recent changes with the extension of the cycle network and the creation of a bike-sharing network, along with the deconstruction of old misconceptions that point to Lisbon as a city too inclined to cycle,[5] have increased the number of women cyclists. Félix also points to the increase: a change from 17% in 2017 to 26% in 2020. Although they represent only a quarter of all users, the path is positive, boosted by tax incentives, for example, but also by the creation of women's cycling groups that support women in their initiation.

[5] Rosa Félix's research showed that 74 per cent of Lisbon's roads have less than a 5% gradient.

7.2.3 Recognition, not Invisibility

Who is missing from our urban spaces? Toponymy as a measure of equality

An interesting lens through which to uncover the historical and cultural imprint of women in urban spaces can be observed through their toponymy. Often, city landscapes are adorned with names that commemorate significant figures, events, or characteristics, reflecting societal values and power structures, a practice that started in Portugal at the beginning of the eighteenth century. This practice favored men, exposing that the presence of women in toponymy was frequently sparse, mirroring historical biases and gender inequalities. Yet, the scarcity of female-centric toponyms underscores broader societal disparities.

Between 1999 and 2008, the academic journal *Faces de Eva* produced a dossier called "Toponymy in the Feminine" about Lisbon. It was based on the distribution of toponyms by parish, with the Gabinete de Estudos Olisiponenses (Lisbon Studies Office) as its main source. One of the main conclusions of this study points to the presence of more women's names in areas of recent urban expansion or in popular neighborhoods. In 2018, the journalist Rita Costa conducted a similar study and concluded that only 15% of toponyms attributed to urban places to women, mostly religious or aristocratic women, such as saints, queens, and princesses.

In 2021, data scientists Narciso and Lemos launched a multimedia platform called "Gendered Streets",[6] which presents an analysis of the toponymy of Porto, already revealed in Silva's work (2012). The results are identical to previous studies: 44% of the streets are named after men, 4% after women, and the rest mention things or concepts. An interesting analysis by the researchers shows the difference in the length of the streets, showing that not only are there more of them, but the length is much greater when the streets have male names: the longest street with a male name is Av. Fernão Magalhães, with 3.84 km; in turn, the longest with a female name is St. de Santa Catarina, with a length of 1.43 km. The data scientists added that, on average, the length of streets named after men is 14 times longer than that of streets named after women.

More recently, in 2022, Banza (also a data scientist) launched a study on Lisbon's toponymy with a gender focus, analyzing not only the city's streets, but squares, parks, and public facilities, such as schools and hospitals. Of the 5000 streets examined, only 5% have female toponyms, in contrast to the 44% of male toponyms. The same is true for gardens—7% female, 36% male. For hospitals—11% female, 54% male; and for schools—14% of the names are female, 43% male.

Addressing this imbalance involves deliberate efforts to recognize and commemorate women's achievements through toponymic choices, thereby enriching urban landscapes and fostering a more equitable, inclusive city narrative that celebrates women throughout history. This is not a Portuguese subject, as shown by the current

[6] The website is available at https://ruasdogenero.pt/pt.

project "Mapping Diversity".[7] The platform expresses the diversity and representation in street names across 30 major European cities, located in 17 different countries, being Portugal represented by the city of Lisbon. In the European context only 9% of streets are named after a woman, showing this numbers are low in many other places. The lack of streets, squares, or landmarks named after women echoes historical marginalization and underrepresentation. It prompts reflection on the need for greater gender inclusivity in urban toponymy, highlighting the importance of honoring women's achievements and acknowledging their role in shaping cityscapes.

7.3 Municipal Equality Plans: A Tool for Change, Though not Fully Spatial

As previously mentioned, gender equality contributes to the development of more equitable, inclusive, and cohesive territories, that is, fairer spaces. The persistence of city inequalities resulting from contexts based on gender stereotypes requires political intervention and, as such, local authorities have a fundamental role as entities that instigate local development (Queirós and Birrento 2016). Municipalities though are entities that have privileged contact with the population and therefore are strategic units to fight against discrimination.

City Councils can plan and adapt to specific territorial needs to combat the discrimination that is often inherent in a social structure that is even unconsciously shaped (Wall and Amâncio 2007). For instance, the construction of suitable transportation facilities for children and the elderly is very important for women, as they use them the most (Queirós et al. 2010). Local policies can provide a network response to the needs of women, ensuring their safety, as well as identifying and understanding their main vulnerabilities. But as revealed, spatial planning for public policies does not involve women and men on an equal footing which will, ultimately, be doomed to fail equality goals.

Since 2013, it is a duty of the City Council to *ensure the integration of the gender perspective in all areas of action of the municipality, namely through the adoption of Municipal Plans for Equality* (Law 75/2013, art. 33). Municipal Equality Plans are public policy instruments specifically designed for endorsing equality at the local level, which establish strategies for transforming gender asymmetries revealed by diagnosis carried out at the local level, integrating gender mainstreaming measures and specific actions, and setting objectives, indicators, and targets. They are also part of the "National Strategy for Equality and Non-Discrimination 2018–2030" (ENIND), approved by the Council of Ministers Resolution 61/2018, in line with the 2030 Agenda for Sustainable Development.

Not all Municipal Councils in the country are moving at the same speed in reducing inequalities, and greater political commitment is needed to better root equality issues

---

[7] Information available at https://mappingdiversity.eu.

in the local economic and social fabric (Pinto et al. 2018). An enquiry (textual analysis of contents and measures proposed) of the equality plans in force in the Lisbon Metropolitan Area, accessed via the municipalities' websites,[8] ensured a categorization of the measures with spatial/territorial/urban content. Namely, this analysis was carried out for the municipalities of Amadora, Barreiro, Cascais, Loures, Moita, Montijo, Odivelas, Palmela, Sesimbra, and Setúbal.[9]

Which initiatives with a spatial perspective emerge from the municipal plans for equality? Mainly objectives and measures that aim to (i) integrate gender lens in the areas of spatial planning and territorial management (urban planning and environment) mainly focusing mobility and public spaces; (ii) identify inequalities in the enjoyment of public space, which can be achieved through training initiatives and mapping/cartography; (iii) enhance the visibility of women at the local level through toponymy.

Those objectives and measures reveal an awareness that the local public policies for urban planning often ignore that women and men use space differently due to personal and domestic life arrangements with work, or for sociocultural reasons, presenting different needs and concerns. So, the good news is that if urban space is the most comprehensive level of planning to guarantee inclusive territories, then the domains of spatial planning cannot continue to be relegated to the background. The municipal departments responsible for these areas must be involved, alongside social action, in the first line when the goal is to struggle against discrimination. The success of local gender equality policies depends on their integration at broader levels, such as environmental protection, economic development, and social inclusion. It also relies on the actions of entities that operate in networks, primarily local actors (Pinto et al. 2018).

7.4 Final Remarks

Gender is a key factor in the unequal struggles over power, sense of belonging, social identity, and access to resources and opportunities. Understanding the connections between home and work activities helps recognize participation in the labor force, pay, promotions, and other occupations, as well as opportunities and barriers to accessing what spaces and places offer. Municipalities, particularly city councils, have the tools and frameworks to examine how power is exerted and negotiated in and through space and place, influencing what we think, feel, who we are, and what we can become (Mollett et al., 2018: 191).

[8] The Lisbon Metropolitan Area is made up of 18 municipalities: Alcochete, Almada, Amadora, Barreiro, Cascais, Lisboa, Loures, Mafra, Moita, Montijo, Odivelas, Oeiras, Palmela, Seixal, Sesimbra, Setúbal, Sintra, and Vila Franca de Xira.

[9] According to the CIG, two more municipalities have their Equality Plans approved (Cascais and Seixal), as per the map on the website, which was updated on 30 October 2023. However, they were not included in this analysis as they are not publicly accessible on the municipalities' websites.

In Portugal, local authorities have had the instruments and tools since the early 2000s. Municipal plans are strategic tools that can seriously address how lives are shaped by gender power relations. Our assessment of plans in the Lisbon Metropolitan Area offers hope for establishing a feminist and intersectional approach to urban planning. This progress reflects city councils' inclusive and caring strategies.

Despite advancements in gender public policies, cities remain non-neutral spaces where violence persists, and mobility is uneven. The limited objectives and measures in municipal equality plans may not sufficiently address the complexity of the issue, as people face discrimination based on multiple intersecting factors that shape privileges and disadvantages. Additionally, while some aspects of urban planning are gender-responsive, many issues remain unaddressed, including mixed uses, mobility, accessibility, safety, distribution of commerce and services, community buildings, housing, and social mix (UN-Habitat, 2012). One obstacle is the compartmentalized approach of city council departments in Portugal, each addressing specific challenges and goals with little coordination on spatial and multi-scale work, which is crucial at the metropolitan level. Another issue is the scarcity and reliability of gender-specific statistics at the local/urban level, potentially stemming from a lack of awareness of urban planning's effects on people's lives, well-being, and quality of life.

Urgent measures are needed to prepare urban planners to address the needs of women and disadvantaged groups, including involving them in the planning process—decision-making, design, and implementation—and promoting community engagement, advocacy, awareness, and participation in urban planning challenges. UNHabitat (2012) advises planners to adopt gender-aware planning roles. Those within city councils can work directly with grassroots groups to collect and integrate gender indicators into their strategies and projects. Achieving gender equality through sustainable urbanization requires planning professionals to be better equipped for transformation, as "neutrality" still dominates professional language, the public-private dichotomy persists, and gender mobility constraints and privileges endure.

References

Doan PL (2010) The tyranny of gendered spaces—Reflections from beyond the gender dichotomy. Gend Place Cult 17(5):635–654. https://doi.org/10.1080/0966369X.2010.503121

Félix RM (2019) Barriers and motivators to bicycle in low cycling maturity cities: lisbon case study. Ph.D. Degree in Transportation Systems, Universidade de Lisboa Instituto Superior Técnico, Lisbon

Gaspar J, Queirós M, Marques da Costa N, Henriques EB (2009a) Estudo de Diagnóstico e Criação de Indicadores de Género na Área do Território e Ambiente. CEG, Universidade de Lisboa & CIG, Lisboa, p 175

Gaspar J, Queirós M, Marques da Costa, N, Henriques EB (2009b) Guia para o Mainstreaming de Género na Área do Território e Ambiente. CEG, Universidade de Lisboa & CIG, Lisboa, p 53

Gordon MT, Riger S (1991) The female fear: the social cost of rape. University of Illinois Press, Urbana

Greed C, Davies L, Brown C, Dühr S (2003) Gender equality and plan making: gender mainstreaming toolkit. Royal Town Planning Institute, London. https://www.rtpi.org.uk/media/3518/genderequality-planmaking.pdf
Hayden D (1980) What would a non-sexist city be like? Speculations on housing, urban design, and human work. Signs 5(3):S170–S187
Kern L (2020) Feminist city. Verso Books, London
Lages JP (2020) Mujeres, Casas y Ciudades. Más Allá del Umbral (Zaida Muxí Martínez)—Atualização bibliográfica. Finisterra LV(114):199–201. https://doi.org/10.18055/Finis20386
Madariaga de IS (2004) Urbanismo con perspectiva de género. Instituto Andaluz de la Mujer
Madariaga IS de (2013) From women in transport to gender in transport: challenging conceptual frameworks for improved policymaking. J Int Aff 67(1, fall/winter 2013, The Gender Issue: Beyond Exclusion):43–65
Mollett S, Fluri JL, Whitson R, Oberhauser AM (2018) Conclusion and reflections. In: Mollett S, Fluri JL, Whitson R, Oberhauser AM (eds) Feminist spaces. Gender and geography in a global context. Routledge, London, pp 189–95
Monteiro R, Ferreira V (2016) Guia para a Integração a nível Local da Perspetiva de Género na Segurança e Prevenção da Violência no Espaço Público. CES—Centro de Estudos Sociais da Universidade de Coimbra, Coimbra
Muxí Z (2018) Mujeres, Casas y Ciudades. Más Allá del Umbral. dpr-barcelona, Barcelona
Nobre S (2021) Women´s homelessness and housing exclusion in the Northern Lisbon Metropolitan area: an in-depth exploratory study. Universidade Nova de Lisboa, Lisbon. https://run.unl.pt/bitstream/10362/123033/1/Tese%20versão%20final.pdf
Pain R (2001) Gender, race, age and fear in the city'. Urban Studies 38(5–6):899–913. https://doi.org/10.1080/00420980120046590
Paricio A, Regional B (2019) Manual d'urbanisme de La Vida Quotidiana. Urbanisme Amb Perspectiva de Gènere. Ajuntament de Barcelona, Barcelona
Pascoal IS (2023) Contributo para a Inclusão da perspectiva de Género nas Políticas de Mobilidade em Bicicleta: O Caso da Área Metropolitana de Lisboa. Mestrado em Urbanismo Sustentável e Ordenamento do Território. Universidade Nova de Lisboa, Lisbon
Pestana Lages J, Jorge S (2023) A gendered taxonomy on housing precarity: challenges from Lisbon Metropolitan area during the Covid-19 pandemic. Soc. e Território 35(1):51–72. https://doi.org/10.21680/2177-8396.2023v35n1ID32254
Pinto P, Torres A, Queirós M (2018) Auditoria de Género: um roteiro para avaliar a dimensão de género nas políticas públicas. In: Ferrão J, Paixão JMP (eds) Metodologias de avaliação de políticas públicas. Imprensa da Universidade de Lisboa, Lisboa, pp 411–432
PpDM (2017) Feminismos no centro. Plataforma Portuguesa para os Direitos das Mulheres, Lisbon. https://plataformamulheres.org.pt/site/wp-content/ficheiros/2017/06/segunda-brochura-3maio.pdf
Queirós M, da Costa NM (2012) Knowledge on gender dimensions of transportation in Portugal. Dialogue UniversalismE 3(1):47–69
Queirós M, da Costa NM (2020) Planning mobility in portugal with a gender perspective. In: de Madariaga IS, Neuman M (eds) Engendering cities: designing sustainable urban spaces for all. Routledge, New York, NY
Queirós M, Birrento M (2016) Transversalização e Territorialização das Políticas de Igualdade de Género: das orientações de Política Nacional à Intervenção Municipal em Portugal. In: Queirós M, Louro A, Maciel P (eds) IV Workshop RIDOT: O Desafio Do Planeamento e à Observação Territorial Nos Países Ibero-Americanos Para o Século XXI: Dinâmicas, Processos, Experiências e Propostas. IGOT Universidade de Lisboa, Lisboa, pp 175–188
Queirós M, da Costa NM, Palma P, Caetano G, Vieira PJ (2010) Guia Para o Combate à Discriminação Nos Municípios. Planear Para Todos e Todas, Planear a Diversidade. Estudo Sobre Políticas e Medidas de Igualdade e Combate à Discriminação Nas Autarquias Locais. CIG & IGOT-Universidade de Lisboa, Lisbon

Queirós M (2022) Mobilidades desiguais, no espaço e no tempo. In *Mobilidades desiguais*, edited by Ricardo Barbosa da Silva, Silvana Zioni, and Flávia Ulian. Série Estudos Urbanos. Comissão de Pesquisa, Instituto das Cidades, Unifesp, São Paulo, SP, pp 39–53

Souza A (2019) Mulher, Uma Força Que Caminha—Estudo de Caso Em Brasília e Lisboa. Ph.D. Degree in Transportation, Brasília. Departamento de Engenharia Civil e Ambiental, Faculdade de Tecnologia, Universidade de Brasília, Brazil

UN-Habitat (2012) Gender and urban planning: issues and trends. In: Reeves D, Parfitt B, Archer C (eds). nited Nations Human Settlements Programme, Nairobi, Kenya

Wall K, Amâncio L (eds) (2007) Família e Género Em Portugal e Na Europa. Atitudes Sociais dos Portugueses. Imprensa de Ciências Sociais, Lisbon

Chapter 8
The Segregation of the Nouveau Riche: A Metropolitan Issue—The Case of A Coruña

Diego Cidrás and Inês Gusman

Abstract Urban studies concerning inequality propose that analyses focused solely on the socially disadvantaged sectors are insufficient to fully illustrate the phenomenon. More than a decade after the outbreak of the real estate, financial, and social crisis of 2008, segregation research points toward a structural character of the phenomenon, thereby surpassing the perspectives that have confined it to the impacts of the crisis on the most vulnerable groups. A new urban culture based on socioeconomic differentiation and the separation of lifestyles has emerged, where consumption and leisure activities are assuming an increasingly important role. Simultaneously, hypermobility, emphasis on connectivity and technology, as well as new family models, have reconfigured the role of housing as a classic object of segregation. In this context, this chapter examines the evolution of urban segregation in the metropolitan region of A Coruña during the period 2001–2021, through the study of distribution patterns among the socially and economically privileged classes. To achieve this, a longitudinal multivariable analysis is conducted using census data and the Household Income Distribution Atlas, disaggregated by census tracts. This Galician metropolitan area has stood out in Spain over recent years due to its significant capacity to attract human and economic capital, particularly revolving around multinational companies in the textile sector and petroleum refining. The empirical work yields a territorial interpretation of the so-called "INDITEX Effect" in the study area, where forms of urban segregation are emerging beyond the municipal scope of A Coruña.

Keywords Inequality · Segregation · Urban culture · Housing · INDITEX · Galicia

D. Cidrás (✉) · I. Gusman
Departamento de Xeografía, Universidade de Santiago de Compostela, Santiago de Compostela, Spain
e-mail: diego.cidras.fernandez@usc.es

I. Gusman
e-mail: mariaines.gusmancorreia@usc.es

R. C. Lois-González and J. A. Rio Fernandes (eds.), *Urban Change in the Iberian Peninsula*, The Urban Book Series, https://doi.org/10.1007/978-3-031-59679-7_8

8.1 Introduction

Studies on inequality have traditionally framed the discussion in the context of urban segregation (Nightingale 2012; Shertzer et al. 2022; Vaughan and Arbaci 2011). This phenomenon refers to "the extent of relationships established among different social groups, the diverse ways of appropriating public space and inhabiting the city" (Suárez 2011, p. 43). From the emergence of radical Geography to more contemporary studies, the predominant focus of segregation has been on the marginalization of disadvantaged collectives within cities. In this context, institutional and academic discourses have often converged in identifying poverty and various social and cultural biases as driving factors of segregation, thus centering the explanation of the phenomenon on dissenting agents (Botana 2020; Ruiz-Tagle 2016). Perhaps as a consequence of these approaches, segregation studies have been confined to the strictly urban scale, where a significant portion of spatial phenomena indeed flourishes. Therefore, there exists a certain gap regarding what occurs beyond the diffuse limits of the city, especially from the perspective of segregation among affluent collectives (Blanco and Nel·lo 2019).

In line with the ambition of this collective work, this chapter arises from the call to elevate the phenomenon of segregation to a broader scale: the metropolitan. Such profile of residential segregation typically stems from suburbanization processes lacking territorial planning (Echazarra 2010), although strict planning exercises can also lead to segregation (Lens and Monkkonen 2016). In any case, metropolitan residential segregation can be conceived as an indicator of the socio-spatial distribution of the population on a supra-local scale (Rubiales Pérez et al. 2012). This dynamic is often mediated by practices of seclusion, implying a voluntary withdrawal to suburban spaces with a higher potential for homogenization. Among the socio-territorial attributes of metropolitan segregation, negative externalities related to (hyper)mobility (Sampson and Levy 2020), service provision (K'akumu and Olima 2007), or energy consumption (Grossmann 2019) gain relevance, among other aspects. Moreover, a degradation of public space emerges in this context, as the proliferation of segregated residential areas away from the central core exacerbates the disinterest of higher income community members in caring for and coexisting within shared spaces in the traditional nucleus (Harvey 2013).

Secondly, it's worth questioning why, within this metropolitan logic, adopting the perspective of the more affluent population is necessary. In this context, one of the most significant factors influencing housing access in the Iberian context emerges—the rapid increase in inequality following the real estate and financial crisis of 2008 (Zamora-Kapoor and Coller 2014). Over a decade after the Great Recession, substantial improvements in employment and other indicators have not been accompanied by an increase in equity (González-Pérez et al. 2016; Piñeira-Mantiñán et al. 2018). For instance, since 2008, the income of the top 10% wealthiest in Spanish society has increased tenfold compared to the income of the bottom 10%. According to Organisation for Economic Co-operation and Development (OECD) data, Spain and Portugal rank 5th and 6th, respectively, in terms of income inequality among the

18 considered Euro Area countries, with Gini coefficients of 0.320 and 0.313 in 2021 (OECD 2024). This scenario "indicates the inadequacy of an interpretation of segregation solely centered on the most socially disadvantaged sectors" (Blanco and Nel·lo 2019). This is precisely why this study proposes an examination of residential segregation of the Nouveau Riche from a metropolitan perspective.

Identifying practices of metropolitan residential segregation mediated by affluent collectives is not an especially complex task. In the Iberian context, numerous established cases adhere to this logic (Romero-González and Farinós-Das 2016). Perhaps the most notable include the coastal residential area of Cascais near Lisbon (Tulumello and Colombo 2018) or the exclusive developments located to the north of Madrid (Leal and Sorando 2016), among other instances. Within their morphological and social diversity, most of these cases share a temporal pattern, typically dating back to the latter part of the last century. With the aim of assessing a post-2008 crisis reality, this study chooses to focus on the case of A Coruña (NW Spain), which, while having initiated its metropolitan development some time ago, exhibits notably contemporary indicators of enrichment and elitization. The globalization of the INDITEX Business Group (INDITEX), a leader in the manufacturing and marketing of clothing, with its headquarters located in the metropolitan area, motivates this study to ultimately address the segregation of the new affluent in this peninsula corner. How has the residential segregation of affluent collectives in A Coruña evolved over the past two decades? How does the presence of multinational companies like INDITEX relate to the current spatial residence patterns of A Coruña? These are some of the key questions we will address throughout the study, which sets its primary objective as analyzing and discussing the evolution of this segregation between the years 2001 and 2021.

The chapter is structured as follows. Firstly, a literature review is presented to delve into the complex interrelationship between economic dynamics and the phenomenon of residential segregation in urban spaces, focusing particularly on its causes and socioeconomic manifestations and implications within European cities. Subsequently, the metropolitan reality of A Coruña is introduced, particularized in recent years by the influx of thousands of highly skilled workers who, in most cases, opt to reside in newly developed residential areas. Thirdly, results are developed, mediated by a statistical and cartographic analysis of the attributes defining metropolitan segregation in the case of A Coruña. In the discussion and conclusions section, we use the results of our study to highlight the structural nature of urban segregation and its transformation in the ACMA. We focus on the privileged social and economic classes, shedding light on emerging segregation dynamics influenced by factors such as lifestyle separation and the "INDITEX Effect", redefining the role of housing in this evolving urban landscape.

8.2 The Economic Dynamics and the Urban Segregation Phenomenon

As noted above, one of the main characteristics of the contemporary urbanization process in European cities is urban segregation. Although it is a long-lasting phenomenon and a subject of concern in urban policy for decades (Bolt 2009; Tammaru et al. 2016), the increase of social inequalities and the evolution of real estate markets over the last three decades (Piketty 2014) turned into a hot topic of many academic debates (Nel·lo 2021b). This phenomenon is frequently associated with the globalization of the world economy, the erosion of the welfare state, and the prevailing influence of neoliberal policies. The emergence of these trends was discernible as early as the 1980s, but they have further intensified in the aftermath of the 2008 financial crisis (Nel·lo and Donat Muñoz 2023). In a broader sense, urban segregation is the uneven spatial distribution of social groups within urban areas, influenced by factors such as occupation, income, education, gender, and ethnicity. It is frequently addressed as a residential phenomenon (van Ham et al. 2021).

Massey and Denton (1988) pointed out that residential segregation, at its core, refers to the extent of spatial separation between two or more groups within the urban environment, indicating their distinct living patterns. Among the most significant factors of this spatial separation are income and real estate prices, as they determine the capacity of people to choose their place of residence (Maloutas and Fujita 2016). According to economic theory, housing prices are determined by wages and income on the demand side and by the availability of housing units on the supply side (Florida and Mellander 2010). Although some scholars have conceived spatial segregation as being neutral, as it simply involves the over-representation of certain groups and the under-representation of others as a natural outcome of the economic system, for others it should not be regarded as an impartial process (Piekut 2021). Within this last academic current, there is a strong belief that spatial distance follows social distance (Duncan and Duncan 1955). Moreover, residential segregation might be both a cause and a consequence of unequal access to social resources and public services. Consequently, the rising residential disparities are poised to play a significant role in generating and perpetuating social inequalities (Arundel and Hochstenbach 2020).

Although residential segregation can be assessed through different angles, a significant aspect that has captured the attention of researchers is the isolation of the healthier people. Over the last decades, the rich have become richer, and public policies and state institutions have been unsuccessful in tackling poverty (Atkinson 2015). High-income people can afford to select housing options that are aligned with their preferences, leading to increasing spatial separation between people with high and low incomes. In fact, this phenomenon can be considered the spatial expression of what Ariño and Romero (2016) have coined as the "the secession of the rich". It has become more pronounced than the segregation of the most disadvantaged groups (Nel·lo 2021b). These dynamics can originate gentrification of certain areas of the cities. Gentrification has become a widespread process in the last 30 years and is defined by Hammel (2009) as the displacement of working-class residents of a neighborhood by wealthier professionals. A paradoxical aspect of gentrification, in the short term, is it can diminish the levels of segregation, as high-income groups move into low-income neighborhoods (Tammaru et al. 2020).

Regardless of the adopted perspective to study residential segregation, it seems evident that people with high and low incomes are increasingly living separated in different parts of the city (van Ham et al. 2021). The unequal capacity to choose where to live based on income leads to the concentration of low-income groups in areas with lower prices and poorer urban quality. The high-income groups can afford to live in areas with higher property prices, enjoying a higher quality of urban life, better services, and living among their peers (Harvey 1973; Secchi 2013; Nel·lo 2016). These spatial divisions manifest other dimensions of rising socio-economic inequality. For Fujita and Maloutas (2012), the segregation in the housing market is a consequence of the income inequality produced in the labor market. In fact, it was in the 1990s that the author Sassen (1991) firstly called attention to the consequences that the occupational structure changes were producing in the urban dynamics of the global cities. For this author the best-connected cities, often serving as control centers for international economic activity, attract both high-salaried employees in advanced business services and low-salaried employees in consumer service industries. According to the author, this was contributing to the social polarization of these cities.

Nevertheless, residential segregation is not limited solely to economic classes but is becoming more evident among worker skills. Recent studies reveal a trend where skilled workers are increasingly concentrating in space (Florida and Mellander 2016; Berry and Glaeser 2005). This concentration is impacting housing values by driving up demand in local markets. Berry and Glaeser (2005) observed a growing disparity in human capital levels among different metropolitan regions in the United States. This can be associated with the reconfiguration of the occupational landscape which transitioned from manufacturing-based economies to knowledge-driven post-industrial models resulting in the spatial clustering of high-paid and low-paid workers, often driven by the preferences of the "new middle class" or "creative class" (Boterman et al. 2018).

van Ham et al. (2020) found through their study in London that, while levels of segregation remained relatively stable from 1981 to 2011, the social geography of the city experienced a significant shift. In the 1980s, the rich tended to reside on the outskirts of London, while the poor were concentrated in the center. However, by 2011, this pattern had reversed, with affluent individuals now living in the city center and lower income groups residing on the outskirts. A similar transformation has been observed in the city of Toronto (Hulchansky 2010). These changes are aligned with the identified trend that cities have become consumption hubs. According to Penco (2015), successful cities are capable of attracting knowledge-based enterprises and institutions, such as high-tech enterprises, headquarters, and high-value services of big corporations, creative businesses, universities, and research centers, while they are a hub for knowledge production and consumption. This transformation has given rise to a class of high-income earners who either work and live within the city or commute from outside but opt to reside in the city to enjoy its amenities. As a consequence, many cities and especially their central areas turned focal points of consumption, consumer services, desirable amenities, and attractive destinations for an increasingly affluent workforce (Jayne 2005).

In a research conducted in various cities worldwide, van Ham et al. (2021) found that high-income workers are not solely attracted to city centers but also seek out appealing coastal regions, while lower income groups are pushed toward the peripheries of urban regions. In the Iberian context, and according to Blanco and Nel·lo (2019) and Nel·lo (2021a), the major cities exhibit four distinctive features in their contemporary residential segregation patterns: its structural nature, its social and spatial pervasiveness, its tendency to deepen, and its transition from a local to a metropolitan scale. This spatial polarization is not limited to the city level but occurs across regions, especially in metropolitan areas. Considering these complexities, it becomes imperative to adopt a supramunicipal scale when assessing the residential segregation phenomenon.

8.3 A Coruña: A Territory of Metropolitan Wealth?

The Metropolitan Area of A Coruña, locally known as “Gran Coruña” (Great Coruña), is situated at the northern vertex of the European Atlantic Axis (Fig. 8.1). Despite its colloquial name, this metropolitan area assumes an intermediate scale within Iberian parameters. It encompasses an urban area of approximately 500 km^2, inhabited by nearly 500,000 people, hence representing one-fifth of the Galician population. Defining its precise boundaries is intricate, as A Coruña, much like most Spanish metropolitan spaces, currently lacks a distinct legal entity. In this regard, there have been numerous efforts and proposals to formalize the metropolization of this Galician urban aggregate. Fundamentally, two major approaches have been pursued. On the one hand, a monocentric approach at a contained scale has been articulated by the municipality of A Coruña and 10 surrounding municipalities. This scheme has been—and continues to be—employed by various state, community, and local institutions, as well as in various attempts to establish an administrative unit (AMAC 2023). On the other hand, the Galician administration has proposed a broader framework through its territorial planning, envisioning an even larger urban region. This concept would also encompass the urban area of Ferrol (Xunta de Galicia 2011), a city with functional ties to A Coruña. However, this urban region has never been precisely delineated and has not been subjected to any legal articulation. This study adopts the first approach, not only because it is more established institutionally and socially, but also because it reveals a clearer residential integrity, as we will elucidate further.

The development of this metropolitan area stems from the uniqueness of its central nucleus. Situated on a granite peninsula in the Atlantic waters, A Coruña is the most compact provincial capital in the Spanish state, covering an area of less than 40 km^2. This morphological particularity has led to its urban growth occurring primarily in a radial pattern and through several terrestrial exit vectors from the city (Concello

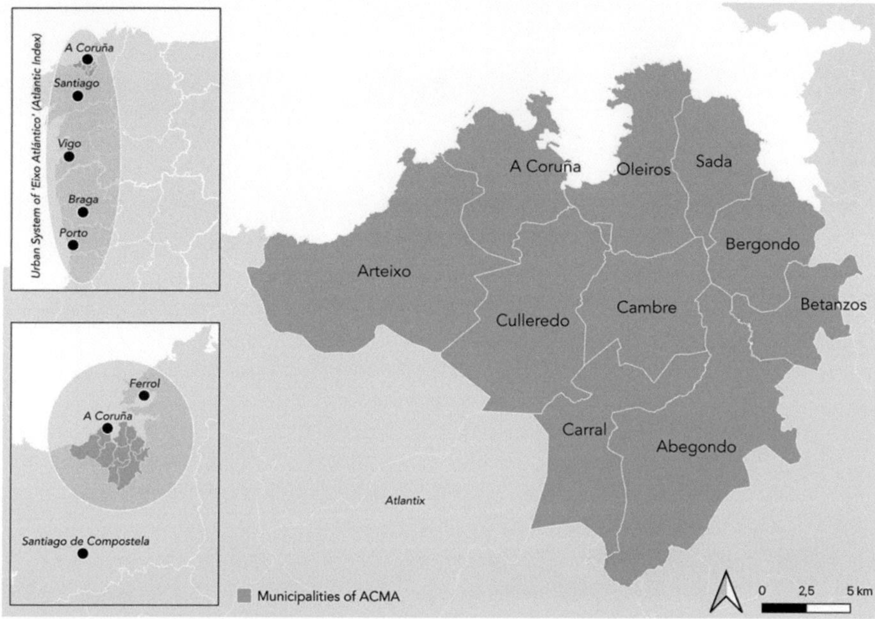

Fig. 8.1 Location of the A Coruña metropolitan area. Map created by the authors based on Información Xeográfica de Galicia and Instituto Geográfico Nacional

da Coruña 2017). Simultaneously, the intensive use of urban space results in the city center having the highest population density in the Iberian Peninsula and one of the highest in all of Europe, although this statistic does not extend to the entirety of the municipality (Fig. 8.2). In any case, these morphological attributes anticipate two problematic issues for the city: firstly, the limitation of its public space, mostly dedicated to infrastructure for private vehicles; and secondly, the impossibility of indefinitely expanding the housing stock, especially in a context where less than 25% of municipal land is classified as rural. To these, we must add a third issue that transcends formal logic: A Coruña is the third Spanish city with over 200,000 inhabitants with the highest percentage of vacant housing (14.7%), a structural issue of the Galician territory (Instituto Nacional de Estadística 2023). All these pressures ultimately point toward a rupture of the municipal scale from which urban policies have historically been developed.

The process of population diffusion at the metropolitan scale in A Coruña dates back to the 1970s and 1980s. In a context of limited urban regulation, factors such as accessibility to developable land, improvements in mobility infrastructure toward the center of A Coruña, and the demand for nature-oriented areas as living spaces led to the formation of a first metropolitan ring in the municipalities of Culleredo, Arteixo, Oleiros, and Sada, mainly. The latter two, located in the easternmost sector of the

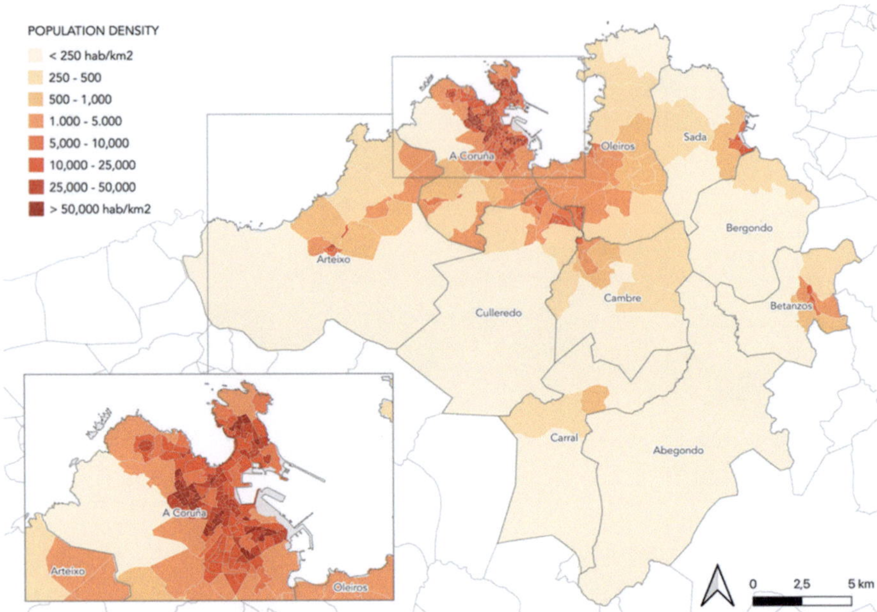

Fig. 8.2 Population density in ACMA, 2021. Map created by the authors with data from the Instituto Nacional de Estadística (2023)

current A Coruña Metropolitan Area (ACMA),[1] initially functioned as destinations for second residences. The consolidation of the metropolization process occurred between the 1990s and 2000s when primarily a generation of young adults navigated the challenging access to housing in A Coruña by moving to the immediate periphery. As Table 8.1 shows, during the period from 1991 to 2021, a significant increase in the population of the ACMA can be appreciated with a rise in the resident population by 20.7%. However, this demographic growth was not evenly distributed across the various municipalities. On the one hand, certain municipalities experienced substantial population growth over this 30-year period, such as Culleredo (110.87%), Oleiros (99.53%), Cambre (99.80%), Sada (84.25%), and Arteixo (83.56%). On the other hand, some municipalities saw a population increase, but the rates were not as remarkable as in the previously mentioned ones. For instance, Bergondo had a growth rate of 26.43%, Betanzos saw a 9.27% increase, and Abegondo experienced a modest growth rate of 0.48%. In the case of A Coruña, despite maintaining its status as a major population hub within this metropolitan area, with a population stabilizing around 245,000 inhabitants, there was a slight decrease of 0.57%. These figures underscore that the population expansion within this metropolitan area can be primarily attributed to the growth of peripheral municipalities rather than the central core. These numbers underscore that the population expansion within this

[1] *Área Metropolitana da Coruña* (AMAC) in Galician language, as it is known locally.

metropolitan area can be primarily attributed to the growth of peripheral municipalities rather than the central core. In fact, the majority of diffuse residential areas emerged, both in the first metropolitan ring and in a second one, featuring municipalities like Abegondo, Betanzos, Cerceda, and Carral. This exodus scenario was countered by the municipal policies of A Coruña through the creation of small to medium pockets of new housing in neighborhoods like Nuevo Mesoiro, Someso, and Los Rosales, as well as through special plans aimed at regenerating degraded neighborhoods. Thus, the city of A Coruña manages to sustain its population consistently around 250,000 inhabitants, while numerous low-density residential developments have continued their metropolitan expansion. However, this process was contained during the decade 2011–2021, leading to a current scenario of balance between A Coruña and its peripheries (Fig. 8.3).

Understanding the economic support that has accompanied this residential diffusion remains essential. In this respect, unlike other metropolitan areas, that of A Coruña is characterized by a polycentric and decentralized economic activity. In fact, eight out of the ten highest revenue companies in the ACMA have their headquarters in municipalities within the metropolitan periphery (Table 8.2). The oil refinery (A Coruña, Arteixo), the Sabón industrial park (Arteixo), and the recent construction of the Outer Port (A Coruña, Arteixo) have situated the largest employment hubs of the ACMA outside the urban center over the past decades. In certain cases, this polarity is reinforced by the tertiary use of industrial land in the outer ring of the city, where several large-scale motor-accessible shopping centers are located, catering to both the city center and the outskirts. Consequently, the city center of A Coruña is primarily dedicated to office activities and service provision. In contrast to this, another distinct feature of this territory is that most of the supralocal facilities—such as major healthcare centers, the university, cultural and leisure facilities, and large markets—are strictly situated within the core of A Coruña. As a result, the existing

Table 8.1 Evolution of resident population in ACMA (1991–2021)

Municipality	1991	2001	2011	2021	Growth rate (%)
Abegondo	5,466	5,694	5,709	5,492	0.48
Arteixo	17,931	22,709	30,482	32,914	83.56
Bergondo	5,392	6,179	6,722	6,817	26.43
Betanzos	11,930	12,423	13,537	13,024	9.17
Cambre	12,330	18,691	23,649	24,635	99.80
Carral	5,229	5,282	6,064	6,581	25.86
Coruña, A	246,953	239,434	246,028	245,541	*– 0.57*
Culleredo	14,631	22,076	29,207	30,853	110.87
Oleiros	18,480	26,886	34,133	36,874	99.53
Sada	8,935	11,351	14,870	16,463	84.25
ACMA	347,277	370,725	410,401	419,194	20.71

Source Instituto Nacional de Estadística

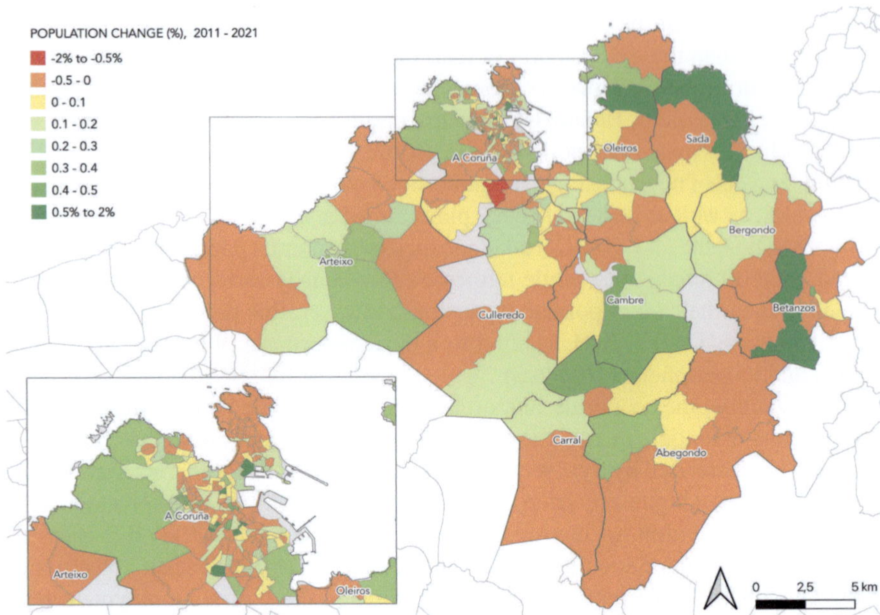

Fig. 8.3 Population change (%) in ACMA, 2021. Map created by the authors with data from the Instituto Nacional de Estadística (2023). In gray, census sections without data due to statistical confidentiality

spatial segregation between residence, work, and leisure in A Coruña generates a significant demand for mobility that is challenging to address.

Understanding the socioeconomic dimension of ACMA requires a specific focus on the role played by its main economic powerhouse: the INDITEX Group. Recognized as a global leader in fast fashion through its own brands like ZARA and PULL&BEAR, this corporate giant is headquartered in Arteixo, western ACMA. INDITEX was founded in 1985 with its flagship brand ZARA and has undergone a remarkable rise to become a global fashion brand. Its journey began with domestic expansion in Spain in the late 1980s and early 1990s, followed by a strategic push into international markets, notably in Europe and the United States in the 2000s. Today, INDITEX operates in 202 markets across five continents, boasting a vast network of over 7,300 stores in 96 markets and a robust online platform. From Arteixo, INDITEX manages essential aspects such as organization, design, planning, and forecasting for the entire group. Thus, a symbol of globalized fashion trends paradoxically addresses its production planning within a relatively compact metropolitan space. While the company's annual reports do not provide segmented labor information (see INDITEX Group 2023), it's estimated that around 5,000–6,000 individuals work directly for one of its companies at the headquarters. According to the last ARDAN report (2023), the economic profile of A Coruña and surrounding municipalities is characterized by a notable emphasis on the textile and apparel sector. This sector holds the top

Table 8.2 The 20 companies with the highest sales volume in the province of A Coruña (2021)

No	Company	2021 sales (€)	Workers	Sector	Municipality
1	INDITEX	11,351,000,000	–	Clothing	**Arteixo**
2	ZARA España	2,143,936,968	13,157	Clothing	**Arteixo**
3	Metalurgia Galaica	1,593,940,299	55	Steel	Naron
4	Bershka España	1,365,999,745	3,482	Clothing	**Arteixo**
5	Pontegadea	1,288,000,000	17	Real state	A Coruña
6	Vego Supermercados	–	6,136	Food retailers	A Coruña
7	Gadisa	1,054,437,398	705	Food products	**Betanzos**
8	ABANCA	1,047,715,000	–	Financial intermediaries	**Betanzos**
9	TEMPE	–	1,680	Clothing	**Arteixo**
10	Fashion Logistics	–	–	Land transport	**Arteixo**
11	Pull&Bear España	944,062,785	2,731	Clothing	Narón
12	Stradivarius España	904,023,423	3,790	Clothing	**Arteixo**
13	FINSA	820,509,198	2,596	Wood industry	Santiago
14	Massimo Dutti	712,479,319	2,520	Clothing	**Arteixo**
15	Hijos de Rivera	573,555,931	890	Beverages and others	A Coruña
16	ZARA Home	555,323,224	491	Clothing	**Arteixo**
17	Mercartabria	532,659,982	2,989	Food retailers	**Betanzos**
18	Conservas Rianxeira	491,685,267	292	Food products	Boiro
19	Aluminios Cortizo	446,884,730	955	Non-ferrous metals	Padron
20	Rinsa del Noroeste	443,983,112	747	Food products	Ribeira

In bold, companies located in areas of the ACMA other than A Coruña ($n = 11$), with the majority of them operating in the textile sector ($n = 7$)
Source Informa, Business Information (website): https://data.humdata.org/dataset/kontur-population-dataset

position in the metropolitan gross value added (GVA), contributing 20.87% to the local economy. Within this sector, INDITEX's dominance is unmistakable, with the company accounting for an impressive 94.37% share of the sector's GVA and employing more than half (53.25%) of the sector's workforce in A Coruña.

In addition to the textile industry, the retail sector for clothing and accessories also plays a significant role, contributing 14.42% to the regional GVA. This sector is heavily influenced by INDITEX as well, with 7 out of 55 businesses belonging to this group, accounting for 99.55% of the sector's GVA in the Comarca. Furthermore, other sectors like business consultancy and management also contribute to the economic landscape of this area. Medium to highly qualified professionals in fields such as international trade, design, and ICT industries collectively transform the metropolitan area of A Coruña into a burgeoning employment hub in the northwest Iberian region.

Certainly,INDITEX's presence in ACMA transcends mere labor dynamics. In this context, a phenomenon popularized in recent years by a local media outlet but difficult to define emerges: the "INDITEX Effect" (Becerra 2017). This denotes a process of social change in A Coruña, as well as an aesthetic transformation of the city, primarily catalyzed by the emergence of a new creative class. Its most immediate expression has been observed in the traditional neighborhoods of the urban center. Here, districts previously impacted by degradation until the early 2000s have undergone a commercial revitalization with an almost underground aesthetic (Rodríguez-Barcón 2020). Fashion hubs, design studios, and business models enriched by the synergies generated by the multinational INDITEX have reshaped the image of the urban center. Nevertheless, pockets of significant vacant housing still persist. This phenomenon has been analyzed and discussed by various scholars, characterizing it as "boutification" or commercial gentrification not necessarily preceded by residential gentrification (Rodríguez-Barcón 2020; Moledo-Mariño 2018). If the new affluent individuals aren't primarily residing in the center of A Coruña, where can we find them?

8.4 Nouveau Rich Segregation in A Coruña: An Issue at Two Scales

The processes of segregation among local economic elites emerge as a result of the distinctive urban evolution within the ACMA. In a regional context marked by deruralization, where the major urban centers of Galicia are often referred to as the beneficiaries of an increasingly urbanized region, the phenomenon of the *Great Coruña* can only be comprehended beyond its municipal boundaries. This expansion ushers in the potential for a bifurcated scenario of segregation: firstly, within the urban core, predominantly encompassing the historic center and the expansion district, traditional neighborhoods become focal points of interest for the creative classes. In line with Rodríguez-Barcón's (2020) case study in central A Coruña, reinvestment in these areas can subsequently trigger processes of gentrification. Secondly, on a broader scale, the consolidation of the shift away from a compact city model—frequently mediated by the development of residential complexes or detached housing units—constitutes another significant domain susceptible to segregation. In this context, our study centers on the socio-residential patterns inherent in the labor profiles of the so-called "INDITEX Effect", specifically focusing on highly paid workers.

In recent decades, the ACMA has seen a percentage increase in highly paid residents.[2] Although the proportion of individuals dedicated to director and manager

---

[2] According to the Instituto Nacional de Estadística, in the year 2021 in Spain, the groups of Directors and Managers and Professional Scientific and Intellectual Technicians had the highest representation in high salaries (2342.2 euros or more). In fact, the average monthly salary of those employed in these two groups was 4418.4 euros for the first group and 3141.1 euros for the second group. So these two groups are where the highest paying professions and occupations are concentrated.

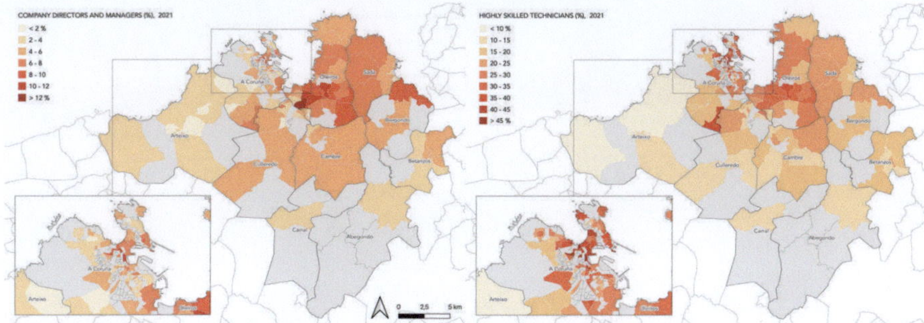

Fig. 8.4 Percentage (%) of highly skilled workers in ACMA, 2021. Map created by the authors with data from the Instituto Nacional de Estadística (2023). In gray, census sections without data due to statistical confidentiality

positions has decreased in the total workforce of the ACMA between 2011 and 2021 (from 5.2 to 4.2%), the number of professional scientific and intellectual technicians has increased and currently represents 21.9% of the total workforce. In the metropolitan ensemble, higher values are observed for both groups compared to the corresponding data for the entire Galicia, where the first group represents 3.9% of the total workforce, while the second group represents 17.9%. However, when considering the sub-municipal scale for analysis and taking into account the data illustrated in Fig. 8.4, a different panorama is detected.

The residential distribution of these professional groups follows certain distinctive patterns. In the case of the group of directors and managers, there is a noticeable inclination toward residing in traditionally affluent areas of the metropolitan area near the Oleiros, traditionally known as having the best living conditions in the area We see that the highest proportions of these professions in the workforce are recorded in the census sections of the municipalities of Coruña, Oleiros, Sada, and Bergondo. However, it is in the municipality of Oleiros where the highest values are reached, especially in the census sections in the western part of the municipality, those closest to A Coruña, where in many census sections the professions and occupations associated with managerial roles represent more than 10% of the total.

Inversely, the group of Professional Scientific and Intellectual Technicians is more concentrated in A Coruña's city center, where they can easily access a range of leisure and cultural amenities. In fact, it is in the municipality of A Coruña where the highest values of workers of this class are reached. In this municipality, there are several census sections where over 45% of employed individuals are engaged in professions and occupations associated with this group. The majority of these sections are located in iconic areas of the city: on the one hand, in its historical zone, and, on the other hand, in one of the most important economic activity hubs of the city, the Orzán and Peixaría area. It is precisely in the municipalities of A Coruña and Oleiros where the highest proportions of these workers are recorded, 25.6% and 26.7% respectively. In Oleiros, the residential areas where there are more significant percentages of these workers coincide with those of the first group. It is

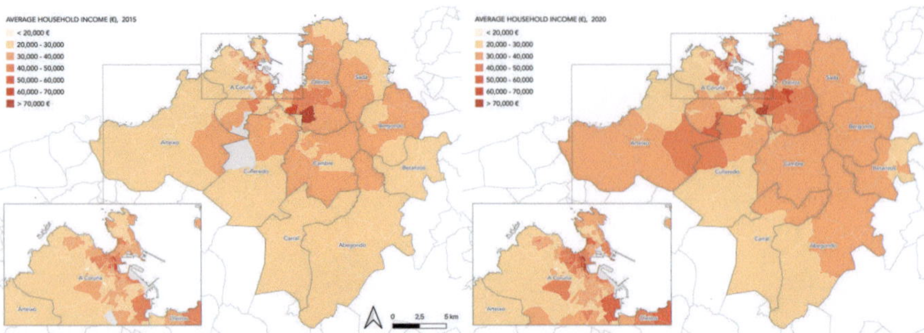

Fig. 8.5 Average household income (€), 2015–2020. Map created by the authors with data from the Instituto Nacional de Estadística (2023). In gray, census sections without data due to statistical confidentiality

noteworthy that, although these two groups follow similar distribution patterns as they are concentrated in the municipalities in the eastern quadrant of the northern metropolitan area, while individuals engaged in management and leadership roles have a higher presence in the coastal territories of the estuary, Professional Scientific and Intellectual Technicians have a greater presence in the central areas of the city of A Coruña.

It is in this metropolitan area where three of the five municipalities with the highest per capita gross income in Galicia in 2020 are located: Oleiros (23,238€), A Coruña (19,409€), and Bergondo (18,335€). However, when considering gross income per household at the sub-municipal scale, as illustrated in the maps in Fig. 8.5, the territory is characterized by different income levels. According to the 2020 data, pockets of wealth in this territory are located in the municipalities of Oleiros and A Coruña. In these municipalities, we can observe a higher number of census sections where gross household income exceeds 60,000 euros. It is worth mentioning that the census sections with the highest income levels are located in the central zone of the municipality of A Coruña and along the coastal strip of Oleiros.

In reality, the territorial and socioeconomic dichotomy illustrated by these images between the municipalities to the east and west of the Municipality of A Coruña can be explained by the different development trajectories of the municipalities. If in the case of Arteixo, the development model followed resulted in a high percentage of industrial land and a significant proportion of high-rise housing, in Oleiros, single-family housing predominates. In fact, along with the municipality of Cambre, Oleiros is the municipality where the percentage of employed individuals who do not commute outside the municipality for work is the lowest (approximately 13%), a percentage very different from what is recorded in Coruña (59.8%), Betanzos (28.8%), and Arteixo (27.2%).

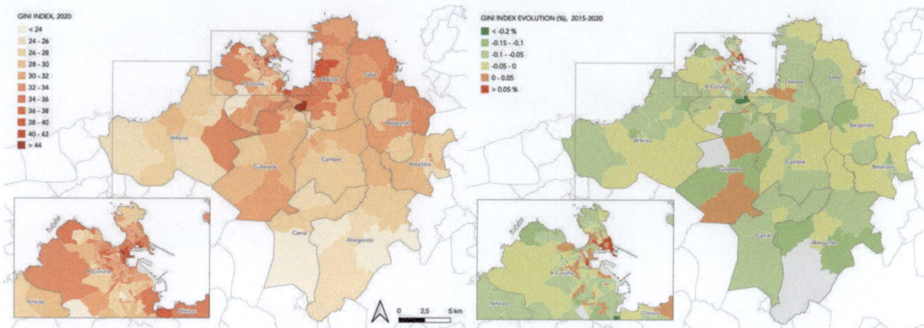

Fig. 8.6 On the left, Gini index, 2020. On the right, Gini index evolution (%), 2015–2020. Map created by the authors with data from the Instituto Nacional de Estadística (2023). In gray, census sections without data due to statistical confidentiality

Areas with the highest concentration of wealth tend to be precisely those where the Gini index,[3] which measures economic inequality, reaches higher levels, as illustrated in Fig. 8.6. The census sections in A Coruña, Oleiros, and Sada that record higher income levels are also the areas where the Gini index is higher. This connection between wealth distribution and economic inequality presents significant challenges in terms of equity and social justice. However, as depicted in the map on the right side of Fig. 8.6, between 2015 and 2020, there has been a trend of decline in this indicator in most of the territory, except for certain areas in A Coruña, Oleiros, and Culleredo.

The housing market has witnessed a consistent increase in prices in recent years, particularly following a brief deceleration period in 2020, which was attributed to the COVID-19 pandemic. The housing prices continue to surge due to a combination of factors. These include the negligible profitability of deposits, a growing demand for more spacious and larger dwellings, the presence of real estate investment funds, and highly favorable real estate financing conditions. This is a trend that has impacted the entire Spanish territory, albeit to varying degrees, including municipalities in the ACMA. Such an increase in housing prices has repercussions on rental prices, which in recent years have reached historic highs in almost the entire country, including the study area. This is one of the initial observations when examining the maps in Fig. 8.7. Between 2015 and 2021, the average rental price per square meter increased in nearly all municipalities, with particularly significant increases noted in A Coruña (20.6%) and Oleiros (20.7%). It is precisely in these two municipalities where census sections with the highest average rental prices are located, with several of them surpassing values of 700 euros.

---

[3] The value of the Gini index ranges from 0 to 100, with zero representing maximum equality (all citizens have the same income) and 100 representing maximum inequality (all income is held by a single citizen).

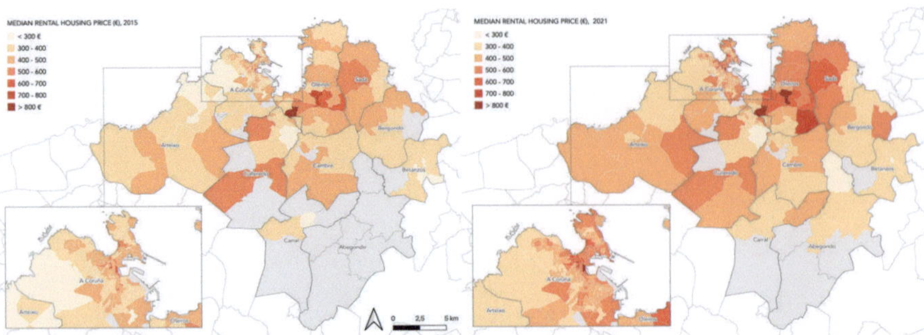

Fig. 8.7 Median rental housing price in ACMA, 2015–2021. Map created by the authors with data from the Instituto Nacional de Estadística (2023). In gray, census sections without data due to statistical confidentiality

8.5 Discussion and Conclusions

The findings from this study reveal that, consistent with prior research conducted in various European contexts (Haandrikman et al. 2023; Nel·lo and Gomá 2018), there are conspicuous and enduring patterns of socioeconomic segregation within the ACMA. This phenomenon extends to both the metropolitan and municipal levels. Notably, residential segregation can be attributed to disparities in the ability to choose one's place of residence, with financial resources playing a pivotal role in shaping housing choices (Hulchansky 2010). Individuals with higher incomes possess the financial means to select housing options that align with their preferences, thereby contributing to an expanding spatial divide between those with substantial incomes and those with lower incomes. As highlighted by the contextual analysis presented in this study, the ACMA has experienced a significant economic transformation driven by the presence of global giants like INDITEX and other corporations. Multinational companies such as INDITEX not only attract highly skilled workers but also stimulate economic activity, resulting in a concentration of executives and managers within the organization and its associated supply chains. Nevertheless, the municipalities housing these economic activities do not consistently align with the housing preferences of higher earning individuals.

Although spatial sorting of residents according to socioeconomic status is an old urbanistic phenomenon (Nightingale 2012), contemporary trends such as increasing economic inequalities and the intensification of urban expansion beyond administrative borders have added layers of complexity to the understanding of urban segregation. Various studies and overviews have furthered the discourse on how to measure segregation and at what scale, illustrating an evolution from mono-centric urban structures to more dispersed models (Tammaru et al. 2016). This transformation exposes a dynamic interplay between economic and social dynamics in inner cores and the surrounding areas, as the results of our study show. Although the municipality of Arteixo is an important hub of economic activity in ACMA, high-income people are concentrated in certain parts of the municipalities of A Coruña and Oleiros. Indeed, even within these two municipalities, high-paying jobs are distributed unevenly across the area. This disparity may suggest the emergence of specialization in economic and social functions within this metropolitan area, with certain areas becoming hubs for economic activities and others primarily designated for residential purposes.

The results allow us to identify two attractive areas for high-income households: first, areas previously recognized as high-income neighborhoods, and second, the city center of A Coruña, the most populated municipality of the ACMA, which is considered a hub for consumption, consumer services, and sought-after amenities. In fact, a central finding of this study highlights the observed residential duality between the group of executives and directors and highly skilled technicians. This contrast is evident at a metropolitan level through a specific residential pattern: the former group predominantly maintains their housing positions in the urban periphery, while the latter gravitates toward compact neighborhoods, typically situated in central areas and near the coastline. In the context of classical approaches to occupational-based segregation (Berry and Glaeser 2005), these results align with the emergence of a new creative middle class (Boterman et al. 2018), a trend accentuated by the presence of INDITEX in the metropolitan space. A Coruña thus amplifies residential interest, likely mediated by variables such as leisure and consumption, as suggested by Moledo Mariño (2018) in their exploration of the central Orzán neighborhood. In this regard, the findings have confirmed an accompanying increase in rental prices in neighborhoods of heightened interest to the emerging creative middle class.

This duality appears to extend to the realm of social inequality. In a context where the ACMA initially had high levels of inequality, the results reveal a decrease in metropolitan inequality over time, with the exception being the neighborhoods of greatest interest to the creative middle class. Therefore, the trajectory of inequality is more problematic in these areas than in traditional elite residential areas. In line with the arguments put forth by Penco (2015), the consumption vibrancy and the emergence of small central businesses linked to the high-tech sector in the urban center of A Coruña are driving a significant revaluation of several central neighborhoods. In this regard, the data indicate a process of gentrification in areas of interest such as the Old Town or specific neighborhoods like the Peixería, particularly in the housing sector.

In conclusion, we can note that the so-called "INDITEX Effect", mediated by the influx of a new creative class to the vicinity of A Coruña, is accentuating a shift in metropolitan residential patterns. The residential segregation practices of the dominant classes that were prevalent until a decade ago, characterized by the development of private urbanizations in the urban periphery, are giving way to a revaluation and elitization of central spaces. This process aligns with other urban examples in the Iberian Peninsula and internationally. While the socioeconomic transformation brought about by the textile sector appears to be a driver of this process, future studies based on specific data related to the business group could shed more light on the elitization and segregation of the key neighborhoods targeted by its associated working creative class. In any case, it is already evidenced that the highest tensions regarding current housing prices are concentrated in the areas of greatest residential interest to the local new creative middle class.

The integration and growing interdependence of the territory are among the key consequences of metropolitan dynamics. However, it is important to note that the extension of these dynamics does not lead to a uniformization of living conditions or income levels across the encompassed territories (Nel·lo and Gomà 2018). Indeed, metropolitan networks tend to promote the specific roles and functions of various areas, and when land and housing markets are integrated at the metropolitan level, this tends to intensify the patterns of residential segregation. Consequently, each neighborhood not only develops unique characteristics based on its municipality but also takes on distinctive traits in connection with the larger metropolitan context to which it belongs. The concentration of high-income workers in certain municipalities, as evident in the ACMA, has implications on urban conditions and the municipal resources, creating a challenging situation where municipalities with fewer needs tend to possess more resources. This resource disparity significantly impedes the ability of local governments to implement urban and social policies aimed at addressing income inequalities and enhancing service accessibility. That is why it is so important to have governing structures functioning at a metropolitan scale. Such metropolitan governance allows for a more equitable distribution of resources and coordinated efforts to tackle these pressing issues effectively, transcending the limitations of individual local administrations.

Acknowledgements The work of Inês Gusman has been supported by the grant UP2021-042 (Ayudas Margarita Salas) funded by Ministerio de Universidades (Spanish Government), Next Generation EU.

References

ACMA—A Coruña Metropolitan Area (2023) Área Metropolitana de A Coruña: Inicio. Available online in: http://www.amac.gal/estrategia-metropolitana-amac/es/area-metropolitana-de-a-coruna?argIdioma=es

Ariño A, Romero J (2016) La secesión de los ricos. Galaxia Gutenberg, Barcelona

Arundel R, Hochstenbach C (2020) Divided access and the spatial polarization of housing wealth. Urban Geogr 41(4):497–523

Atkinson AB (2015) Inequality. Harvard University Press

Becerra J (2017) El efecto inditex transforma A Coruña [The inditex effect transforms A Coruña]. La Voz de Galicia. Available online in: https://www.lavozdegalicia.es/noticia/coruna/coruna/2017/09/24/empuje-inditex-moderniza-ciudad/0003_201709H24C2991.htm

Berry CR, Glaeser EL (2005) The divergence of human capital levels across cities. Pap Reg Sci 84(3):407–444. https://doi.org/10.1111/j.1435-5957.2005.00047.x

Blanco I, Nel·lo O (2019) Políticas públicas contra la segregación urbana. https://www.barcelona.cat/metropolis/es/contenidos/politicas-publicas-contra-la-segregacion-urbana

Bolt G (2009) Combating residential segregation of ethnic minorities in European cities. J Hous Built Environ 24(4):397–405

Botana C (2020) La promoción pública de segregación urbana. Crít Urbana 12:17–20

Boterman WR, Manting D, Musterd S (2018) Understanding the social geographies of urban regions through the socio-economic and cultural dimension of class. Popul Space Place 24(5). https://doi.org/10.1002/psp.2130
Concello da Coruña (2017) Estratexia de Desenvolvemento Urbano Sostible Integrado de A Coruña. Available online in: https://www.coruna.gal/movilidad/es/planes-y-programas/eidus?argIdioma=es
Duncan OD, Duncan B (1955) A methodological analysis of segregation indexes. Am Sociol Rev 20(2):210
Echazarra A (2010) Segregación residencial de los extranjeros en el área metropolitana de Madrid. Un análisis cuantitativo. Rev Int Sociol 68(1):165–197
Florida R, Mellander C (2010) There goes the metro: how and why bohemians, artists and gays affect regional housing values. J Econ Geogr 10(2):167–188. https://doi.org/10.1093/jeg/lbp022
Florida R, Mellander C (2016) The geography of inequality: difference and determinants of wage and income inequality across US metros. Reg Stud 50(1):79–92. https://doi.org/10.1080/00343404.2014.884275
Fujita K, Maloutas T (2012) Residential segregation in comparative perspective making sense of contextual diversity. Routledge
González-Pérez JM, Lois-González RC, Piñeira-Mantiñán MJ (2016) The economic crisis and vulnerability in the Spanish cities: urban governance challenges. Procedia Soc Behav Sci 223:160–166
Grossmann K (2019) Energy efficiency for whom? A conceptual view on retrofitting, residential segregation and the housing market. Sociol Urbana Rurale 119
Grupo Inditex (2022) Memoria Anual Grupo Inditex. Available online in: https://static.inditex.com/annual_report_2022/pdf/Memoria-anual-grupo-inditex-2022.pdf
Haandrikman K, Costa R, Malmberg B, Rogne AF, Sleutjes B (2023) Socio-economic segregation in European cities. A comparative study of Brussels, Copenhagen, Amsterdam, Oslo and Stockholm. Urban Geogr 44(1):1–36. https://doi.org/10.1080/02723638.2021.1959778
Hammel D (2009) Gentrification. In: Kitchin R, Thrift N (eds) International encyclopedia of human geography. Elsevier, pp 360–367
Harvey D (1973) Social justice and the city. Edward Arnold
Harvey D (2013) The political economy of public space. In: The politics of public space. Routledge, pp 23–188
Hulchansky D (2010) The three cities in Toronto: income polarization among Toronto's neighbourhoods, 1970–2005. University of Toronto
Instituto Nacional de Estadística (2023) Censo de Población y Viviendas 2021. Resultados sobre Hogares y Viviendas. Available online in: https://www.ine.es/prensa/censo_2021_jun.pdf
Jayne M (2005) Cities and consumption. Routledge
K'akumu OA, Olima WH (2007) The dynamics and implications of residential segregation in Nairobi. Habitat Int 31(1):87–99
Leal J, Sorando D (2016) Economic crisis, social change and segregation processes in Madrid. In: Socio-economic segregation in European capital cities: east meets west, pp 214–237
Lens MC, Monkkonen P (2016) Do strict land use regulations make metropolitan areas more segregated by income? J Am Plann Assoc 82(1):6–21
Maloutas T, Fujita K (2016) Residential segregation in comparative perspective: making sense of contextual diversity. Routledge
Massey DS, Denton NA (1988) The dimensions of residential segregation. Soc Forces 67(2):281–315. https://doi.org/10.1093/sf/67.2.281
Moledo Mariño N (2018) Proceso de gentrificación en el barrio del Orzán (A Coruña): percepción vecinal y análisis crítico de un fenómeno de transformación urbana. Available online in: https://ruc.udc.es/dspace/bitstream/handle/2183/22288/MoledoMarino_Noa_TFG_2018.pdf?sequence=2

Nel·lo O (2016) Desigualdad social y desegregación urbana en la región metropolitana de Barcelona. In: García DB, del Río Lafuente MI, Matos GM (eds) Estudios de geografía urbana en tiempos de crisis: Territorios inconclusos y sociedades rotas en España. Biblioteca Nueva, pp 95–120
Nel·lo O (2021a) Efecto Barrio. Segregación residencial, desigualdad social y políticas urbanas en las grandes ciudades ibéricas. Tirant lo Blanch
Nel·lo O (2021b) Urban movements and the challenges of the European city. In: Social movements and public policies in southern European cities. Springer International Publishing, pp 15–32
Nel·lo O, Gomà A (2018) Geographies of discontent: urban segregation, political attitudes and electoral behaviour in Catalonia. City, Territory Architect 5(1). https://doi.org/10.1186/s40410-018-0099-z
Nel·lo O, Donat Muñoz C (2023) Segregación residencial y recursos locales en el área urbana de Valencia: la necesidad de gobernanza metropolitana. Cuad Geogr Univ València 110:289. https://doi.org/10.7203/CGUV.110.25423
Nightingale CH (2012) Segregation: a global history of divided cities. University of Chicago Press
OECD (2024) Income inequality (indicator). https://doi.org/10.1787/7f420b4b-en. Accessed 20 May 2024
Penco L (2015) The development of the successful city in the knowledge economy: toward the dual role of consumer hub and knowledge hub. J Knowl Econ 6(4):818–837. https://doi.org/10.1007/s13132-013-0149-4
Piekut A (2021) Re-theorising spatial segregation: a European perspective. In: Urban inequality and segregation in Europe and China. Springer, pp 13–38
Piketty T (2014) Capital in the twenty first century. Harvard University Press
Piñeira-Mantiñán MJ, Durán-Villa FR, Taboada-Failde J (2018) Urban vulnerability in Spanish medium-sized cities during the post-crisis period (2009–2016). The cases of A Coruña and Vigo (Spain). Urban Sci 2(2):37
Rodríguez-Barcón A (2020) El papel de la clase creativa en la gentrificación comercial del centro histórico de A Coruña. Finisterra 55(113):89–116
Romero-González J, Farinós-Das J (2016) Cities and urban and metropolitan regions in Spain: a new agenda in a global context. In: Urban governance in southern Europe
Rubiales Pérez M, Bayona J, Pujadas I (2012) Patrones espaciales de la segregación residencial en la Región Metropolitana de Barcelona: Pautas de segregación de los grupos altos. Scripta Nova. Rev Electrón Geogr Cienc Soc 16(423)
Ruiz-Tagle J (2016) La segregación y la integración en la sociología urbana: revisión de enfoques y aproximaciones críticas para las políticas públicas. Rev INVI 31(87):9–57
Sampson RJ, Levy BL (2020) Beyond residential segregation: mobility-based connectedness and rates of violence in large cities. Race Soc Probl 12(1):77–86
Sassen S (1991) The global city: New York, London, Tokyo. Princeton University Press
Secchi B (2013) La città dei ricchi e la città dei poveri. Laterza, Bari
Servicio de Información Empresarial ARDÁN (Consorcio de la Zona Franca de Vigo) (2023) ARDÁN GALICIA 2023. Informe Económico y de Competitividad
Shertzer A, Twinam T, Walsh RP (2022) Zoning and segregation in urban economic history. Reg Sci Urban Econ 94:103652
Suárez AL (2011) Segregación residencial en la región metropolitana de Buenos Aires [en línea]. In: Balian de Tagtachian B, Suárez AL (coords) Pobreza y solidaridad social en la Argentina: aportes desde el enfoque de las capacidades humanas. Educa
Tammaru T, Marcińczak S, van Ham M, Musterd M (2016) Socioeconomic segregation in European capital cities: east meets west. Routledge
Tammaru T, Marcińczak S, Aunap R, van Ham M, Janssen H (2020) Relationship between income inequality and residential segregation of socioeconomic groups. Reg Stud 54(4):450–461. https://doi.org/10.1080/00343404.2018.1540035
Tulumello S, Colombo A (2018) Inclusive communities, exclusionary city, planning n/a? Mapping condomínios fechados semi-quantitatively in Lisbon, Cascais (and Barreiro). In: Changing societies: legacies and challenges, vol 1, pp 481–507

van Ham M, Uesugi M, Tammaru T, Manley D, Janssen H (2020) Changing occupational structures and residential segregation in New York, London and Tokyo. Nat Hum Behav 4(11):1124–1134. https://doi.org/10.1038/s41562-020-0927-5

van Ham M, Tammaru T, Ubarevičienė R, Janssen H (2021) Urban socio-economic segregation and income inequality. In: van Ham M, Tammaru T, Ubarevičienė R, Janssen H (eds) Springer International Publishing

Vaughan L, Arbaci S (2011) The challenges of understanding urban segregation. Built Environ 37(2):128–138

Xunta de Galicia (2011) As Directrices de Ordenación Territorial (DOT): Descarga dos documentos das directrices da ordenación do territorio. Available online in: https://cmatv.xunta.gal/seccion-organizacion/c/CMAOT_Instituto_Estudos_Territorio?content=Direccion_Xeral_Sostibilidade_Paisaxe/Directrices_Ordenacion_Territorio/seccion.html&std=descarga-dot.html

Zamora-Kapoor A, Coller X (2014) The effects of the crisis: why southern Europe? Am Behav Sci 58(12):1511–1516

Part III
Finnance and Real Estate

Chapter 9
Urban Change in the Iberian Peninsula: A 2000–2030 Perspective—Housing, Real Estate, and the Economy in Portuguese Cities

Luís Carvalho

Abstract Urban economic geography must acquire an increasing role in the study of the city and its growth processes. This is what is done in this chapter reserved for Portugal. On the one hand, the main cities and metropolitan areas have concentrated on the most competitive companies and economic sectors while growing. On the other hand, urban space has become a privileged place for business and income accumulation, based on real estate activities. Shops, department stores, and homes are products that are bought, sold, and exchanged in a voracious market, a market that has attracted large sums of increasingly internationalized capital. The financialization of the Portuguese city has been consummated in recent decades, and the period of acute economic crisis at the beginning of the 2010s has only intensified this phenomenon. Large fortunes and investment funds acquire a growing volume of houses, apartments or entire neighbourhoods in Lisbon and Porto, to develop the lucrative business of speculation. At the same time, the tourist attraction generated by these cities, and all of Portugal, translates into the proliferation of tourist rentals, which contribute to the expulsion of traditional residents from urban centres. All these processes, in an orderly manner, and distinguishing the metropolitan realities of Lisbon and Porto from the rest, are studied for Portugal in this chapter.

Keywords Housing crisis · Supply scepticism · Zoning · Financialization · Porto · Portugal

> To be clear, it is difficult to see how the programme will create new housing in an effective and fast manner. (…) [I don't see it coming out of] forcing landlords to rent, nor from [retrieving back] short term rentals [properties] (…) nor in the total absence of parliamentary political consensus. (…) [It is] an example of a bad start to deal with a need that became dramatic, crucial, and very urgent (…)

L. Carvalho (✉)
School of Economics and Management, Centre of Economics and Finance (CEF.UP), Centre of Studies in Geography and Spatial Planning (CEGOT), University of Porto, Porto, Portugal
e-mail: lcarvalho@fep.up.pt

R. C. Lois-González and J. A. Rio Fernandes (eds.), *Urban Change in the Iberian Peninsula*, The Urban Book Series, https://doi.org/10.1007/978-3-031-59679-7_9

(Marcelo Rebelo de Sousa, portuguese president, on vetoing a new national policy package to deal with the housing crisis, 21.08, 2023)

9.1 Why Have Housing and Real Estate Become so Important Recently?

In August 2023, the Portuguese president Marcelo Rebelo de Sousa vetoed a legislation package called "*Mais Habitação*" ("*More Housing*"), prepared by the national government, to deal with the ramping housing crisis affecting major Portuguese cities (Rebelo de Sousa 2023). Independently of his (political) reasons, this veto illustrates two major facts. On the one hand, like in other countries (e.g. Rodriguez-Pose and Storper 2020), housing shifted from being a local concern to becoming a national issue of major political relevance. On the other hand, it highlights the difficulty in reaching a political consensus on the most appropriate solutions to deal with the problem.

But why has housing become such a key issue in national Portuguese politics? It is difficult to downplay the role that housing always had in cities, in general, and in sustainable urban development, in particular. The provision and access to housing—a constitutional right in many countries, like Portugal—is closely related to real estate markets, as well as to the rights and interests associated with it (Mendes and Tulumello, 2022). For many households, owning or renting a dwelling is the largest expenditure and investment they make in a lifetime (Costa, 2009); housing provides shelter, security, privacy, and identity for individuals and families, and is closely related to social and emancipation, in particular for young people. Hence, housing quality and affordability has always had significant impacts on the well-being, health, and economic productivity of cities (Glaeser, 2008). Likewise, in Portugal and elsewhere, housing is also a factor shaping urban social inclusion (or exclusion) and environmental quality, as it influences the distribution and accessibility of land and buildings, sprawl, and access to employment, among many others.

Yet, the interest in housing—both from a political and academic perspective—has only recently grown significantly in Portugal. To be sure, housing had been a major issue in Portugal in the aftermath of the "Carnation Revolution" of 1974 but became much less so after the country joined the European Union (in 1986) and the Eurozone (in 2002), which came with low interest rates, a boost in new construction. The situation contours started to change in the aftermath of the 2008–12 financial crisis, namely as the Portuguese housing market became affected by the increasing mobility of people and investment, both domestically and internationally (e.g. Carvalho et al., 2019; Rodrigues, 2022). The rising attractiveness of the country's major cities as touristic and knowledge "hubs" led to heightened flows of visitors, students, professionals, and investors, and had an impact on the demand and supply of housing, as well as on housing prices and rents. Also, technological change played a role, especially the emergence of online platforms that facilitate short-term rental of housing units, such as Airbnb (Fernandes et al. 2018, 2019). These platforms have created new opportunities and challenges for the housing market, as they affect the profitability, competition, and regulation of different types of housing, and had profound impacts in cities like Porto and Lisbon (Jover and Cocola-Gant, 2023; Chamusca et al., 2019).

And still, the aftermath of the COVID pandemic raised the housing issue in Portugal to the status of a "crisis". Like in many other European countries, the post-pandemic economy came with higher construction costs (linked to the scarcity of materials), labour shortages (as many workers left the construction industry), overall inflation, and thus the substantial rise of interest rates and increasing difficulties in access to credit (e.g. Rodrigues, 2022). The COVID pandemic has also changed the patterns of living and working, increasing the importance of staying at home (Mendes, 2020), as well as the demand for more space and amenities in housing units—at least for the ones that can afford it. Likewise, the explosion of remote work brought Portugal onto the map of so-called "digital nomads", thus creating new housing pressure in amenity-rich and internationally well-connected Portuguese cities (e.g. Borges et al., 2022).

The remainder of this chapter analyses how the housing affordability crisis revealed itself and grew in Portugal over the last decade—in Porto and Lisbon in particular—reviewing what its main causes appear to be. Then, it investigates the variegated policy perspectives that have emerged to (try to) deal with it in the country, framing these responses in relation to the international scientific debate that opposes "housing-as-opportunity" and supply-side perspectives with demand curbing and "supply scepticism" viewpoints. The chapter concludes with a reflection on what the implications are for cities of the new national positioning of the housing debate in Portugal.

9.2 The Reasons Behind the Crisis: Speculation or New Market Fundamentals?

Perhaps the simplest and more telling visualization of the Portuguese urban housing crisis is the one in Fig. 9.1. It illustrates how the prices of houses have increased significantly above the income levels of most households over the last decade. As described in a recent study by FFMS (2023), this has led to a deterioration of housing affordability, both for purchase and rental. The reminder of this section largely draws on this study to review some of the most significant evidence of such a crisis, and what it looks like.

One of the indicators that shows the evolution of house prices is the real house price index, which adjusts the nominal prices for inflation. According to FFMS (2023), and based on official national statistics, this index has risen by more than 30% since 2013, reaching a historical high in 2020, despite the economic contraction caused by the covid pandemic. The increase in house prices has been more pronounced in Lisbon and Porto, where the index has grown by more than 50%

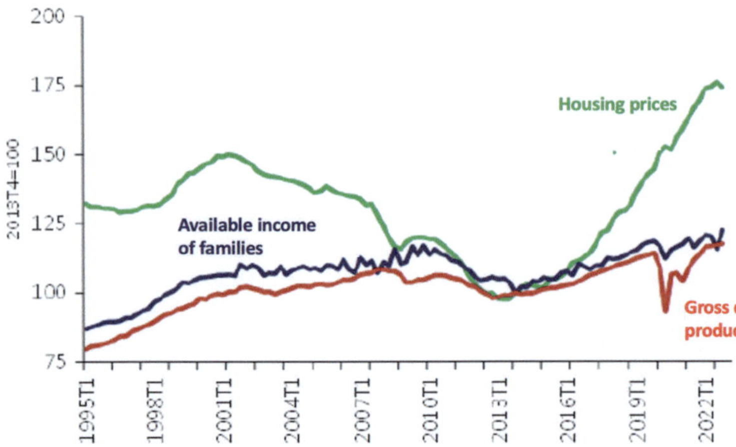

Fig. 9.1 Gross domestic product, available income, and housing prices. *Source* FFMS—Fundação Francisco Manuel dos Santos (2023)

and 40%, respectively, in the same period. Another indicator that reflects the deep social impact of house prices is the income required to buy or rent a median house in Porto and Lisbon. This indicator measures the minimum percentile of the income distribution that a household needs to have to afford the housing costs, assuming a maximum effort rate of 40%. According to FFMS (2023)—based on the national survey of household income, and on local housing price statistics—for a household with two active members, the income required to buy a median house in Lisbon or Porto increased considerably between 2017 and 2020, reaching the 60th and 70th percentiles, respectively. This means that only the households in the (rather) upper part of the income distribution can afford to buy a house in these cities. The situation is even worse for households with one active member—for example, a single parent, or a young professional—where the income required is above the 90th percentile in both cities (FFMS, 2023). Beyond the numbers, it follows from this that the housing crisis in Portuguese cities is likely having serious social and economic implications, for lower income groups, or people entering the housing market for the first time.

Some authors argue that a major explanation for this state-of-affairs in Portugal—and in Porto and Lisbon in particular—is housing financialization (Cordeiro Santos, 2019; Drago, 2020). This would mean that a large chunk of the housing stock in the most appealing urban districts in Porto and Lisbon would have become speculative assets, being traded and "changing hands" frequently, seeking for ever-growing returns, and not actually available in the market for "real" use. Over the last decade, some anecdotal examples of famous properties allegedly being "financialized" and/or used for very short occupation became popular, being frequently mentioned by pundits and commentators. FFMS (2023) qualitatively tested this hypothesis, employing a model that considers various economic factors—to measure so-called *market fundamentals*—that can influence housing prices, such as disposable income, interest rates, loan stock, residential investment, foreign investment, and unemployment rates. The authors found that, over the last decade, there have indeed been periods in which housing prices in Porto and Lisbon cannot be explained by those variables alone, most notably at the end of 2020 and 2022, suggesting speculative behaviour. However, they also found that more structural changes in the market fundamentals played by far a more decisive role, pointing towards a relevant qualitative change in the real estate markets involving Portugal´s major cities—as actually has been pointed out by many others over the last years (e.g. Rodrigues, 2022; Mendes and Tulumello, 2022).

From the side of housing demand, three major changes are usually referred to. First, the Portuguese population has undergone demographic changes in recent years; there has been a decrease in the number and size of households due to ageing, low birth rates, and increased divorces (Azevedo, 2020), and this has given rise to a substantial mismatch between available and desired housing stock, notably in terms of typology (e.g. smaller housing), location, and quality. Second, as it has been extensively analysed elsewhere (e.g. Cocola-Gant and Gago, 2019; Cordeiro Santos, 2019), Portuguese real estate has attracted the attention of foreign investors and buyers, especially from European citizens benefiting from free capital movements, "golden visas", and other sorts of special residence permits. While the magnitude of this effect is up to debate, such external demand has at least partially contributed to the appreciation of housing prices, especially in tourist and historic areas (FFMS, 2023), and this effect has been shown to be very relevant in relation to the new demand for occupation of properties in the short-term rental market, as early studies in the major Portuguese cities suggest (Fernandes et al., 2018, 2019; Franco and Santos, 2021). Third, it has become increasingly difficult to buy a house through bank loans. Access to housing credit in Portugal has become more challenging in recent years due to rising home prices, employment instability, more restrictive credit conditions—still a consequence of macro-economic prudency after the financial crisis—and worsening interest rates (FFMS, 2023). In this setting, families need to have high incomes, significant savings, and/or family inherited wealth (see Piketty and Zucman, 2015) to be able to purchase a median-priced home, especially in the metropolitan areas of Lisbon and Porto, making younger and more vulnerable families most affected by this situation (FFMS, 2023).

On the supply side, many commentators—and more liberal economists in particular (e.g. Brinca and Duarte 2022; Rodrigues, 2022; Santos Ferreira, 2023)—stress the rigidity of housing supply and the difficulties associated with building more and faster. Housing is known to be a particular type of economic good whose supply is rather rigid in short term, being unresponsive to price changes, and only mildly responsive in medium run (O'Sullivan 2011; Costa, 2009). In the Portuguese context, excessive regulation, bureaucracy in licencing processes, lack of stability in real estate regulation ("scaring out" long-term investors), high tax burden in the sector, the requirement for a substantial amount of equity capital, and unpredictability in assessment processes due to a multitude of norms and interpretations are often pointed as culprits (e.g. Santos Ferreira, 2023). While it is unlikely that decreasing these costs would right away result in lower housing prices and in the development of more affordable housing typologies (Costa, 2009), it is hard to deny that, at least, they pose obstacles on the supply of new housing. In fact, according to the national housing statistics, only 150 thousand new homes have been built in Portugal, which is less that 25% of the number of homes built the decade before (FFMS, 2023). Part of it has to do with the focus on urban rehabilitation, namely in large cities, but also resonates with the lack of labour force and a shrinking construction industry in the aftermath of the financial and covid crises (Santos, 2023). And, on the top of it, with a few exceptions (in Porto), the supply of social housing is widely considered to be marginal in the country.

9.3 How to Deal with It: The Entrenchment of Opposing Views

While most commentators and political parties in Portugal could agree with the key aspects of such a diagnostic, a deeper disagreement emerges when it comes to the most desirable solutions to deal with it. As mentioned—and not without consequences—such a debate has somehow moved away from the pragmatism of the local level to the more ideological context of national politics. And, also in academia and think tanks, there are growing barricades between the proponents of supply side and private-led solutions, on the one hand, and the proponents of deeper interventions in the market, together with the limitation of excessive demand (in particular of international buyers and renters), on the other hand. To be sure, such a divide is not a Portuguese specificity, but one that has grown more markedly in recent years.

One of the broadly proposed strategies to address the growing housing shortages is to increase the supply of new housing, both at "affordable" and at higher market rates, as well as more social and public housing. Many arguments are put forward in favour of such an approach (e.g. Manville et al., 2022, for a review; Gyourko and Molloy 2015). First, their proponents contend that land can be used more intensively to allow for more housing (e.g. by building higher, and in now more protected places, relaxing local zoning regulations), while programmes such as inclusionary zoning—as being tried out in Porto and Lisbon (Alves, 2021; Pinto, 2020)—can ensure that market-rate, expensive housing also includes some affordable units at the lip stream, contributing to social mix in more expensive city areas. Second, they assert that new housing—even if targeting higher market segments, can still lower prices and rents in other submarkets; this would take place by alleviating competition and provoking trickle-down effects, under which older housing—now more affordable—would "filter down" to medium and low market segments (O´Sullivan 2011). While worldwide evidence for filtering is rather scant and results vary substantially, some studies seem to show that building new housing is associated with lower displacement though competition phenomena and citywide price moderation (Manville et al., 2022). Third, building new houses is also seen a driver of economic growth and job creation, which is seem as a particularly relevant argument in semi-peripheral European countries like Portugal (Guimarães, 2023).

From this perspective, while rental subsides to tenants are seen as acceptable in the short run (namely, as it may raise overall market prices), the key levers to be pulled at the side are, on the one hand, zoning de-regulation and licensing bureaucracy reduction and, on the other hand, fiscal stability, low taxation, land provision, and subsidies to real estate developers so that less economically appealing housing are also built in the most expensive cities. Not surprisingly, this approach is largely favoured by market players and their constituencies, namely real estate developers, investors, and the overall industry, and can be seen as a form of *public policy for housing*, instead of *public housing policy*. From this perspective, policies limiting housing and real estate demand, whether from locals or international citizens, tourists, or investors are seen as undesirable (on economic and normative grounds), and ineffective (on legal and regulatory grounds) (see, for example, FFMS, 2023).

Other scholars, commentators, and political parties—notably at the "left" spectrum—contest this vision, highlighting many trade-offs, controversies, and uncertainties. Overall, these arguments can be gathered around what has been called internationally the "supply scepticism" school of thought (Been et al., 2019). First, supply sceptics don't see filtering happening on a significant scale, not in the short run, and not at all for high-end properties in desirable urban locations (e.g. Zuk and Chapple, 2016; Rodriguez-Pose and Storper 2022); they fear that new housing will actually induce more demand, especially from wealthier outsiders, who will bid up prices and offset any possible supply effect. Second, they highlight that more housing across the board—and concomitantly less zonnig and urban environmental regulation—would endanger urban quality, the distinctive amenities, and environmental values that make for the economic attractiveness of many cities (Storper, 2010). Third, they worry that new housing will have other negative spillover effects in neighbourhoods, such as increasing rents, fuelling gentrification, and ultimately displacing existing residents.

More recently, also economic geographers jumped into this debate, casting additional doubts on de-regulation agendas as a means to deal with housing crises in economically dynamic cities (Rodriguez-Pose and Storper, 2020, 2022). Likewise, they point that there is little evidence that the construction of housing in desirable areas of dynamic cities has a trickle-down effect on the rest of the housing market—which can be counteracted by gentrification, financialization, and speculation. Also, they stress that beyond de-regulation, the actual and fundamental determinants of housing prices are land values, economic structure, income distribution, and the location of urban amenities—see also Carvalho (2022a, b), for the case of the Portuguese case. They demonstrate that housing prices are driven by the demand for land in high-productivity and high-income areas, and that density alone does not ensure affordability, and is thus unlikely to have a significant positive effect on those most affected by housing crises.

9.4 Conclusion

Ultimately, the policy package "Mais Habitação" was approved following the presidential veto of August 2023—but still only with the favourable votes by its proposing political party. The package mixes many different types of housing supply and demand side measures, and some policies with a strong interventionist flavour, such as additional taxation and bans for new short-term rental property, forced rental of abandoned dwellings, as well as international demand curbing (notably, the end of "golden visas"). To be sure, these measures might never be enforced—at the time of this writing, new national elections have been called out by the president. Yet, it is clear that real estate and housing issues have gained a much more preeminent role in Portugal over the last years and are most likely to keep having it in the future. Hence, it was not only the nature of the real estate market fundamentals that changed in Portugal and its major cities: the housing debate—and the implications of its shortage and lack of affordability—moved from being a local affair to having a prime, highly contested, and an ideological spot in Portuguese politics.

As mentioned, the design and implementation of housing policy are subject to debates and controversies which, ultimately, also reflect the diversity and complexity of the urban housing market. As argued by Rodriguez-Pose and Storper (2022), the jury is still out on which types of policies work better—while they are sceptical of deregulation and supply-side measures, at least in isolation. However, it can be argued that a balanced policy mix is increasingly required to deal with housing issues in cities, and that the most appropriate policies are context-dependent—which is a hard pill to swallow for the proponents of one-size-fits-all, more ideologically infused and extreme recipes. While places can certainly learn from one another, what works in Copenhagen or North America will probably not be the most effective recipe in Portugal or Spain.

However, while housing politics became national, it would be problematic just to wipe this discussion out of the local level. Ultimately, the effects, and (many of) the causes of housing shortage and affordability issues, boil back to cities. Hence, there is a latent risk of miscoordination between national and local levels when it comes to policy design and implementation. In this sense, both national and local policymakers concerned with urban housing issues would need to combine strong knowledge about how real estate markets work, with a deep place sensitivity and integrated perspectives on the (changing) links between housing, the economy, and other dimensions that make up for sustainable urban development.

References

Alves S (2021) Novas tendências do Planeamento e Gestão Urbanística (New trends in Urban Planning and Management). Confidencial Imobiliário, Research —Academic papers. May 2021

Azevedo A (2020) Como vivem os portugueses? População e famílias, alojamentos e habitação (How do the Portuguese live? Population and families, housing and accommodation). Fundação Francisco Manuel dos Santos, Lisboa

Been V, Ellen IG, O'Regan K (2019) Supply skepticism: housing supply and affordability. Hous Policy Debate 29(1):25–40

Borges I, Brás S, Machado A, Leite S, Costa E, Mota S (2022) Digital nomads: a growing trend in hospitality in Portugal. In: Advances in tourism, technology and systems: selected papers from ICOTTS 2021, vol. 2. Springer Nature Singapore, Singapore, pp 549–559

Brinca P, Duarte (2022) The importance of housing in Portugal: a macroeconomic perspective. In Rodrigues PMM (2022) The real estate market in Portugal: prices, rents, tourism and accessibility. Fundação Francisco Manuel dos Santos, Lisboa

Carvalho L, Chamusca F, Fernandes JAR, Pinto J (2019) Gentrification in Porto: floating city users and internationally-driven urban change. Urban Geogr 40(4):565–572

Carvalho L (2022a) O nexo entre a (nova) economia e o imobiliário nas cidades portuguesas (traduzido em castelhano). In Lois R, Fernandes J, Esposito E, O mundo visto desde as cidades da Península Ibérica e do Brasil. Tirant lo Blanch, Madrid

Carvalho L (2022b) O nexo entre a (nova) economia e o imobiliário nas cidades portuguesas (The Nexus between the (New) economy and real estate in Portuguese Cities). In Lois R, Fernandes J, Esposito E (eds) O mundo visto desde as cidades da Península Ibérica e do Brasil. Tirant lo Blanch, Madrid

Chamusca P, Rio Fernandes J, Carvalho L, Mendes T (2019) The role of Airbnb creating a "new-old" city centre: facts, problems and controversies in Porto. Bol De La Asoc De Geógrafos Españoles 83(2820):1–29

Cocola-Gant A, Gago A (2019) Airbnb, buy-to-let investment and tourism-driven displacement: a case study in Lisbon. Environ Plann A: Econ Space 0308518X19869012

Cordeiro Santos A (ed) (2019) A nova questão da habitação em Portugal (The New Housing Question in Portugal). Atual Editora, Coimbra

Costa J (2009) Mercado fundiário e habitação (Land Market and Housing). In: Costa JSPN (coordenação), Compêndio de Economia Regional, vol.1, cap. 17, Princípia Editora, Cascais, pp 607–638

Drago A (2020) Depois da financeirização da cidade: A crise urbana na AML e os meios de a esconjurar (After the financialization of the city: the urban crisis in AML and ways to Conjure It). In Reis J (Org.) Como reorganizar um país vulnerável?. Conjuntura Actual Editora, pp 377–417

Fernandes J, Carvalho L, Chamusca P, Mendes T (2018) O Porto e a Airbnb (Porto and Airbnb). Book Cover, Porto

Fernandes J, Carvalho L, Gago A, Chamusca P, Mendes T (2019) Lisboa e a Airbnb (Lisbon and Airbnb). Book Cover, Porto

FFMS (2023) A crise de habitação nas grandes cidades—uma análise (The Housing Crisis in Major Cities—An Analysis). Policy Papers da Fundação. Fundação Francisco Manuel dos Santos, Lisboa

Franco SF, Santos CD (2021) The impact of Airbnb on residential property values and rents: evidence from Portugal. Reg Sci Urban Econ 88:103667

Glaeser EL (2008) Cities, agglomeration and spatial equilibrium. Oxford University Press, Oxford

Glaeser E (2012) Triumph of the city: How our greatest invention makes us richer, smarter, greener, healthier, and happier. Penguin Books, New York, NY

Guimarães R (2023) Editorial—Alojamento local é economia local (Local lodging is local economy). Confidencial Imobiliário 382. Abril 2023, Porto

Gyourko J, Molloy R (2015) Regulation and housing supply. Handb Reg Urban Econ 5:1289–1337

Jover J, Cocola-Gant A (2023) The political economy of housing investment in the short-term rental market: insights from urban Portugal. Antipode 55(1):134–155

Manville M, Lens M, Monkkonen P (2022) Zoning and affordability: a reply to Rodríguez-Pose and Storper. Urban Stud 59(1):36–58. Inclusive zoning

Mendes L (2020) How can we quarantine without a home? Responses of activism and urban social movements in times of COVID-19 pandemic crisis in Lisbon. Tijdschr Econ Soc Geogr 111(3):318–332

Mendes L, Tulumello S (2022) Habitação e dinâmicas sociodemográficas no Portugal contemporâneo: Contributo para uma discussão sobre 'Habitação e natalidade—Enquadramento e políticas públicas'" (Housing and sociodemographic dynamics in contemporary Portugal: Contribution to a discussion on 'Housing and natality - Framework and public policies'). Fórum Sociológico 40

O´Sullivan A (2011) Urban Economics, McGraw-Hill Higher Education

Piketty T, Zucman G (2015) Wealth and inheritance in the long run. In Handbook of income distribution, vol. 2. Elsevier, pp 1303–1368

Pinto (2020) Novo PDM do Porto quer incentivar construção de habitação acessível (New Porto Master Plan aims to encourage the construction of affordable housing). J O Público 20 July 2020

Rebelo de Sousa M (2023) Carta ao Presidente da Assembleia da República, 20 de Agosto de 2023 (Letter to the President of the Assembly of the Republic, August 20, 2023). Presidência da República, Lisboa

Rodrigues P (2022) The real estate market in Portugal: prices, rents, tourism and accessibility. Fundação Francisco Manuel dos Santos, Lisbon

Rodríguez-Pose A, Storper M (2020) Housing, urban growth and inequalities: The limits to deregulation and upzoning in reducing economic and spatial inequality. Urban Stud 57(2):223–248

Rodríguez-Pose A, Storper M (2022) Dodging the burden of proof: a reply to Manville, Lens and Mönkkönen. Urban Stud 59(1):59–74

Santos (2023) Construção: problema da falta de mão de obra "tenderá a agravar-se" (Construction: labor shortage problem "will tend to worsen"). Idealista news, 21 July 2023

Santos Ferreira H (2023) 2024: Que habitação queremos ter em Portugal? (2024: What housing do we want to have in Portugal?). J Público imobiliário 13 Dec 2023

Storper M (2010) Review of Edward Glaeser, "Triumph of the city." J Econ Geogr 11(6):1079–1082

Zuk M, Chapple K (2016) Housing production, filtering and displacement; untangling the relationships. University of California, Institute of Governmental Studies, Research Brief, Berkeley, CA

Chapter 10
The City Building Business in Spain

María José Piñeira Mantiñán and Ramón López Rodríguez

Abstract An indisputable characteristic of Spanish cities and metropolises has consisted, for almost a century, in presenting themselves as enormously attractive spaces for the speculative business of housing and the purchase and sale of land. The city is identified less as a pleasant place to live for its citizens and more as a privileged area of capitalist business. This observation encourages the analysis in this chapter, where housing and the struggle for control of space in the main Spanish cities becomes the focus of the exhibition. The processes of appropriation, zoning and generation of capital gains in urban land are explained. The major actors and groups that promote this process are listed. But, at the same time, emphasis is placed on innovative practices and social content in housing management based on unique local experiences such as Barcelona, Cádiz or Pamplona in a short period. The debate on the right to the city remains alive and housing costs continue to present themselves as the highest that a household must face throughout its life. However, since the 2008 crisis, the obsession with accessing home ownership and an expanding rental market, still not effectively regulated, make up a panorama that will be analysed in these pages for the whole of Spain.

Keywords Urban sprawl · Metropolianisation processes · Unsustainable growth model · Artificialisation · Madrid · Spain

10.1 Introduction

In this chapter we want to address the result of the practices that have been developed to produce urban space, focusing on the Spanish case. This case presents particular characteristics that have intensified the process of urbanisation. It has been guided by

M. J. P. Mantiñán (✉) · R. L. Rodríguez
University of Santiago de Compostela (USC), Santiago, Spain
e-mail: mariajose.pineira@usc.es

R. L. Rodríguez
e-mail: ramonlopez.rodriguez@usc.es

R. C. Lois-González and J. A. Rio Fernandes (eds.), *Urban Change in the Iberian Peninsula*, The Urban Book Series, https://doi.org/10.1007/978-3-031-59679-7_10

a neoliberal logic, where the financialisation of the economy and real estate investments have been the bet of the economic growth model and generator of employment in the country. Growth subordinated to the tertiarisation of the economy has led to the dematerialisation of industrial activity in favour of services and consumer activities in large cities and their metropolitan areas (Méndez 2014). These decentralisation dynamics have been referred to by various authors as the consolidation of traditional metropolitan areas (Pons 2012; Olazabal and Bellet 2017; Delgado 2018).

It was from the 1970s onwards that these major urban transformations took place in the largest Spanish cities. A deconcentration of the population in the central cities, accompanied by a dispersion of urbanisation in nearby small and medium-sized municipalities, was characterised by low building density and intense land occupation (Nel-Lo 2007). This residential expansion in the peripheries of the central cities broke with the traditional compact city model that had been developed up to that time. But it was from the 1980s onwards that Spanish urban agglomerations in general underwent important changes in their physiognomy, evolving towards a model of a city dispersed throughout the territory that increased land occupation (Troitiño 2003). This expansion of built space along the peripheries of the consolidated city was favoured by the redefinition of the labour and residential market (Feria 2010), an improvement in accessibility and connectivity conditions, as well as a land supply at a better price. A metropolitan reality was generated, in which urban forms were increasingly extensive, lax and complex.

A process that became more aggressive from the end of the 1990s, when we witnessed the so-called brick fever (Gaja 2017), was characterised by hyperproduction in the real estate sector, which was to change the character of urbanisation (Burriel de Orueta 2008). This was the beginning of a period that many authors have called 'the prodigious decade of Spanish urban planning' (Bellet Sanfeliu and Andrés López 2021; Burriel de Orueta 2008; Górgolas 2019). Urban land is developed according to the rules of the real estate market: it is bought and sold for its economic and financial value (Gaja 2008; Lois et al. 2016). The result is a sprawl of the city beyond the immediate peripheries—reaching 50–60 km in large cities and 10–20 km in medium-sized cities (Lennert et al. 2012; Zoido et al. 2013)—and the growth/consolidation of some nuclei that in the medium term are going to become medium-sized cities. Thus, the process of metropolitanisation in Spain, characterised by some authors as a post-Fordist phenomenon on a global scale, will be the result of an economic model based on construction. Consequently, there was an intense artificialisation of the territory (Fig. 10.1) with the support of an urban planning that was the product of a territorial policy with a marked economicist (mercantilist) character (Górgolas 2017).

This model of economic growth and urban transformation continued for four decades. Betting on an expansive and deregulated growth along the urban peripheries, which responded to economic and speculative interests (Rullán Salamanca 2011; Lois et al. 2016; Olazabal and Bellet 2017). The link between the growth of cities and the different stages of economic expansion, and even the role played by housing in the circuits of capital (López and Rodríguez 2010), is not something that authors such as Marx or Harvey have not explained before. In fact, the latter explains very well

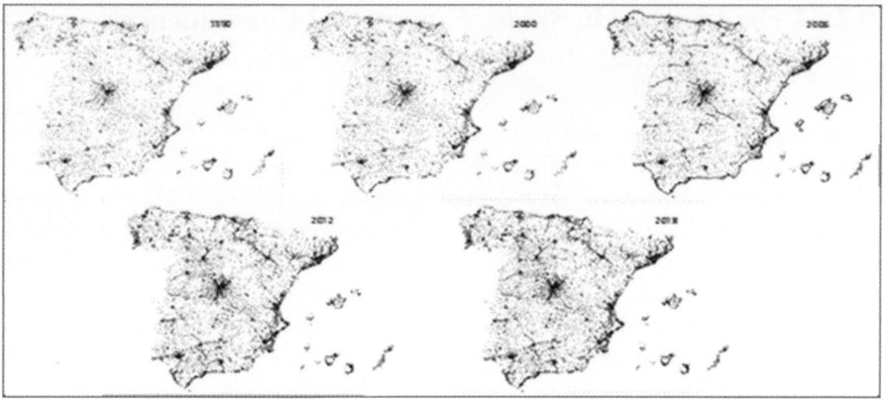

Fig. 10.1 Evolution of artificial land in Spain. (1990, 2000, 2006, 2012, 2018). *Source* Own elaboration based on Corine Land Cover

the relationship of dependency that has been generated between urbanisation and capitalism. The former needs the surpluses generated by capitalism to be realised. (Harvey 2012) Therefore, the production of urban space has been and will be the result of agents acting within the framework of the capitalist system. And as Capel (1974: 19) put it: 'In a capitalist society, the city and space in general do not belong to its inhabitants and are not shaped according to their interests, but according to the interests, sometimes contradictory, of a series of agents' (owners, developers, companies and public bodies).

Since approximately the middle of the twentieth century, we can affirm that there has been an expansion of the urban, which has not been uninterrupted, not even during periods of crisis. But the financial crisis of 2008 and its effects (company closures, unemployment, evictions, cuts in basic services) brought the construction process to a standstill and in turn the expansion of the urban as it had been done. Construction is focused on the interior reform of the consolidated urban fabric and the promotion of actions on unconsolidated urban land (Bellet Sanfeliu and Andrés López 2021), a model that, with the outbreak of the economic crisis, which has had enormous social repercussions, has been rejected by part of the population. The cases of political corruption associated with urban planning (Romero 2010) that came to light on a daily basis led to the emergence of new social movements, as well as the mobilisation of citizens. This was due to the population's rejection of both the model of governing cities and the management of their growth.

10.2 Urban Sprawl in Spain, Factors and Consequences of an Unsustainable Urban Development Model

The processes of metropolitanisation completely changed the landscapes, economic dynamics and social relations of the affected areas. The crisis of the compact city—with clear boundaries—gave rise to complex spaces, with imprecise limits, in which the population and economic activities were distributed over an extensive territory. During the first phase, rural landscapes coexisted with new urbanisations and rural ways of life with urban ones. However, in the medium term, the urban was liquidating the rural, which was replaced by clearly urban residences, businesses and infrastructures (Zoido et al. 2013). The expansion of built space reached unprecedented levels of occupation, thanks to the construction of a dense network of transport infrastructures (Alguacil 2000). This favoured the transition from monocentric to polycentric metropolitan structures (Marmolejo et al. 2012).

The growth experienced by urban spaces since the 1990s has turned them into the backbone of the territory. They are home to more than 72% of the Spanish population. Madrid and Barcelona are the metropolitan regions par excellence (Fig. 10.2), with the largest populations, followed in importance by Valencia, Sevilla, Zaragoza, Bilbao and Málaga. All of them have been configured in recent decades based on central cities with strong growth and which, at present, with the exception of Bilbao, have between 500,000 and 1,000,000 inhabitants and have acquired a notable supra-regional prominence and centrality. Thirdly, there are various growth axes and conurbations, which constitute the most dynamic spaces in their respective Autonomous Communities, such as the Galician Atlantic urban axis, the Asturian Y, the Valladolid-Burgos axis, the Alicante-Murcia conurbation or the island metropolises. Finally, there are those metropolitan areas that are less central in their regional scope and with a smaller demographic volume than the previous ones, such as the Bay of Cadiz, the coast of Almeria, the Plana of Castelló, the Tarragona-Reus-Costa Dourada conurbation or Santander-Torrelavega (Fig. 10.3).

In the process of metropolitanisation (1990–2000), artificialisation mainly affected Madrid, the Mediterranean coast and the provincial capitals (Figs. 10.1 and 10.2). But from then on, the process became more generalised and much more intense, especially in some sectors: on the one hand, the coast, as some communities have more than 75% of the land adjoining the sea as urban or developable. The pressure on the coast is particularly severe in municipalities such as Cartagena, Marbella, Calviá, Ciutadella, San Josep de Sa Talaia, Denia, San Javier, Estepona and Jávea, (Fig. 10.3) which concentrate a large amount of artificial land in the strip that covers the first 500 m of coastline (Prieto and Ruiz 2013). On the other, the medium-sized cities south of the Guadalquivir, the Ebro axis, the peripheries of the main urban agglomerations (Madrid, Seville, Bilbao, Granda, Valladolid, Vigo, A Coruña), small cities in Castilla-La Mancha and Castilla y León—which act as regional headwaters—as well as urban agglomerations articulated by medium-sized cities (Cuenca, Lleida, Mérida, León), which went from being compact and continuous to becoming lax and extensive in the territory (Olazabal and Bellet 2017).

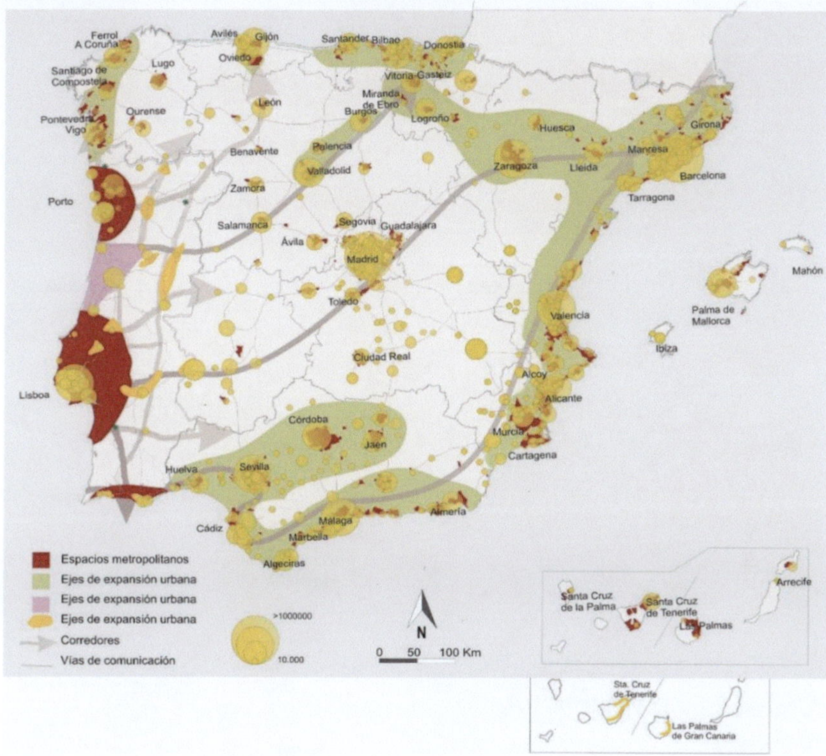

Fig. 10.2 Iberian urban system. *Source* Lois and Piñeira (2017)

There have been multiple economic, regulatory, urban planning and social factors that coincided and favoured the process of metropolitanisation (Fig. 10.4).

Among the former, it is worth highlighting an exceptional economic situation, which was affected by: a general economic boom (1996–2007), unusually low financial rates (which boosted the volume of mortgage debt), the introduction of the euro (which forced the laundering of hidden capital (due to its opacity, the real estate sector is and has traditionally been a target sector for capital in need of 'surfacing'), and the absence of investment alternatives comparable to real estate, given the volatility of the stock markets (Alonso and Furió 2010). Construction thus became the most dynamic sector of the Spanish economy, both in terms of its contribution to GDP, which rose from 6.9% in 1995 to 10.4% in 2006, and in terms of employment, which rose from 8.9% in 1996 to 13.7% in 2006.

The reduction in interest rates changed the behaviour of credit providers and borrowers. The former reoriented their strategies from attracting deposits to granting loans and credits. The latter became accustomed to credit-based financing, with a variable interest rate and repayment over a long period of time (up to 50 years) (Alonso and Furió 2010). These conditions encouraged people to buy a home, even if they were not very creditworthy (Fernández 2016). Thus, it was the demand for

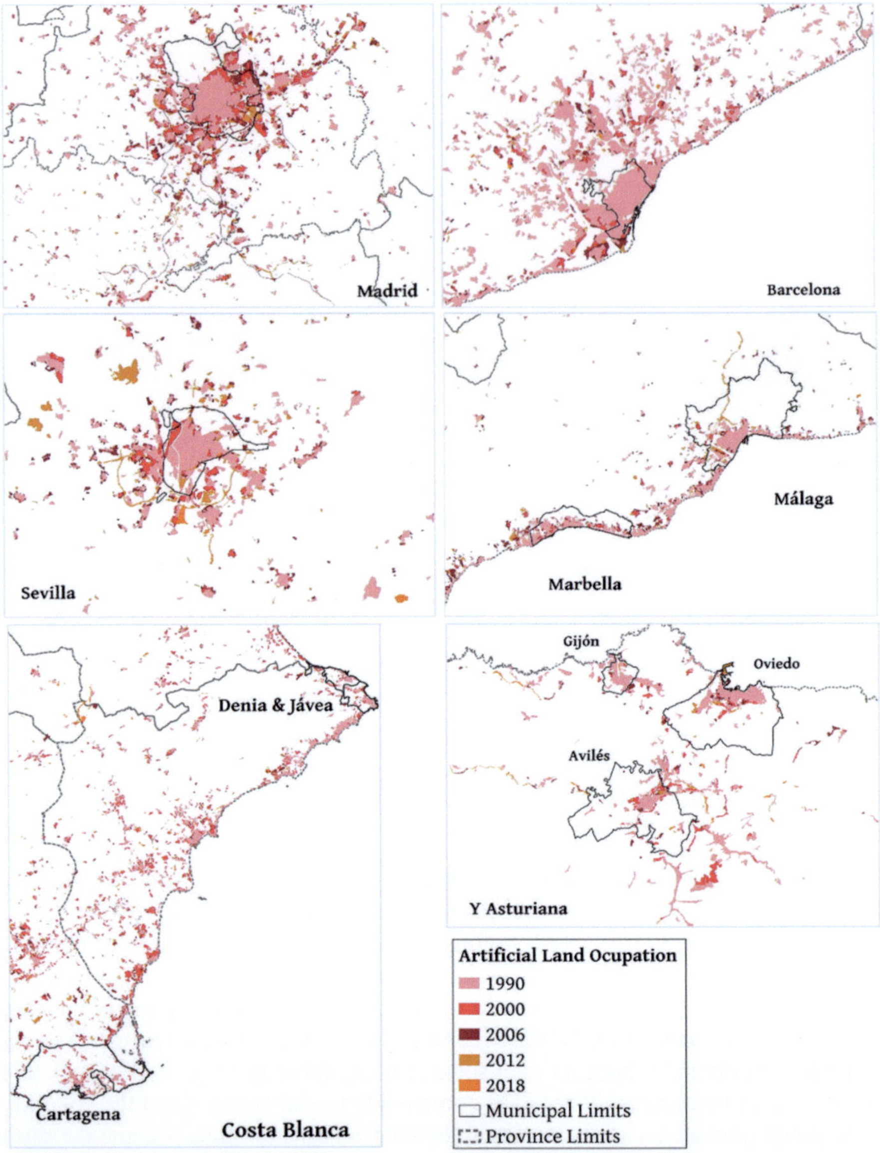

Fig. 10.3 Artificialisation of the land in some of Spain's main cities and urban areas. *Source* Own elaboration based on Corine Land Cover

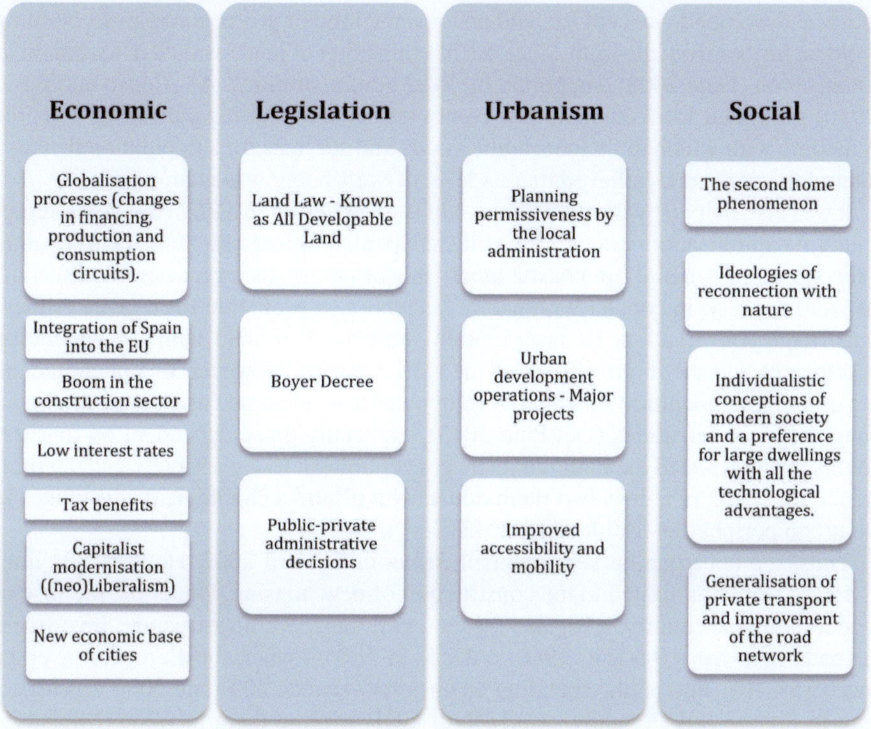

Fig. 10.4 Drivers of metropolitanisation processes. *Source* Own elaboration

housing that sustained the expansion of the construction sector, to which a speculative component would later be added. The increase in the price of land for development and housing led to housing going from being a basic good to being considered a financial asset, whose rising value was guaranteed, and which reached its zenith in 2007. In that year, a square metre of housing cost 2085.5 euros, 133.4% more than in 2000, when it was 893.3 euros. If we add to this the tax advantages derived from paying a mortgage and another series of incentives such as home savings accounts for the under 35 s (Rodríguez Alonso 2009), we can understand the splendour of real estate activity, which went from registering 234,583 new housing starts in 1995 to 664,923 in 2006. There is no doubt that over-financing was the key to the process, with 1.9 million real estate mortgage transactions recorded in 2006, 70% of which were for housing (OSE 2014).

From the regulatory point of view, the Boyer Decree of 1985 underpinned the stimulation of the construction sector, encouraging the purchase of housing by means of a 17% tax credit for the purchase of a primary residence, the abolition of the forced extension of urban lease contracts and the elimination of obstacles to foreign investment. The construction sector also found support in the 1998 Land Law, which sought to lower land and housing prices by opening up the possibility that all land could be developed—except for land under some kind of protection regime. Nothing could be further from the truth, since the liberalisation of land resulted in a tsunami of urbanisation (Gaja 2008), supported by local administrations. As Alonso and Furió (2010) point out, local councils saw commissions and building permits as a source of income with which to finance public goods and services for a population that also entered the game and believed that its level of well-being was guaranteed.

It was also helped by the phenomenon of second homes, which in the 1980s represented a capitalisation strategy for affluent Spanish households. For them, owning a second home—mainly in coastal areas- meant materialising a desirable lifestyle, far from the city, in contact with nature, as well as a profitable form of investment or heritage conservation. Its proliferation went hand in hand with the generation of transport and the improvement of the road network. However, this phenomenon became the monoculture of urban development and, in some cases, even the brand image of some territories (Del Pino 2015), especially those dependent on tourism. To this must be added the individualistic conceptions of today's society and the preference for living in a detached main house with all the technological advantages in the urban peripheries (Zoido et al. 2013).

Thus, the construction sector remained booming until 2007. However, its lines of action were not limited to the construction of new housing. Under the premise of modernising the country, it began a race to build transport infrastructure. Investment in roads rose from 0.6% in 1980 to 1.8% in 2004—when the European average was 0.85%-1%; and public spending on airports between 2000 and 2009 was higher than in France and Germany, at 0.18% of GDP, tripling them in value (Ministerio de Fomento 2011; OECD 2011). This is how Spain came to have 15,965 km of expressways, 50 airports and 2900 km of high-speed railways (ADIF; Lois et al. 2016). However, there were miscalculations and some projects turned out to be failures: the Toledo-Albacete high-speed train line had to be closed due to lack of users; the airports of Ciudad Real and Castellón became ghost spaces; or the motorways from Cartagena to Vera (Almería)—with less than 2000 vehicles per day compared to the 9000 projected—or the one from Madrid to Toledo, with only 996 vehicles per day compared to the 17,000 planned.

In this context, researchers such as Nel.lo (2001) affirm that the dispersed city, the ideal urban planning model until the crisis, supported by administrations, territorial agents and citizens, is not a city, but a mere urbanisation. Under its apparent quality of life, it hides a multitude of disadvantages, such as dependence on vehicles, lack of services (educational and health centres) and facilities, functional specialisation, social segregation and a large number of environmental impacts, which make it inefficient, unsustainable and unsupportive. The economic crisis not only paralysed the processes of urban expansion, but also brought to light the economic, social and environmental costs of the dispersed growth model (Fig. 10.5). The golden age of the Spanish economy gave way to what Marín (2010) has defined as a superimposition of two crises: an external one (the international and financial crisis) that overlapped with another of internal origin, derived from the exhaustion of the growth pattern of previous years.

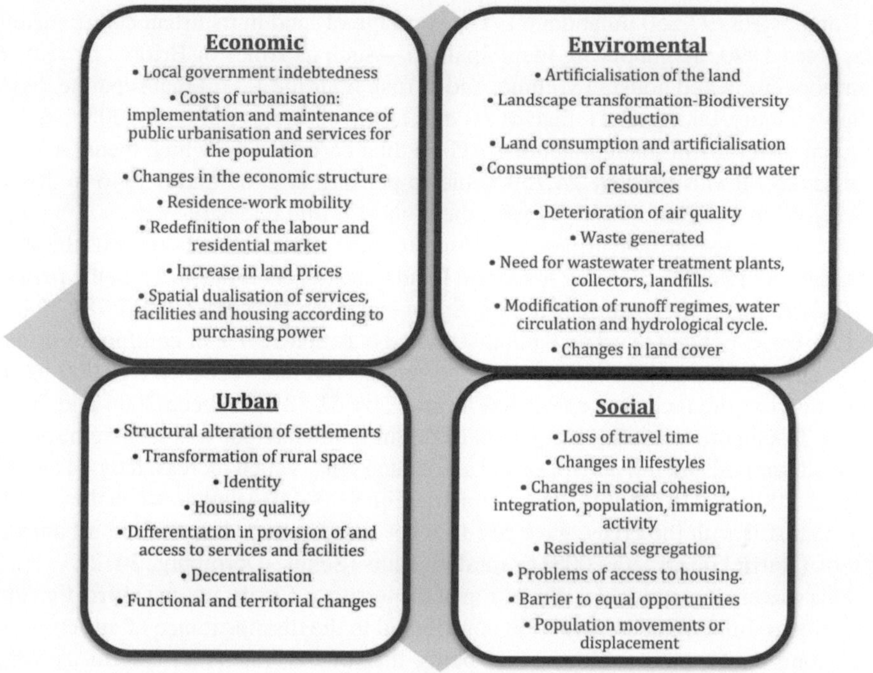

Fig. 10.5 Economic, social, urban planning and environmental costs of sprawl. *Source* Own elaboration

Local administrations were aware of the consequences of their level of indebtedness. As a result, in 2010, 30% of Spanish municipalities went into receivership and began a phase of austerity policies and redundancy plans. It is estimated that municipal revenues fell by 30–70%, making it difficult for them to pay for the most basic services, and they had to resort to confirming with financial institutions to be able to guarantee them (Jiménez 2011). In absolute terms, Madrid was the city that accumulated the most debt (6.4 million/€), followed by Barcelona (1.2 million/€), Valencia (890 million/€), Zaragoza (748 million/€), Malaga (743 million/€) and Seville (454 million/€).

But as we have already mentioned, part of the success of dispersed growth was based on the proliferation and improvement of extensive transport networks, which favoured the use of private vehicles. This was confirmed by the Movilia survey (Ministerio de Fomento 2008), referring to the year 2006, which recorded that journeys to work in Spanish metropolitan areas accounted for 30.3% of total daily mobility, and were mostly made by private transport (59%) as opposed to public transport (25%). This, together with a decline in average vehicle occupancy (Cameron, Lyons and Kenworthy 2004), led to the overloading of many city access roads. Traffic jams and wasted time became the daily torment of hundreds of citizens living in the peripheral areas, not only of large cities, but also of smaller ones, such as Santiago de Compostela (97,260 inhabitants). The shortage of land in its urban centre meant that, since 1990, neighbouring municipalities—such as Ames or Brión—increased their population and housing volume, and at rush hour the 12 km that separate them collapsed daily, taking an average of 30 min to travel. It was not until 2008 that the regional government inaugurated the AG-56 dual carriageway, which meant a 73% time saving. It was used by 24,759 vehicles per day in 2009 and 31,946 in 2018, making it the Galician motorway with the highest traffic intensity.

One of the most critical aspects resulting from the construction boom was the artificialisation of land. According to Corine Land Cover data (Table 10.1), in the period 1990–2006 the artificial surface area in Spain increased by 52% to 1,017,784.30 ha. 62.17% corresponded to urban land, distributed between 28.7% of continuous urban fabric and 33.41% of discontinuous urban fabric. The latter reached 340,000 ha in 2006, and despite the crisis continued to grow, by 61.38% between 2006 and 2012 and 11.74% in the following six-year period. Similar behaviour was seen in the areas destined for road and rail networks and associated land, which increased by 447.41% in 1991–2006, and in the areas under construction (328.41%) which, when they came to a standstill with the crisis, gave rise to what some authors have called urbanised deserts (Burriel de Orueta 2008) or modern ruins (Schulz-Dormburg, 2012).

This was an unsustainable occupation of the territory. Firstly, we can affirm that the extension of horizontal construction contributed to the disappearance of agricultural areas, one of the most serious cases being that of the Valencian Huerta. A strip of peri-urban cultivation, with a high landscape and symbolic value, but of which 1600 ha were destroyed between 1991 and 2011 (Soriano 2015; Hermosilla 2007). The causes: speculation linked to urbanisation or expectations of changes in land use, the construction of road infrastructures, the lack of protection and the lack of coordination of municipal territorial policies and plans.

On the other hand, the construction of residential complexes and infrastructures led to a reduction in the earth's capacity to absorb CO_2 by destroying numerous natural ecosystems (Plataforma Tecnológica Española de Eficiencia Energética 2016); waste production, electricity and water consumption increased. Several studies carried out in the metropolitan region have shown the relationship between dispersed population and energy consumption (López 2018) and water consumption (Sauri and Cantó 2008; Vera 2006). Thus, it has been found that between 1992 and 2012 electricity consumption grew by 75%, at a higher rate than the population, and that municipalities such as Sitges, Gallifa, Campins, Sant Vicenç de Montalt or Graner have a higher consumption than the rest. In terms of water consumption, consumption is multiplied by 2 or 3 in municipalities that are committed to urban development planning and residential accommodation. This is the case of Torrevieja, which went from a consumption of just over 2 million m3 in 1981 to 5´8 million in 1991 and 9´7 million in 2001. In addition to household consumption, there is the maintenance of gardens—which in the metropolitan region of Barcelona has been estimated at around 36% of total household consumption (Domene et al. 2005)—and swimming pools, of which there are around 52,000 in the outskirts of Barcelona, distributed along the coast and in small and medium-sized municipalities (Vidal 2007).

Table 10.1 Evolution of artificial surfaces in Spain (Ha)

		1990	2000	2006	2012	2018
Urban fabric	Continuous urban fabric	253.42,646	271.79,853	292.72,685	215.36,885	215.21,860
	Discontinuous urban fabric	230.59,769	284.80,137	340.00,620	491.07,939	548.71,688
Commercial, industrial and transport areas	Industrial or commercial areas	76.62,437	120.73,579	146.61,564	244.45,138	266.28,235
	Road and rail networks and associated land	3.86,995	7.46,144	21.18,448	39.15,754	51.06,530
	Port areas	8.22,838	9.91,629	10.75,853	8.77,632	9.31,119
	Airports	14.58,864	15.51,664	19.02,856	19.71,318	22.03,439
Mining, landfill and under construction areas	Mining extraction areas	46.06,330	64.53,728	75.59,704	81.19,569	82.77,970
	Landfills and dumps	6.09,837	7.07,084	5.90,681	13.10,214	13.60,551
	Areas under construction	16.62,886	31.94,116	71.24,008	86.01,289	27.90,906
Non-agricultural artificial green areas	Urban green areas	4.51,534	5.93,516	6.67,388	17.87,382	18.49,371
	Sports and leisure facilities	8.96,387	20.28,141	28.04,623	46.12,058	46.69,916
National total		669.60,523	839.99,591	1.017.78,430	1.262.85,178	1.302.11,585

Source Own elaboration based on Corine

These spaces lack territorial and social cohesion. The proliferation of private and gated residential complexes has been generating a physical fragmentation of urban space, an increase in social segregation and a change in citizen relations and behaviour (Villar and García 2016). Urban developments such as those of La Moraleja, Soto del Real or El Encinar in Madrid; or those of Montcabrer, La LLobera, Can Tolrà and Sant Crist in the metropolitan area of Barcelona, have favoured a territorial duality, in which income levels and the predominant typology of urban development go hand in hand. In the case of Madrid, there is a clear division between the northern metropolitan sector (Pozuelo de Alarcón, Boadilla del Monte, Majadahonda, Alcobendas Las Rozas) with an average income between 50,000 and 70,000 € (INE 2017) and where urban development's proliferate, and the south, where middle-class working-class neighbourhoods predominate (Fuenlarada, Getafe or Parla) where income ranges between 21,000–25,000 € (Fig. 10.5). The same happens in Barcelona, where the northern sector forms the first industrial belt of the metropolitan area (Hospitalet, Badalona, Terrassa, Sabadell, Granollers), with lower rents (€23,000–27,000) compared to Matadepera, San Just Desvern, San Cugat de Vallès, Sant Vicen'de Montalt, which double the average rent (Fig. 10.6).

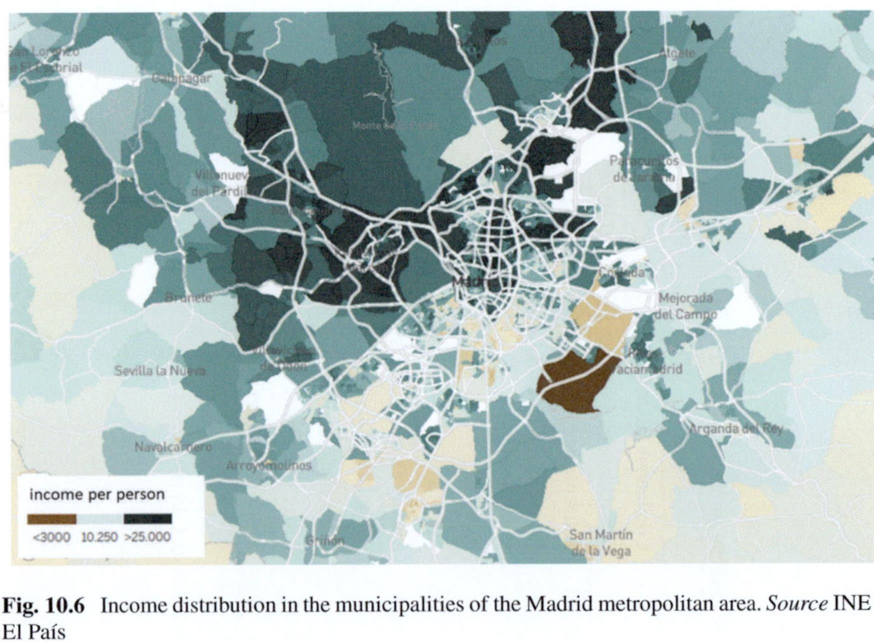

Fig. 10.6 Income distribution in the municipalities of the Madrid metropolitan area. *Source* INE & El País

10.3 The Process of Land Artificialisation in the Urban Area of Madrid

The macrocephalic Spanish urban system prevailing until then is diluting and moving towards a polycentric model (Fig. 10.6). The best example is Madrid, the urban area with the largest population size in the country (Ministerio de Fomento 2018). In 1991 it was the head city of its Autonomous Community, as only six cities (Móstoles, Leganés, Alcalá de Henares, Fuenlabrada, Alcorcón and Getafe) of its first peripheral crown had more than 100,000 inhabitants. However, in 2006, the latter already numbered nine with Parla, Alcobendas and Torrejón de Ardoz—which registered demographic increases of over 30%—while Móstoles and Alcalá de Henares exceeded 200,000 inhabitants and Fuenlabrada was on the verge of reaching the 200,000 mark. (Fig. 10.7) In this way, a city of cities (Nel.lo 2002) was articulated in which compact areas converged with other dispersed ones, new centralities with marginal areas and urbanised spaces with expectant ones (Olazabal and Bellet 2017).

One of the main characteristics of Madrid is that it has managed to transform its metropolitan model faster and more intensely than any other Spanish city or urban area (Valenzuela Rubio 2011). In the 1960s, coinciding with the years of industrial development in the country and the arrival of labour in the cities, it favoured the development of working-class dormitory neighbourhoods outside the city of Madrid, mainly in the southern municipalities of Alcorcón, Leganés, Getafe, Fuenlabrada and Parla and the municipalities to the east, the Corredor del Henares (Coslada, San Fernando de Henares, Torrejón de Ardoz and Alcalá de Henares) (Ayuntamiento de Madrid 2014). This is the beginning of the metropolitan condition, where spaces outside the compact city begin to be occupied.

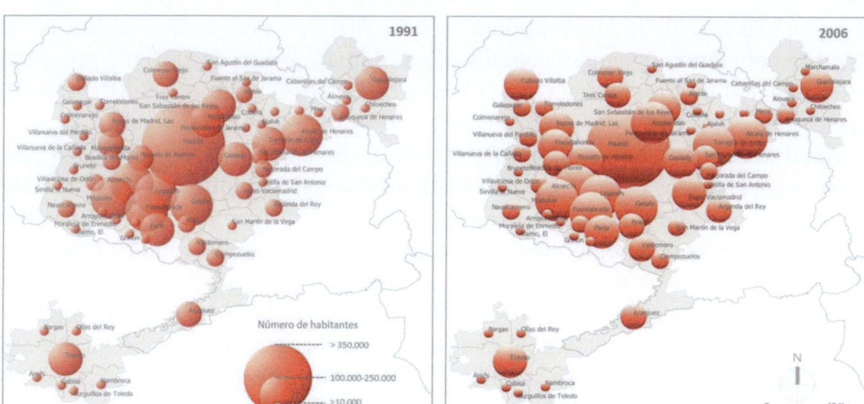

Fig. 10.7 Population growth in Madrid and its peripheral areas. *Source* 2001 Population Census and 2006 Register of Inhabitants

At the beginning of the 1980s, the urban configuration of Madrid had the characteristics of a monocentric metropolitan area that concentrates the higher functions, with the central city being the destination of labour flows from a ring of dormitory towns with insufficient facilities (Valenzuela Rubio 2011). An urbanising phenomenon began in Madrid of rapid urban growth and decentralisation of tertiary activities towards municipalities on the periphery. This urban growth will lead to the Metropolitan Area delimited in 1963, extended years later to 27 municipalities, being surpassed.

There were two expansionary cycles between 1985–1992 and 1995–2007, where the transformation of the urban and territorial structure was of great magnitude (González López et al. 2018). According to López de Lucio (2003), between 1991 and 2001 the main elements of urban territorial transformations in the case of Madrid have been public investment in communication infrastructures (road network and public transport improvements) with a notable improvement in high-capacity roads, commuter trains or the metro network. Likewise, the legal support provided by urban planning (local and regional) has led to notable changes in the land integrated or destined to form part of the artificial landscape in the community. This has favoured the expansion of the urban sprawl in a short period of time. A period in which local or metropolitan governments promoted growth through city-marketing strategies to achieve global visibility (Méndez 2012) by promoting large urban projects in the capital and attracting or holding events such as Madrid's candidacy for the 2012 Olympic Games. These actions have been reinforced by more limited interventions in the city's neighbourhoods or in other municipalities of the Metropolitan area.

During this period there has been a process of growing suburbanisation towards the south and east, largely driven by the need to find affordable housing, and towards the north and west, in a predominantly middle-class movement towards housing estates (Díaz Orueta 2007). The centre of the region is losing relative weight in favour of the municipalities of the metropolitan crown. The high growth of housing is not related to population dynamics, but the price of housing is related to the processes of socio-spatial differentiation (Díaz Orueta 2007; López de Lucio 2003; Méndez 2014).

With the arrival of the financial crisis in 2007, the unfeasibility of this expansive growth model became clear. This will entail the reversal of multiple dynamics of the territorial urban reality. Before the crisis, the metropolitan areas were the ones that brought together population growth, urban development, activities and employment. After the crisis, it was the centre that was less affected, with lower increases in unemployment. However, the socio-spatial segregation of Madrid's urban area can be seen, with higher unemployment rates in the south-east, where lower income and qualification groups live, unlike what happened in the north-west, where investment in knowledge-intensive services was the focus (Méndez 2014).

10.4 Conclusions

The process of metropolitanisation of the last forty years in Spain has favoured the transition from a macrocentric model to a polycentric one, in which new multifunctional centres are gaining prominence, integrating housing with areas of activity and employment, consumption and leisure, R&D&I institutions and specialised training facilities. Since the mid-1980s there has been extraordinary urban growth, which quickly led to a change in the metropolitan model. Since 2000, the rate of growth has slowed down in the municipalities of the nearby crown in favour of more distant municipalities.

This territorial complexity is the ideal starting point to move towards better cohesive, managed and planned metropolitan spaces. It is time to organise the territory from a systemic approach in order to try to understand the functioning of urban systems from a holistic and integrating perspective. Governing spaces as complex as metropolitan ones is no easy task, but it is a reality that some urban spaces have surpassed their municipal, provincial and even regional administrative limits, becoming true axes of urban growth.

Policies and initiatives aimed at managing this territorial complexity have been relegated to the background in the face of the impacts of the crisis and the citizen unrest caused by cases of corruption and vulnerability. The 2015 municipal elections marked a turning point, with new political forces emerging from the citizens' movement coming to local power. Among their promises were to implement a new model of city-making, stop corruption and address vulnerability. For citizens, they represented an alternative to the previous growth model. The priority was no longer for the city to expand, but to take care of the neighbourhoods. Political forces in recent years have focused on 'taking care' of disadvantaged intra-urban spaces, ignoring the problems that afflict the urban peripheries, especially in those areas where urban skeletons and housing estates proliferate and where, although housing construction was completed, services were not provided for the population.

When analysing the evolution of artificialisation by category, the largest increases since the crisis in absolute terms are in Urban Discontinuous, the result of the less dense urban growth associated with the construction boom. This growth has been maintained despite the crisis, but the data also show a drastic decrease since 2012 in areas under construction, which may be linked to the completion of projects that were started earlier. But in this post-crisis period, the biggest increase in artificialisation is due to commercial and industrial areas and transport infrastructures (except ports and airports) and to urban green areas and sports and leisure facilities. This may be associated with a transformation of consolidated urban space to the development and implementation of certain services and facilities in urban growth outside the central city.

Due to the length of this work, the differentiation of the categories has been done in a general way and it is therefore risky to make a confirmation of what is really happening with this process of land artificialisation. In the future, this should be done by cities, municipalities or urban areas, breaking them down into levels of artificiality, which would allow us to appreciate their distribution and the possibility of identifying the network of sub-centralities that make up the territory. It would also allow us to check whether these make up a more balanced territory that is less dependent on the centre.

References

ADIF. Líneas de alta velocidad. www.adif.es

Alguacil J (2000) Calidad de vida y modelo de ciudad. Boletín CF+S > 15—Calidad de vida urbana: variedad, cohesión y medio ambiente. http://polired.upm.es/index.php/boletincfs/article/view/2237/2319

Alonso M, Furió E (2010) La economía española. Del crecimiento a la crisis pasando por la burbuja inmobiliaria. Cah de Civilisation Espagnole Contemporaine 6. http://journals.openedition.org/ccec/3212

Ayuntamiento de Madrid (2014) Dimensión territorial y socioeconómica de la región metropolitana De Madrid, Monográfico en Barómetro de Economía de la Ciudad de Madrid, 39, 1° TRIMESTRE

Bellet Sanfeliu C, Andrés López G (2021) Urbanización, crecimiento y expectativas del planeamiento urbanístico en las áreas urbanas intermedias españolas (1981–2018). Inv Geográficas (76):31–52. https://doi.org/10.14198/INGEO.18054

Burriel de Orueta EL (2008) La década prodigiosa del urbanismo español (1997–2006). Scripta Nova. Rev Electrónica de Geografía y Cienc Soc XII/270(64)

Cameron I, Lyons TJ, Kenworthy JR (2004) Trends in vehicle kilometers in world cities. 1960–1990: underlying drivers and policy responses. Transp Policy 11:287–298

Capel H (1974) Agentes y estrategias en la producción del espacio urbano español. Rev de Geogr 19–56

Corine Land Cover. Centro nacional de información geográfica. http://centrodedescargas.cnig.es/CentroDescargas/catalogo.do?Serie=SIOSE

Del Pino JA (2015) Estructuras residenciales y movilidad: más allá de la segunda residencia. Centro de Investigaciones Sociológicas, Madrid

Delgado C (2018) Los procesos de metropolitanización dispersa: Castro Urdiales (Cantabria) en la región urbana de Bilbao. Bol De La Asoc De Geófrafos Españoles 78:474–517

Díaz Orueta F (2007) Los grandes proyectos de desarrollo urbano y la reconfiguración socio-espacial de las ciudades: El Barrio De Lavapiés (Madrid). en Cuad Urbano 6:169–194

Domene E, Saurí D, Parés M (2005) Urbanization and sustainable resource use: the case of garden watering in the Metropolitan Region of Barcelona. Urban Geogr 26(6):520–535

Feria JM (2010) La movilidad residencial y los procesos de urbanización metropolitanos en España. En Feria JM, Albertos JM (Coord.) La ciudad metropolitana en España: procesos urbanos en los inicios del siglo XXI. Civitas Thomson Reuters. Cizur Menor. pp 23–47

Fernández D (2016) La crisis económica española: Una gran operación especulativa con graves consecuencias. Estud Internacionales (Santiago) 48(183). https://doi.org/10.5354/0719-3769.2016.39883

Gaja F (2008) El tsunami urbanizador en el litoral mediterráneo. El ciclo de hiperproducción inmobiliaria 1996–2006. Scripta Nova: Rev Electrónica de Geografía y Cienc Soc XII/270(66)

Gaja F (2017) Planning and urban growth what to do with urbanized vacant áreas in the land of Valencia. In: En Brebbia CA, Miralles JL (eds) Environmental and economic impact on sustainable development. WIT Press, Boston, pp 104–112

González López J, Benito Moreno M, Pozueta Echavarri J (2018) La metrópolis tras la burbuja inmobiliaria. Transformación del suelo y formas de crecimiento urbano en Madrid. en II Congreso

Internacional ISUF-H Zaragoza 2018, 13–14 de septiembre de 2018, Universidad de Zaragoza. pp 135–146
Górgolas P (2017) Burbujas inmobiliarias y planeamiento urbano en España: una amistad peligrosa. Cuadernos de Investigación urbanística 111:5–64
Górgolas P (2019) Del «urbanismo expansivo» al «urbanismo regenerativo»: directrices y recomendaciones para reconducir la herencia territorial de la década prodigiosa del urbanismo español (1997–2007). Aplicación al caso de estudio del litoral andaluz Ciudad y Territorio. Estud Territoriales 199:81–100
Harvey D (2012) Ciudades rebeldes. Del derecho de la ciudad a la revolución urbana. Akal.
Hermosilla J (2007) El patrimonio hisráulico del Bajo Turia: L'Horta de València. Generalitat Valenciana. Valencia
Instituto Nacional de Estadística (INE) (2017) Atlas de distribución de renta de los hogares. Estadística Exp. www.ine.es
Jiménez A (2011) Las Instituciones locales en tiempo de crisis. Una nueva estructura institucional ajustada a las necesidades en tiempo de crisis. QDL., 25. Fundación Democracia y Gobierno Local
Lennert M, Rozenblat C, Van Hamme G, Smetkowski M (2012) Orientaciones para el futuro de las ciudades: estudios recientes y nuevas perspectivas para las ciudades europeas. In: En Montaner JM, Subirats J (eds) Repensar las políticas urbanas: apuntes para la agenda urbana. Diputación de Barcelona
Lois RC, Piñeira MJ (2017) El modelo territorial y urbano resultante. Un sistema urbano metropolitano y policéntrico más formal que funcional activo. In: En Farinós J, Olcina J (eds) Geografía Regional de España. Tirant Humanidades. Valencia
Lois RC, Piñeira MJ, Vives S (2016) El proceso urbanizador en España (1990–2014): una interpretación desde la geografía y la teoría de los circuitos de capital. Scrita Nova Rev Electrónica de Geografía y Cienc Soc XX (539)
López J (2018) Consumo doméstico de electricidad y forma urbana en la región metropolitana de Barcelona. Bol De La Asoc De Geógrafos Españoles 76:329–357. https://doi.org/10.21138/bage.2525
López I, Rodríguez E (2010) Fin de ciclo: financiarización, territorio y sociedad de propietarios en la onda larga del capitalismo hispano (1959–2010) Traficantes de Sueños. Madrid
López de Lucio R (2003) Transformaciones territoriales recientes en la región urbana de Madrid. en Urban 8:124–161
Marín Quemada JM (2010) La crisis financiera internacional, las dos crisis de la economía española y algunas recomendaciones de política económica. Rev del Inst de Estud Económicos. Lecciones de la crisis 1
Marmolejo C, Chica E, Masip J (2012) ¿Hacia un sistema de metrópolis españolas policéntricas?: evolución de la influencia de los subcentros en la distribución de la población. ACE: Architecture. City Environ 18:163–190
Méndez R (2012) Globalización, neoliberalismo y dinámicas metropolitanas en madrid, en Documentos Y Aportes En Administración Pública Y Gestion Estatal (DAAPGE) 1(19):29–49
Méndez R (2014) Expansión y crisis del modelo neoliberal en Madrid. In: Hidalgo R Janoschka M (eds) La Ciudad Neoliberal. Gentrificación y exclusión en Santiago de Chile, Buenos Aires, Ciudad de México y Madrid. Geolibros 19, Santiago de Chile, pp 217–232
Ministerio de Fomento (2008) Movilia 2006/2007. https://www.mitma.gob.es
Ministerio de Fomento (2011) Informe 2009–2010 sobre el sector de las autopistas de peaje en España. http://www.fomento.gob.es/
Ministerio de Fomento (2018) Áreas urbanas en España 2018. Constitución, Cuarenta años de las ciudades españolas
Nel.lo O (2002) Cataluña, ciudad de ciudades. Ed. Milenio. Barcelona
Nel.lo O (2001) Las areas metropolitanas. In: En Gómez Mendoza J, Gil Olcina A (coords.); Geografía de España. Ariel. Barcelona

Nel.lo O (2007) La tercera fase del proceso de metropolitanización en España. In: En Artigues A et al. (coords.); Los procesos urbanos postfordistas. Actas del VIII Coloquio y Jornadas de Campo de Geografía. Universitat de les Illes Balears y Asociación de Geógrafos Españoles. Palma

Observatorio de la Sostenibilidad (OSE) (2014) Observatorio de la Vivienda y Suelo. Informe Anual. Ministerio de Fomento. Madrid

OECD (2011) Trends in Transport Infrastructure Investment 1995–2009. International Transport Forum. Statistics Brief. París. http://internationaltransportforum.org

Olazabal E, Bellet C (2017) Procesos de urbanización y artificialización del suelo en algomeraciones urbanas 1987–2011. Cuad Geográficos 57(2):189–210

Plataforma Tecnológica Española de Eficiencia Energética (2016) Uso eficiente de la energía en las ciudades. Foro de las Ciudades, Madrid

Pons JJ (2012) Metropolitanización: procesos y estructuras demográfico–territoriales. En La población en clave territorial. Procesos, estructuras y perspectivas de análisis. Asociación de Geógrafos Españoles y Universidad de Cantabria. Santander. pp 21–26

Prieto F, Ruiz JB (2013) Costas Inteligentes. Estudio realizado para Greenpeace España. Madrid. https://archivo-es.greenpeace.org/espana/Global/espana/report/costas/Costas%20Inteligentes%20F.%20Prieto.pdf

Rodríguez Alonso R (2009) La política de vivienda en España en el contexto europeo. Deudas y Retos. Boletín CF+S, 47/48, 125–172

Romero J (2010) Construcción residencial y gobierno del territorio en España. De la burbuja especulativa a la recesión. Causas y consecuencias. Cuad Geográficos de la Univ de Granada 47(2):17–46

Rullán Salamanca O (2011) La regulación del crecimiento urbanístico en el litoral mediterráneo español. Ciudad Territorio 42(168):279–297

Sauri D, Cantó S (2008) Ambiente Integración de políticas sectoriales: Agua y urbanismo Fundación Nueva Cultura del Agua. Panel científico-técnico de seguimiento de la política de aguas. Universidad de Sevilla. Sevilla

Schulz-Dormburg J (2012) Ruinas modernas. Una topografía de lucro. Àmbit. Barcelona

Soriano V (2015) La huerta de Valencia: Un paisaje menguante. Universidad de Valencia, Valencia

Troitiño MA (2003) Renovación urbana: dinámicas y cambios funcionales. Perspectivas Urbanas 2:1–11

Valenzuela Rubio M (2011) Los Procesos de Metropolización. Madrid, una región metropolitana entre la dispersión y el policentrismo. In: Humbert MyV (coord.) en España en la Unión Europea: Un cuarto de siglo de mutaciones territoriales. Casa de Velázquez 121, Madrid, pp 211–253

Vera JF (2006) Agua y modelo de desarrollo turístico: la necesidad de nuevos criterios para la gestión de recursos. Bol De La Asoc De Geógrafos Españoles 42:155–178

Vidal M (2007) Urbanització dispersa i nous usos de l'aigua: El cas de les piscines a la Regió Metropolitana de Barcelona Unpublished Masters Thesis. Bellaterra, UAB, Institut de Ciència I Tecnologia Ambientals

Villar A, García M (2016) Ciudad segregada en España: urbanizaciones cerradas en Valencia y Sevilla. Rev INVI 31(86). https://doi.org/10.4067/S0718-83582016000100006

Zoido F et al (2013) Diccionario de urbanismo: geografía urbana y ordenación del, territorio. Cátedra, Madrid

Chapter 11
Platform Economy and Uneven Financial Geographies: The Case of Fintech in Portugal

Mário Vale and Daniela Ferreira

Abstract This chapter examines Portugal's fintech ecosystem within the global platform capitalism context, examining its role in urban development and regional disparities. The study addresses the theoretical implications of the global fintech revolution, emphasizing digital transformation and financial inclusion. It details Portugal's fintech evolution, regulatory challenges, and investment trends, noting the concentration of fintech activities in urban centers like Lisbon and Porto. The chapter concludes by discussing the opportunities and challenges facing Portugal's fintech sector, including regulatory navigation, cybersecurity, and the potential for economic inequality.

Keywords Fintech · Platform economy · Digital economy · Portugal

11.1 Introduction

This chapter delves into the evolving Portuguese financial technology (fintech) ecosystem, examining its integration within the broader contours of platform capitalism and the uneven spatial dynamics it engenders. Through an exploration of how fintech innovations are influencing economic patterns, urban development, and regional disparities, we uncover the intricate ways in which technology and geography intersect in the context of Portugal's economic development. This analysis not only contributes to our understanding of Portugal's geographical landscape and urban change but also highlights the pivotal role of fintech in driving forward the nation's economic frontiers, making a significant contribution to the discourse on the Geography of Portugal. Structurally, the chapter is organized to first outline the fintech

M. Vale (✉) · D. Ferreira
Centro de Estudos Geográficos (CEG), Instituto de Geografia e Ordenamento do Território (IGOT), Universidade de Lisboa, Lisbon, Portugal
e-mail: mario.vale@edu.ulisboa.pt

D. Ferreira
e-mail: danielaferreira2@edu.ulisboa.pt

R. C. Lois-González and J. A. Rio Fernandes (eds.), *Urban Change in the Iberian Peninsula*, The Urban Book Series, https://doi.org/10.1007/978-3-031-59679-7_11

landscape, followed by an examination of its economic impacts, and concluding with a discussion on the spatial distribution and urban planning implications, providing a comprehensive overview of the fintech phenomenon within Portugal's geographic context.

11.2 Transforming Finance and Urban Change

The interplay between fintech and urban development narrative unfolds across two distinct but interconnected realms. On the one hand, it elucidates the global fintech revolution and the profound implications of platform capitalism on the financial sector and urban landscapes. Following this, the discussion shifts toward the tangible impacts of fintech on urban change, scrutinizing both its potential to foster more inclusive, efficient, and sustainable cities and the critical challenges it presents. Through this examination, we aim to uncover the multifaceted role of fintech in shaping and challenging the future of urban living.

11.2.1 The Global Fintech Revolution and Its Theoretical Implications

The rapid advancement of technology and widespread internet adoption have fundamentally altered our lifestyles and perceptions of environments. This transformation extends beyond personal life, reshaping urban planning and economic sector growth (Vale et al. 2024). Notably, the financial sector has significantly evolved through technological innovation. In this digital era, it is recognized that we are entering a novel economic phase, often termed "platform capitalism" or "the platform economy."

This platform logic stems from the creation and integration of companies on digital platforms and is characterized by the promotion of business models open to the participation of different types of external actors connected in horizontal networks. These actors represent a varied range of resources that provide companies with the possibility of developing new products and services that would not have been possible before. The designation for this new context is still not consensual among various researchers, perhaps due to the fact that the topic is an object of interest in very diverse scientific fields (including, but not limited to, geography, economics, sociology, information science, management, marketing, or humanities), which also highlights its social, economic, and even political relevance. Several terms are emerging, such as the platform economy, platform capitalism, the gig economy, the sharing economy or the fourth industrial revolution (McKinsey and Company 2015; Srnicek 2017; World Economic Forum 2015). Interestingly, it is, on the contrary, consensual that, more than a new context, we are facing a new economic paradigm marked by important technological transformations, the increasing use of digital technologies in companies, and also by a strong transformation of the user's role as a participatory actor in product and service development processes (van Dijck 2013; Ramaswamy and Ozcan 2014; Yoo et al. 2016; Srnicek 2017).

The platform economy is based on a business model, which is considered in some studies to be a new, more digital model (Jabłoński and Jabłoński 2020; Wirtz 2019), and in which digital platforms are the fundamental structures. These are characterized by operating under an open and flexible logic to interaction and exchange of resources between different types of actors (Hoyer et al. 2010; Prahalad and Ramaswamy 2004; Ramaswamy and Ozcan 2014).

Within the emergence of digital platforms and in this new context of the platform economy, the term "Fintech," emerges. Fintech has profoundly changed how enterprises and consumers access and use financial services. Fintech is defined by a set of innovations that come from digital technologies for the financial sector and that improve the financial industry by offering new exclusively online-based products and services (Millian et al. 2019; Langley and Leyshon 2020; Wójcik 2021). This concept or model, although directly linked to the financial sector, can target different end markets and companies of various sizes that wish to implement innovative solutions in their service offering.

In this sense, the financial sector has been revolutionized by a wide range of new services that allow faster and easier access to certain financial tools that previously required stricter requirements for access. These new financial platforms facilitate, on the one hand, bank transfers and payment for products and services at lower costs, greatly facilitating the processes of buying and selling products and services using credit cards, prepaid cards, or mobile phones at very low costs. On the other hand, they facilitate access to credits and the possibility of making investments in a more simplified way. Furthermore, it also facilitates banking operations and offers new opportunities for individuals to create savings accounts, making it easier to manage bank accounts. There are also Fintech services more focused on cryptocurrencies that make it easier for their purchase, sale, and storage, using blockchain technology. Finally, among the various services available, there are also insurance Fintech, also known as insurtech, which use innovative technologies designed to extract maximum efficiency from the current insurance business model. All these services become just a smartphone away, as long as the user has internet access, as FinTech removes the need to go to a physical branch.

With the emergence of Fintech, we have seen a greater ease of access to banking operations. This has been discussed thoroughly in the literature, which highlights how Fintech is a driver for promoting the financial inclusion of individuals, especially those who are marginalized from the conventional banking system. According to the World Bank (2017), a greater access to Fintech can significantly contribute to the reduction of poverty and social inequalities, given that such technologies allow lowering costs and make financial services more user-friendly in general. But it is also true that Fintech platforms and their mobile financial services have also been understood as an instrument of financialization in the poorest countries. From a geographical point of view, Fintech can be framed as an emerging field that leverages a sort of data-based colonialism, if we take into account that most platforms and investors are located in the Global North, while there is a growing number of users in the Global South. In this sense, while the potential of Fintech platforms to promote financial inclusion has been widely recognized (Lai and Samers 2020; Maya Declaration 2018; GPFI 2020), we must underline the Janus face of this recent phenomenon.

11.2.2 Fintech and Urban Change

The relationship between fintech and urban evolution presents a complex narrative that, while showcasing potential for more inclusive, efficient, and sustainable cities, also unravels significant challenges that need a critical examination. Fintech's ability to revolutionize urban commerce, banking accessibility, and smart city development underpins its role in shaping future urban landscapes. Fintech might also supports economic decentralization, empowering local businesses and communities. However, this optimistic view must be balanced with a critical understanding of the nuanced implications of rapid fintech growth.

The concerns surrounding fintech extend beyond its potential to widen the digital divide among urban populations, which is a critical element of the contemporary urban life (Ferreira et al. 2021). The reliance on user data for fintech solutions presents privacy and security risks, highlighting the need for strong mechanisms to protect sensitive information, also another critical element for contemporary societies (Kitchin 2014). Moreover, the displacement of traditional financial institutions raises questions about the future of local economies and employment, suggesting that the fintech emergence could lead to economic dislocations and job reduction in the financial sector, straining urban labor markets. A critical approach must also scrutinize fintech's role in urban development through the lens of sustainability and inclusivity. While fintech offers tools for environmental and social improvements, it also faces the risk of perpetuating existing inequalities and introducing new forms of exclusion.

While fintech offers significant potential to transform urban environments, it also presents several persistent challenges. Effectively addressing these challenges is vital for leveraging fintech's positive impact on urban development while minimizing its negative consequences. This requires the establishment of a robust regulatory framework and the implementation of policies designed to protect socially disadvantaged individuals and underprivileged urban areas. Such measures are essential to ensure that the benefits of fintech are equitably distributed and contribute to the sustainable and inclusive growth of urban spaces.

11.3 The Rise of Portuguese Fintech Ecosystem

In Sect. 11.2, we explored the theoretical underpinnings of fintech's disruptive potential on global financial systems, emphasizing concepts such as platform capitalism, digital transformation, and financial inclusion intercepting urban change. These frameworks serve as a lens through which we can analyze the empirical data presented in this section, based on the report from Portugal Fintech (2003), focusing on Portugal's evolving fintech ecosystem. By examining the growth of fintech startups, regulatory adaptations, and their impact on urban and regional economic development, we can contextualize these phenomena within the broader theoretical discussions. For instance, the emergence of platform-based financial services in Portugal exemplifies the practical manifestation of platform capitalism, while initiatives to enhance financial inclusion through fintech innovations reflect the theoretical expectations about technology's role in democratizing access to financial services. This direct correlation between theory and practice not only strengthens our argument but also enriches our understanding of fintech's multifaceted impact, particularly in the geographical and economic landscape of Portugal.

11.3.1 Evolution of the Fintech Ecosystem

The fintech sector globally has witnessed an unprecedented surge, reshaping the financial services landscape with innovative technological solutions. Portugal, mirroring this global trend, has emerged as a dynamic player in this area. The evolution of the Portuguese fintech ecosystem—especially conceived in Lisbon headquarters—is a testament to the country's progressive adoption of digital technologies, fostering a conducive environment for financial innovation. This introduction sets the stage for understanding how Portugal, with its unique economic and technological context, has nurtured a thriving fintech scene, contributing significantly to the broader narrative of digital transformation in financial services.

The regulatory environment plays a critical role in shaping the fintech ecosystem, influencing both the entry of new players and the expansion of existing ones. It also ensures that the evolution of fintech services aligns with broader financial and economic objectives. The Portuguese government has implemented supportive policies and regulatory frameworks that have been instrumental in fostering a conducive environment for fintech growth. These policies aim to balance innovation promotion with financial stability and consumer protection. Despite the supportive environment, there are major challenges in navigating the complex regulatory landscape, which includes compliance with both national and European Union regulations.

Portugal's fintech sector has experienced rapid growth in recent years (Fig. 11.1), characterized by an increasing number of startups and diversification in services offered. This growth is reflective of a broader global trend but is particularly notable in the Portuguese context due to its scale and speed. The market has seen significant

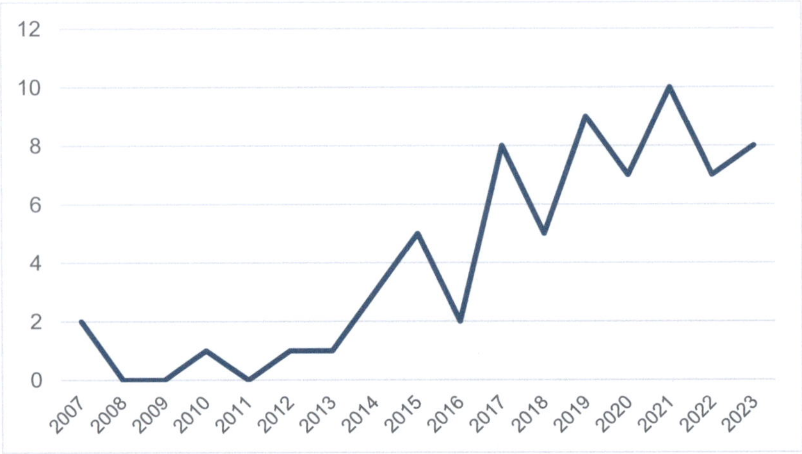

Fig. 11.1 Creation of fintech companies in Portugal, 2007–2023. *Source* Based on *Portugal Fintech Report 2023*

evolution in various segments like digital banking, payments, and investment solutions. These segments have not only grown in number but also in the sophistication of the services offered, driven by customer demand and technological advancements. A key factor in this growth has been the supportive environment for innovation and entrepreneurship. Initiatives by both government and private sectors have played a crucial role in providing the necessary resources and platforms for fintech to thrive.

The Portuguese fintech (Fig. 11.2) ecosystem has experienced a remarkable upsurge in venture capital and funding, reflecting a strong investor confidence in its potential and sustainability. This positive investment trend is substantiated by the impressive total funding of over €1.1 billion euros (€1,144,000,000) to date, as reported in the Portugal Fintech Report 2023. This influx of capital, coming from both domestic and international investors, has not only injected financial resources but also brought in valuable expertise and access to global markets. It serves as a catalyst for the sector's continuous growth, enabling Portuguese fintech startups to expand their operations and innovate in their product and service offerings, thus solidifying Portugal's position as a dynamic player in the global fintech landscape.

The fintech ecosystem has been at the forefront of adopting advanced technologies such as Artificial Intelligence (AI), blockchain, and machine learning. These technologies have revolutionized various aspects of financial services, from customer service automation to enhanced security protocols. The integration of these technologies has significantly improved service efficiency and opened up new avenues for product innovation. This has allowed Portuguese fintech companies to offer more personalized and efficient services, catering to a diverse range of consumer needs and have been able to provide more tailored financial solutions and a better understanding of customer preferences.

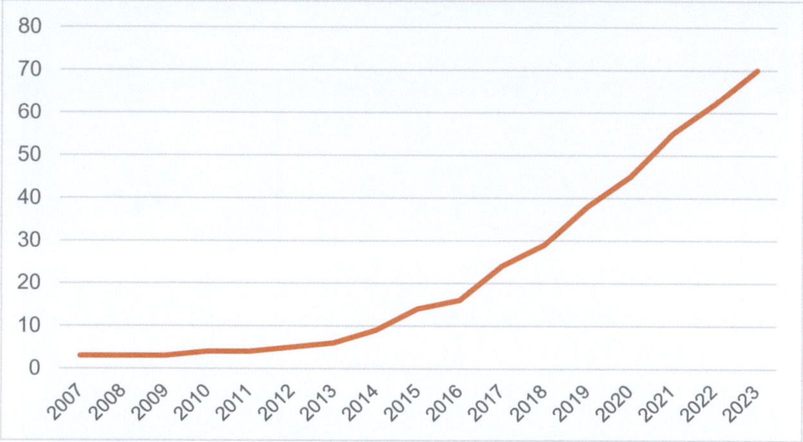

Fig. 11.2 Accumulated number of fintech companies in Portugal, 2007–2023. *Source* Based on *Portugal Fintech Report 2023*

11.3.2 Fintech's Geography

Figure 11.3 clearly illustrates the spatial concentration of fintech companies in Portugal, with a significant clustering in the country's two largest urban centers: Lisbon and Porto. This distribution pattern is indicative of a common trend in the technology and startup sectors, where companies gravitate toward major cities due to several advantageous factors. Lisbon and Porto offer robust ecosystems that are conducive to the growth of fintech firms, including access to a skilled workforce, advanced infrastructure, and supportive regulatory environments. Moreover, these cities—and particularly Lisbon—are hubs of financial activity, providing fintech companies with proximity to potential clients and partners in the traditional banking and finance sectors.

The concentration in these urban areas also suggests network effects, where the value of being part of a dense, interconnected community increases with every new company that joins. This creates a virtuous cycle, attracting more fintech startups and contributing to the innovative dynamism of the sector. In contrast, the more peripheral regions show significantly fewer fintech firms, which may reflect challenges such as smaller local markets, less access to capital, and fewer networking opportunities. In this disused and low-density context, it is noticeable a potential association of little concentration in Braga and Aveiro to the presence of thriving university and research environments.

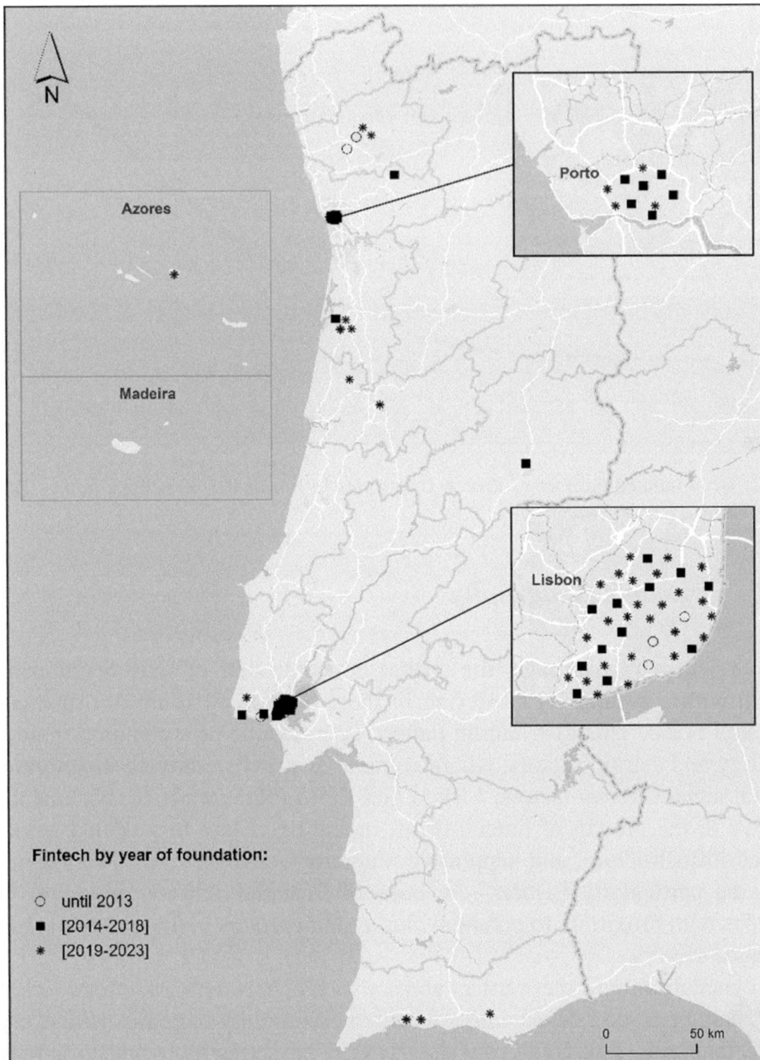

Fig. 11.3 Spatial distribution of Fintechs in Portugal, 2007–2023. *Source* Based on *Portugal Fintech Report 2023*

11.4 Conclusion

The Portuguese fintech ecosystem, characterized by its rapid growth, technological innovation, and supportive regulatory environment, stands as a beacon of potential within the global financial technology landscape. With an investment exceeding €1.1 billion, the ecosystem's vibrancy and the confidence it inspires herald a future of continued expansion and innovation. This optimism is grounded in Portugal's ability to adapt to market conditions, leverage technological advancements, and foster collaboration, underscoring its significant contribution to both the national economy and as a model of resilience in the financial technology sector.

However, this situation is not without its challenges. Navigating complex regulatory frameworks, ensuring compliance, and managing cybersecurity risks are paramount for maintaining momentum. Furthermore, the rapid pace of technological advancement necessitates a continuous effort to stay at the forefront of innovation. Despite these hurdles, the opportunities for growth and expansion are vast, bolstered by supportive governmental policies, interest from international investors, and potential collaborations with traditional financial institutions.

Yet, as we delve deeper into the implications of fintech's rise, critical considerations come to the forefront. The blend of fintech into urban settings also introduces contingencies that challenge traditional urban economic structures, demanding careful consideration of how these technologies might exacerbate disparities, alter employment landscapes, and require adaptive regulatory frameworks to ensure equitable urban transformation. In Portugal, the concentration of fintech activities in urban centers and a very special focus in Lisbon highlights a growing digital divide, raising concerns over equitable access and the potential marginalization of much of the country. The rapid development of fintech, while driving economic innovation, also necessitates a thoughtful examination of its impact on employment, privacy, and the traditional banking sector. As Portugal navigates this fintech revolution, striking a balance between innovation and social equity becomes crucial. Ensuring that the benefits of technological advancements are distributed widely and justly across the country is imperative.

In light of these considerations, fintech activity has become a pivotal area of focus for policymakers and stakeholders focusing urban change. This highlights the critical need for strategies that support not only more balanced growth but also incorporate a geographic dimension, promoting a more inclusive form of urban development.

References

Ferreira D, Vale M, Carmo RM, Encalada-Abarca L, Marcolin C (2021) The three levels of the urban digital divide: bridging issues of coverage, usage and its outcomes in VGI platforms. Geoforum 124:195–206

GPFI (2020) G20 2020 financial inclusion action plan. Global Partnership for Financial Inclusion

Hoyer W, Chandy R, Dorotic M, Krafft M, Singh SS (2010) Consumer cocreation in new product development. J Serv Res 13(3):283–296

Jabłoński A, Jabłoński M (2020) Social business models in the digital economy. Palgrave Macmillan Cham

Kitchin R (2014) The data revolution: big data, open data, data infrastructures and their consequences. SAGE Publications

Lai K, Samers M (2020) Towards an economic geography of FinTech. Prog Hum Geogr 45(4):720–739

Langley P, Leyshon A (2020) The platform political economy of FinTech: reintermediation, consolidation and capitalization. New Polit Economy 26(3):376–388

Maya Declaration (2018) Maya declaration progress report today's targets, tomorrow's impact. Alliance Fin Incl
McKinsey and Company (2015) Global media report, 2015: global industry overview. Glob Media Entertainment Pract. https://www.mckinsey.com/~/media/McKinsey/dotcom/client_service/Media%20and%20Entertainment/PDFs/McKinsey%20Global%20Report%202015_UK_October_2015.ashx
Millian E, Spinola M, Carvalho M (2019) Fintechs: a literature review and research agenda. Electron Commer Res Appl 34. https://doi.org/10.1016/j.elerap.2019.100833
Portugal Fintech (2003) Portugal Fintech Report 2023. (available online:portugalfintech.org)
Prahalad CK, Ramaswamy V (2004) Co-creation experiences: the next practice in value creation. J Interact Mark 18(3):5–14
Ramaswamy V, Ozcan K (2014) The co-creation paradigm. Stanford University Press
Srnicek N (2017) Platform capitalism. Polity Press
Vale M, Ferreira D, Rodrigues N (eds) (2024) Geographies of the platform economy. Critical perspectives. Springer, Berlin
van Dijk J (2013) The culture of connectivity: a critical history of social media. Oxford University Press
Wirtz BW (2019) Digital business models. Springer Nature, Cham. https://doi.org/10.1007/978-3-030-13005-3
Wójcik D (2021) Financial geography II: the impacts of FinTech—Financial sector and centres, regulation and stability, inclusion and governance. Prog Hum Geogr 45(4):878–889
World Bank (2017) Global findex database. World Bank Group, Washington
World Economic Forum (2015) Industrial internet of things: unleashing the potential of connected products and services. New York. http://www3.weforum.org/docs/WEFUSA_IndustrialInternet_Report2015.pdf
Yoo KH, Sigala M, Gretzel U (2016) Exploring tripAdvisor. In: Egger IR, Gula I, Walcher D (eds) Open tourism. Tourism on the Verge. Springer, Berlin

Chapter 12
The Economic and Financial Capitals of Spain: Their Significance in the Structuring of the Urban System

Rubén C. Lois González

Abstract The concept of the global city has enjoyed enormous prestige in the academic world in recent decades. Faced with a restrictive interpretation of this name, global cities have proliferated across all continents, to refer to large cities that play a central role of economic command and decision-making in a significant region of the world. For this reason, this expression is analysed for the Spanish case, based on an obvious question: Can Madrid and Barcelona be authentic global cities due to their central role in certain financial, business and creative activities? Undoubtedly, these two capitals have served as a reference node for the banking, energy, creative or telecommunications sectors throughout Latin America and the Mediterranean world. Both concentrate a large part of the headquarters of the main Spanish transnational companies and are essential to define the economic-financial capitals of the country. However, a review of the headquarters of the largest Spanish companies also shows the importance of regional metropolises such as Bilbao, Valencia or Seville, which have consolidated their position as secondary centres of economic power in Spain. The analysis of this Urban Geography of Capitalism of the largest nation in the Iberian Peninsula will occupy the attention of this chapter.

Keywords Global city · Economy · Financial geography · Wealth · Metropolises

12.1 Introduction

The primary transformation observed in the Geography of Spain over the last half-century is the consolidation of the urbanisation process. The term 'consolidation' is employed due to the significant urban growth experienced throughout the country since the 1960s and 1970s. These urbanisation dynamics initially focused on compact cities, later extending to their peripheries, and more recently, evolving into a process

R. C. Lois González (✉)
Departamento de Xeografía, University of Santiago de Compostela, Santiago de Compostela, Spain
e-mail: rubencamilo.lois@usc.es

R. C. Lois-González and J. A. Rio Fernandes (eds.), *Urban Change in the Iberian Peninsula*, The Urban Book Series, https://doi.org/10.1007/978-3-031-59679-7_12

of metropolitanisation (Capel Sáez and Vilà Valentì 1970; Nel.lo Colom 2001; Lois González and Piñeira Mantiñán 2017). This process entails the concentration of demographic and economic activities around major urban centres or regions. In Spain, these centres are primarily Madrid and Barcelona, which together account for over a third of the country's total population when considering their metropolitan regions. Additionally, five major cities and metropolises (Valencia, Sevilla, Málaga, Bilbao and Zaragoza), with populations ranging from 600,000 to just over one and a half million inhabitants, function as regional metropolises (Feria Toribio and Albertos Puebla 2010; Méndez et al. 2015; Albertos Puebla and Sánchez Sánchez 2017; Lois González and Piñeira Mantiñán 2017).

These primary urban centres define the economic and financial capitals of Spain, although the Atlantic Urban Axis of Galicia, the central city of Asturias, the Valladolid-Burgos axis, Bajo Segura, or the islands of Gran Canaria and Mallorca play also complementary roles to a lesser extent (Albertos Puebla and Sánchez Sánchez 2017; Lois González and Piñeira Mantiñán 2017). On the one hand, the seven main metropolitan areas concentrate more than half of the total population and significantly contribute to national wealth creation, as indicated by available GDP data. These metropolitan areas also attract the wealthiest social groups. On the other hand, they act as magnets for most major economic corporations in the country, particularly those listed on the IBEX 35, the selective indicator of the Spanish stock market that includes the 35 most prominent companies based on business volume and capitalisation. Among these corporations, financial entities such as Banco Santander, BBVA and CaixaBank wield direct influence in shaping major economic centres on the international scale. This centrality in a globalised world must also be measured in the role of these cities as communication hubs. In essence, the Spanish economy has rapidly transformed, introducing substantial changes in territorial organisation through an intense urbanisation process and shaping significant economic capitals. This newfound economic centrality and power will be the focus of our work.

Among the major Spanish metropolitan areas, Madrid stands out, leveraging its status as the capital of the nation to reinforce its economic importance. Madrid has benefited from various processes of centralisation, coupled with a regional policy of lower taxes, with the objective of attracting capital. Madrid serves as a prime example of the accumulation of economic, political and financial power, aiming at the creation of a hegemonic urban system on a global scale. Moreover, the city of Madrid no longer confines all central directive activities within its borders, with the organisation of the space resulting in the creation of a metropolitan region with over six million inhabitants (Méndez Gutiérrez del Valle 2011; Sanz Berzal 2014; Gutiérrez Puebla 2014). Barcelona's metropolis follows closely, with metropolitan area comprising over five million residents. Although the number of multinational companies based in this city is lower, and there is no capacity to capitalise on the advantages of being the political capital of the Autonomous Region, the city's industrial tradition, its status as a major Mediterranean port, the development of the creative economy, and the rise of tourism explain its position as the second major reference city in the Spanish urban system (Casellas et al. 2010; Pareja-Estaway 2010; Revista Económica de Catalunya 2011). While more modest in size, Valencia, Sevilla (metropolitan areas of 1.5 million inhabitants each) and Bilbao (one million) also qualify as major economic centres. In Zaragoza, a highly favourable logistical situation contributes to the formation of a significant city in a sparsely populated region, and in Málaga, tourism, culture and associated innovation domains help explain a population similar to that of Sevilla (Alonso Logroño et al. 2006; Caravaca et al. 2013; Torres Enjuto 2013; Salóm Carrasco et al. 2015).

12.2 An Approach to the Economic Wealth of Spanish Regions and Metropolises

As previously indicated, gaining a deeper understanding of economic centralisation processes in Spain is of considerable interest, and starting with data published in the *Anuario Estadístico de España* (INE 2019) (Statistical Yearbook of Spain) (Yearbook, 2022) proves highly valuable (Table 12.1). Among these data, our focus has been on fundamental information concerning productive capacity in each territory, as detailed in the Gross Domestic Product by Autonomous Region table. Using these values, the total production amount can be calculated. While the extrapolation of Autonomous Region data to metropolitan areas is not automatic, this information provides an initial baseline for the contribution of the primary urban and metropolitan areas to the overall wealth creation in the country.

In accordance with the most recent data, estimated for the year 2021, a total of five Autonomous Regions (Madrid, Cataluña, Andalucía, Valencia and the País Vasco) account for 67.04% of the total GDP value. This reaffirms our notion that wealth generation is concentrated in a limited number of metropolitan regions. The Autonomous Region with the highest percentage participation in GDP is Madrid, reaching 19.4%, closely followed by Cataluña with 19.0%. It is worth noting that in Madrid, there is a significant alignment between the metropolitan region and the Autonomous Region, while Cataluña comprises a broader territorial scope than the Barcelona metropolis. Together, these two Regions make up 38.4% of the Spanish GDP, a figure surpassing the 33–34% represented by the Madrid and Barcelona regions in the total demographic population. This indicates a higher concentration of wealth in large metropolises relative to the population (Albertos Puebla and Sánchez Sánchez 2014, 2017).

Regarding other territories, it is important to highlight that the significance of Andalucía is undoubtedly attributable to its demographic weight and the diversity of its territorial resources. Furthermore, two major metropolitan areas, Sevilla and Málaga, are located in this region, both attracting numerous innovative economic sectors (Lois González and Piñeira Mantiñán 2017). Following at a certain distance from Andalucía is the Comunitat Valenciana, which also has a prominent metropolitan centre. Valencia exclusively contributes a substantial portion of the regional GDP, with the economic development of the Alicante coastline, the network

Table 12.1 Gross Domestic Product at market prices by Autonomous Region. Current prices. Thousands of euros

	2016	2017	2018	2019	2020 (P)	2021 (A)
National total	1,114,420,000	1,162,492,000	1,203,859,000	1,245,513,000	1,117,989,000	1,206,842,000
01 Andalucía	148,401,635	155,397,322	160,586,830	164,929,489	148,844,318	160,747,479
02 Aragón	34,210,384	35,640,507	36,867,512	38,049,621	35,463,884	37,964,347
03 Asturias, Principado de	21,692,554	22,595,763	23,225,369	23,711,269	21,270,976	23,441,444
04 Balears, Illes	29,826,671	31,387,148	32,740,793	33,935,302	26,186,402	30,321,427
05 Canarias	42,007,287	44,209,198	45,829,590	47,182,560	38,604,715	42,655,988
06 Cantabria	12,735,884	13,228,614	13,743,992	14,180,759	12,838,938	13,846,024
07 Castilla y Leon	54,764,564	56,058,853	58,545,871	59,798,553	54,822,777	58,119,051
08 Castilla—La Mancha	38,314,671	39,897,062	41,480,976	42,398,854	39,342,356	42,286,077
09 Cataluña	212,655,821	221,482,095	229,083,098	238,308,749	212,437,867	229,418,251
10 Comunitat Valenciana	103,220,394	107,799,459	111,743,287	115,606,552	104,671,205	112,603,499
11 Extremadura	18,504,867	19,538,313	20,091,410	20,478,928	18,890,603	20,117,062
12 Galicia	58,271,372	60,397,896	62,425,422	64,220,115	58,638,430	63,230,043
13 Madrid, Comunidad de	212,360,938	222,566,193	231,423,389	242,093,235	218,468,468	234,639,003
14 Murcia, Region de	29,366,951	30,540,937	31,094,181	32,319,250	29,720,024	32,205,157
15 Navarra, Comunidad Foral de	18,743,492	19,528,931	20,063,495	20,873,397	18,941,875	20,378,804
16 País Vasco	67,114,320	69,660,724	71,921,485	74,074,980	66,422,227	71,706,061
17 Rioja, La	8,013,970	8,295,967	8,602,306	8,822,400	8,113,815	8,614,076
18 Ceuta	1,636,557	1,660,133	1,720,445	1,774,927	1,662,130	1,760,058
19 Melilla	1,506,146	1,520,137	1,577,931	1,625,086	1,522,182	1,609,185

P Provisional estimate, *A* advance estimate
Source National Institute of Statistics, Regional Accounting of Spain, Series 2016–2021

of small and medium-sized industrial cities, and a privileged position in the Mediterranean corridor further reinforcing its slightly over 9.3% contribution to the value of national production (Salóm Carrasco et al. 2015). Lastly, the País Vasco combines its modest demographic weight (approximately 2.2 million inhabitants currently) with significant economic relevance (5.9% of the Spanish GDP). In this case, we find a territory that has been affluent and industrialised since the early contemporary period, which has successfully transitioned to research-intensive and advanced service sectors for companies in recent decades. This reinforces its overall economic weight, particularly concerning Greater Bilbao, while also boasting a well-balanced and integrated urban system (Ruíz-Urrestarazu and Urrutia 2008). Alongside these prominent territories, regional GDP is more modest but still relatively significant in Galicia, Asturias, the two Castillas, Aragón or Canarias, showing the existence of complementary urban growth axes with those at the country's core.

A clear indicator of the dynamic under analysis is the distribution of average household income and per capita income at the municipal level (Table 12.2). Over the past decade, the *Agencia Tributaria* (Tax Agency) and the Basque and Navarre tax offices have annually published statistics on wealth per person. Examining the list of municipalities with over 10,000 inhabitants and the highest wealth in Spain reveals the following. Firstly, a marked concentration of these municipalities in the metropolitan regions of Madrid and Barcelona (19 and 14 territories, respectively, out of a total of 60, considering income per household) and in the País Vasco (9 localities), with only a few well-identified exceptions, such as Canarias' example of Santa Brígida, linked to wealth generated by tourism, Oleiros in Galicia, a residence for executives of the multinational ZARA-INDITEX, and affluent suburbs of Valencia such as Godella or Pamplona's Zizur Mayor and Aranguren. This localisation reflects a concentration of advanced tertiary activities, innovation, services to companies, and financial activities that have experienced a privileged situation in these areas. Secondly, while Barcelona, Bilbao or Valencia do not appear on this list, and San Sebastián and Madrid do, it is because while several of Spain's wealthy major cities hover around 40,000 euros in family income and 15,000 per person, they do not enter this select group dominated by clearly smaller cities, medium-sized cities and highly valued residential suburbs in the vicinity of the capitals. Thirdly, a more detailed geographical analysis reveals how the northwest periphery of Madrid, a series of discontinuous territories in the metropolitan area of Barcelona and municipalities near Bilbao and San Sebastián constitute the chosen territories for higher-income groups. In this regard, Pozuelo de Alarcón, Boadilla del Monte, Torrelodones, Majadahonda and Alcobendas in the Madrid vicinity stand out in the ranking. In the vicinity of Barcelona, Sant Cugat del Vallés and San Just Desvern hold relevant positions on the list, while the Basque municipalities of Getxo, Mungia and Sopela, near Bilbao, also appear.

In this map of affluent and metropolitan Spain, the presence of executive positions and professional offices of major firms is significantly higher in Madrid, Barcelona, and the Basque and Navarre cities compared to the rest of the country. Indeed, these three examples coincide in concentrating a substantial number of advanced service

Table 12.2 Spanish municipalities with more than 10,000 residents, sorted by per capita income. Ranking of the top 60

Municipality	Average income per person	Average household income	Residents
Pozuelo de Alarcón	31,670	83,377	87,080
Boadilla del Monte	26,009	82,973	59,000
Torrelodones	20,874	67,763	24,057
Majadahonda	20,672	65,059	72,173
Sant Cugat del Vallés	21,304	64,848	93,937
Las Rozas de Madrid	20,605	63,653	96,294
Alella	20,913	59,709	10,049
Villanueva de la Cañada	17,295	59,240	22,453
Sant Just Desvern	21,511	57,804	19,279
Tres Cantos	19,863	57,055	40,726
Villaviciosa de Odón	18,547	56,991	28,016
Alcobendas	18,684	56,448	116,895
Sant Quieze del Vallés	18,885	54,275	20,117
Paracuellos del Jarama	17,065	53,671	26,293
San Agustin de Guadalix	16,417	52,092	13,626
Getxo	20,293	51,647	81,260
Algete	15,917	50,456	20,704
Villanueva de Pardillo	15,203	50,302	17,266
Premià de Dalt	17,096	48,959	10,412
Godella	16,317	47,826	13,131
Aranguren	15,697	47,826	11,726
Sant Andreu de Llavaneres	17,319	47,120	11,123
Villalbilla	15,647	46,475	15,046
Rivas-Vaciamadrid	15,441	46,389	93,272
Vilassar de Mar	17,368	45,967	20,970
Moralzarzal	15,323	45,812	13,597
Arrroyomolinos	14,359	45,430	33,757
Daganzo de Arriba	14,594	45,260	10,528
Santa Brígida	15,745	45,064	18,272
Sant Joan d'Alacant	16,119	44,980	24,367
Colmenar Viejo	15,028	44,384	52,465
Mungia	16,885	44,367	15,848
San Sebastián de los Reyes	15,709	44,359	90,958
Eliana, l'	15,566	44,109	18,362
Marratxi	14,449	44,052	38,357

(continued)

Table 12.2 (continued)

Municipality	Average income per person	Average household income	Residents
Castelldefels	15,782	43,718	66,866
Donostia/San Sebastián	18,569	43,696	187,849
Oleiros	16,403	43,578	36,922
Sopela	17,133	43,550	12,242
Zumaia	16,768	43,515	10,153
Palllejà	15,681	43,388	11,640
Roca del Vallés, La	15,899	43,298	10,660
Masnou, El	16,572	43,266	23,803
Argentona	15,436	43,092	12,634
Sant Joan Despí	15,814	43,031	34,072
Madrid	17,059	43,003	3,277,451
Griñon	13,889	42,864	10,489
Oñati	18,075	42,863	11,515
Hondarribia	17,627	42,836	16,904
Alpedrete	15,011	42,819	15,178
Sitges	17,588	42,710	30,257
Zizur Mayor/Zizur Nagusia	15,123	42,634	15,198
Corbera de Llobregat	15,125	42,603	15,029
Sabta Cruz de Bezana	15,237	42,511	13,482
Durango	16,862	42,436	29,234
Zarautz	17,306	42,287	23,101
Cabanillas del Campo	14,300	42,232	11,000
Galapagar	14,382	41,920	34,653
Molins de Rei	15,912	41,845	26,123
Lliça d'Amunt	14,227	41,727	15,654

Source Agencia Tributaria and Basque and Navarre Tax Offices

activities, driving the so-called knowledge economy, and hosting a significant number of prestigious professional offices.

In Madrid, these common elements are supplemented by the prominence of high-ranking officials from the central administration of the State and various public institutions. Additionally, the radial structure of communications in Spain reinforces the role of the metropolitan region as a major logistics hub, hosting numerous executive positions. Within the territory of the Madrid Region, large-sized companies (around 20% of the total Spanish) have been established, along with corporate cities of multinational giants such as Banco Santander, BBVA or Telefónica. All these elements define a remarkable metropolis in the European context, with numerous skilled jobs, outlining an axis of wealth that emanates from the centre, extends along the Castellana, Chamberí and the Salamanca district, and continues beyond the city towards Alcobendas and the West-Northwest towards Pozuelo, Boadilla or Las Rozas (Sanz Berzal 2014; Gutiérrez Puebla 2014; Méndez Gutiérrez del Valle and Sánchez Moral 2014). In Barcelona and its metropolitan area, the privileged economic position stems from a significant industrial past. This industrial base has recently transformed, and, coupled with strategic sectors (automotive, pharmaceutical, optics or publishing), a knowledge-intensive tertiary sector, and numerous activities related to tourism, has made it possible to concentrate a population with a high purchasing power in certain neighbourhoods and nearby towns. Among the areas where this new economy has been located, the 22@ area in Poble Nou, the neighbourhoods of La Sagrera or the Forum, various technological parks in L'Hospitalet and the Vallès, and the Zona Franca of the port stand out. The highest-income groups in the metropolis continue to reside in Sarrià or Pedralbes, although San Cugat del Vallès, San Just Desvern or Sitges have become highly sought-after territories (Generalitat de Catalunya 2010; Revista Económica de Catalunya 2011; Molas and Parellada 2011). Finally, Bilbao and the other Basque-Navarre capitals also benefit from a notable industrial past, which has been redirected through a determined regional policy focusing on innovation. The strength of the network of high-tech parks and companies created is remarkable. Well-paid positions have increased, and the most appreciated residential areas continue to be some classics like Getxo, San Sebastián and Hondarribia, along with other municipalities with enviable coastal locations or excellent connectivity (Ruíz-Urrestarazu and Urrutia 2008; Alberdi 2010).

12.3 Major Economic Corporations and Their Territorial Distribution

The most direct method to locate the major private economic centres of a country involves identifying the headquarters of its leading corporations. In the case of Spain, these firms are associated with an index that comprises the 35 largest companies by market capitalisation, the IBEX. It can be argued that the Spanish economy demonstrates a significant dependence on the prominence exerted by this exclusive group of firms. This group of corporations also explains Spain's economic projection abroad, providing elements for analysing cities or metropolises that aim to play a prominent role in the global urban system.

The IBEX 35 includes the leading companies measured by their stock market liquidity, and typically the most important in the country. However, its sectoral composition is far from balanced and has a sectoral structure which diverges from the overall Spanish economy (Albadalejo García 2014). Out of the 35 existing firms, a total of 8 fall within the powerful energy sector (one including paper production), 6 in finance (5 banks and one insurer) and 3 in telecommunications, leaving the remaining productive sectors with only one or two companies (Table 12.3). Indeed, this data already shows that the IBEX includes sectors with greater strengths. In fact, finance represents 32.95% of the total capitalisation, energy 20.49% and telecommunications 8.4%, which, along with the 10.34% from the textile sector represented exclusively by Inditex, accounts for nearly three-quarters of the combined business volume of the 35 firms. Therefore, there is a pronounced concentration in the most classic and influential sectors of contemporary capitalism, organised into large multinational corporations, generating a massive volume of employment, with significant wage disparities between executives and numerous subordinate personnel.

When it comes to the location of major economic and financial centres of the country, the headquarters of the IBEX 35 companies exhibit a strong polarisation around Madrid. In fact, a total of 23 of these companies are domiciled in the metropolitan region of the Spanish capital, mostly in the city centre and the remaining in the north and northwest periphery (with a concentration in Alcobendas). It can be argued that a good percentage of these companies are present throughout the national territory, but the tendency towards an increased concentration of power in Madrid and the privileged logistical position created for the capital explain this location (Bel 2010). Some notable exceptions to this norm are Banco Santander and BBVA, which maintain their headquarters in the cities where they originated but with the vast majority of their leadership positions in Madrid. In Barcelona, only one company remains after the induced migration of the headquarters of Catalan corporations due to the independence process. However, even though entities such as CaixaBank, Banco de Sabadell, Naturgy, Inmobiliaria Colonial and Cellnex Telecom, formally relocated to Madrid, Valencia or Alicante, they retain much of their central activity in the Catalan capital. Regarding other locations, Greater Bilbao stands out again with four firms, while the rest of the companies remain in their original places such as Inditex in the vicinity of A Coruña, Melià in Palma de Mallorca and Viscofán in Tajonar (Navarra), with the striking exception of ArcelorMittal located in the tax haven of Luxembourg. Once again, the areas that stand out for their wealth generation and income dominance appear dominant in this list, reaffirming the idea that in Spain, the existence of truly prominent economic centres is evident.

In addition to the information related to 2020, it is also worth mentioning that this index has undergone some minor changes since that date. Two new pharmaceutical companies, Almirall and PharmaMar, have been added, and the energy company Solaris replaced MásMovil in the prestigious index. For the purpose of this study, the first has its headquarters in Barcelona, and the other two are based in Madrid. Therefore, in an index of 37, the primacy of Madrid is reaffirmed (gaining two companies and losing one), and Barcelona seems to reappear slightly after the discussed change of headquarters for many of its major companies following the independence crisis.

Table 12.3 Companies belonging to the IBEX 35, 2020

Ticker	Company	Headquarters	Sector	ISIN	Allocation
ANA	Acciona	Alcobendas	Infrastructure	ES0125220311	0.7
ACX	Acerinox	Madrid	Metalworking	ES0132105018	0.55
ACS	Grupo ACS	Madrid	Infrastructure	ES0167050915	2.41
AENA	Aena	Madrid	Transport	ES0105046009	2.75
AMS	Amadeus IT Group	Madrid	Tourism	ES0109067019	6.97
MTS	ArcelorMittal	Luxembourg city	Metalworking	LU1598757687	1.14
SAB	Banco Sabadell	Alicante	Finance	ES0113860A34	1.65
SAN	Banco Santander	Santander	Finance	ES0113900J37	14.97
BKIA	Bankia	Valencia	Finance	ES0113307062	1.35
BKT	Bankinter	Madrid	Finance	ES0113679I37	1.48
BBVA	Banco Bilbao Vizcaya Argentaria	Bilbao	Finance	ES0113211835	7.62
CABK	CaixaBank	Valencia	Finance	ES0140609019	4.88
CLNX	Cellnex Telecom	Madrid	Telecommunications	ES0105066007	0.86
CIE	CIE Automotive	Bilbao	Steel, petrochemical	ES0105630315	0.43
ENG	Enagás	Madrid	Energy	ES0130960018	1.16
ENC	ENCE	Madrid	Energy and pulp/paper	ES0130625512	0.29
ELE	Endesa	Madrid	Energy	ES0130670112	1.6
FER	Ferrovial	Madrid	Transport	ES0118900010	2.79
GRF	Grifols	Barcelona	Healthcare	ES0171996087	2.11
IAG	IAG	Madrid	Transport	ES0177542018	3.22
IBE	Iberdrola	Bilbao	Energy	ES0144580Y14	8.41
ITX	Inditex	Arteijo	Textiles	ES0148396007	10.34
IDR	Indra Sistemas	Alcobendas	Consulting	ES0118594417	0.34
COL	Inmobiliaria Colonial	Madrid	Real estate	ES0139140174	0.84
MAP	Mapfre	Majadahonda	Finance	ES0124244E34	1
MAS	MasMovil Ibercom	Alcobendas	Telecommunications	ES0184696104	0.32
TL5	Mediaset España Comunicación	Madrid	Media	ES0152503035	0.35

(continued)

Table 12.3 (continued)

Ticker	Company	Headquarters	Sector	ISIN	Allocation
MEL	Meliá Hotels International	Palma de Mallorca	Tourism	ES0176252718	0.36
MRL	Merlin Properties	Madrid	Real estate	ES0105025003	1.15
NTGY	Naturgy	Madrid	Energy	ES0116870314	1.92
REE	Red Eléctrica Corporación	Alcobendas	Energy	ES0173093024	1.98
REP	Repsol	Madrid	Energy	ES0173516115	5.34
SGRE	Siemens Gamesa Renewable Energy	Zamudio	Energy	ES0143416115	0.95
TEF	Telefónica	Madrid	Telecommunications	ES0178430E18	7.22
VIS	Viscofan	Tajonar	Food	ES0184262212	0.59

Source CNMV

12.4 Approach to the Financial Geography of Spain

In this brief study on the economic capitals of Spain, it is worthwhile analysing the location of banks, the main financial entities in the country, which have undergone a major concentration process in recent years. In fact, the crisis that unfolded in 2008 virtually wiped out the entirety of *Cajas de Ahorro* (Savings Banks). Moreover, business volume today is substantially concentrated around five or six entities, which hold a dominant position. This simplification process of the financial structure has translated into a strengthening of Madrid's leadership position, far ahead of Barcelona, while some capitals in the North or Levant still retain corporate headquarters of financial institutions, but more as a result of specific decisions or consistent with the tradition of the banks themselves.

This statement is confirmed in Table 12.4. The banking sector has notably concentrated around Banco Santander (close to 20% of the total in Spain), CaixaBank and BBVA. The three entities together account for over 60% of the financial business in Spain, measured in terms of the assets declared by these institutions. Behind, and at quite a distance away, are Bankinter, Sabadell and Banca March, along with the business volume that some foreign entities and investment funds have in Spain. One of the significant novelties of the post-2008 crisis recovery is the emergence of so-called private banking in the general financial business of the country. Private banking, which manages large assets through investment or pension funds, has little physical presence in cities, and lacks an extensive branch network, but channels a growing proportion of capital volume. Additionally, it has allowed foreign firms to establish a strong foothold in Spain (Andbank España and Deutsche Bank rank 7th and 8th in the banking ranking by assets). From a geographical perspective, this landscape has led to a certain dematerialisation of the banking business, whose location is more challenging to trace than when they had large central headquarters and an extensive branch network to serve customers (Jácome Estrella et al. 2022). A second consequence of the process is the increased centrality of Madrid, where much of the business management of traditional banks such as Santander, BBVA or Bankinter is concentrated, as well as the central offices of private banking. Barcelona plays a similar but more modest role, and other cities like Valencia or even Bilbao seem to have lost importance.

Having presented the locational dynamics and headquarters of major Spanish banks, we now delve into their territorial code. Starting with Banco Santander, its spatial organisation is evident from its symbolic centre in the city from which it takes its name and its actual centre in Madrid, based in its financial city of Boadilla del Monte, and employing 7000 workers. The institution maintains a dense network of branches throughout the country, reinforcing the commanding role of Madrid where most control and management functions are located. Internationally, Santander is among the world's leading banks, with a significant presence in Portugal, Brazil, the United Kingdom and Poland, alongside a noteworthy presence in France, Germany, the Netherlands, the United States and Canada, among others. Santander's geography reflects Madrid's potential to become a global city. CaixaBank, on the other hand, retains most of its presence in the metropolis of Barcelona, even though its formal headquarters moved to Valencia. It has a significant footprint across Spain and Portugal, where it acquired BPI. Its international projection is much smaller than Santander's, although its notable shareholding in major companies such as Naturgy, Telefónica, Suez or Cellnex should be highlighted. CaixaBank's role in generating economic power in Barcelona is indisputable. Meanwhile, BBVA is also centred in

Table 12.4 Ranking of Spanish banking by assets (in millions of euros)

Position 2019	Position 2018	Group	Capital 2022	Capital 2018	Growth 2022–2018 (%)
611	1	Santander Private Bank	125,161	103,849	20.5
2	3	CaixaBank Private Bank	116,660	64,912	79.7
3	2	BBVA Private Bank	100,952	68,350	47.7
4	4	Bankinter Private Bank	51,127	35,649	43.4
5	5	SabadellUrquijo Private Bank	32,543	26,953	28.2
6	7	March Bank	23,500	14,790	58.9
7	9	Andbank España	18,805	9549	96.9
8	10	Deutsche Bank	18,000	9150	96.7
9	–	Renta 4	16,398	10,183	61.0
10	–	Singular Bank	13,330	–	–
Total private banking			516,476	343,385	50.4

Source Spanish Banking Association (AEB) and Funds People, 2020

Madrid, not forgetting its Basque origins. Like the two aforementioned entities, it maintains a large territorial presence with thousands of branches. Internationally, BBVA participates in the financial systems of Argentina, Mexico, Peru, Venezuela and Colombia, as well as in the United States and Turkey, following a strategy similar to Santander's. At a distance in terms of business volume, BANKINTER is located in Madrid and reaffirms the financial primacy of this city over the entire Iberian Peninsula, as the institution has a significant presence in Portugal. Lastly, Banco de Sabadell joins CaixaBank in showing Barcelona's banking power, albeit with its headquarters relocated to Alicante. Sabadell has consolidated its position by acquiring small banks and has a significant presence throughout Spain. It is also present in Mexico and the United Kingdom. In conclusion, it is evident that Madrid and Barcelona act as increasingly important financial capitals with a global outlook, beginning with their influence in Latin America.

12.5 Final Considerations: Major Urban Centres in Spain: The Aim of Becoming Global Cities?

In the preceding pages, the notion that the Spanish urban system is increasingly centralised around the metropolises of Madrid and Barcelona has been reiterated. The País Vasco, particularly the Greater Bilbao area, appears not to have lost a prominent economic position, while Valencia, Sevilla, Málaga and Zaragoza serve more as centres only at a regional scale. In this current map of Spanish cities, it can be concluded that a larger metropolis corresponds to higher economic capacity and population well-being. From this point, the focus has been on observing the locations of major corporations and leading banks in the country. The conclusion is clear: Madrid attracts the majority of headquarters, without disregarding the economic significance of Barcelona and Bilbao. What is truly significant about this observation is that many of these spatially polarised corporations have substantial economic tentacles in both Portugal and Latin America (Lois González 2019; Rio-Fernandes et al. 2022). Simply by mentioning examples like Telefónica, Endesa or Banco Santander, it becomes evident that from Madrid (or Barcelona), indisputable economic dominance is exerted over key sectors in Latin American countries, including banking, energy or telecommunications. Hence, the title of this section, 'The aim of becoming global cities?'

An essential component of a global city involves benefiting from high levels of air connectivity, functioning as a genuine hub in this type of transport. Madrid and Barcelona have substantially expanded their terminals and runway lengths at Barajas and El Prat airports. With regard to their international connections, both cities have chosen to prioritise two regions: the whole of Western Europe and Latin America. The geopolitical and geoeconomic code of the two major Spanish cities is clear: to consolidate an increasing number of large multinational firms, promote the growth and internationalisation of their businesses, and act as a natural bridge between Latin

America and the major centres of the European Union. This objective motivated the construction of the spectacular T4 and a logistics centre in Barajas, and the new terminal, close to the business and industrial zone of the port, in El Prat. In this competition for the status of a global city, there is a third reference on the Iberian Peninsula, Lisbon (with an almost collapsed airport), dominating connectivity with Brazil and some African countries. The airport in Sondica, Bilbao, has also enhanced its connectivity with the rest of Europe. Meanwhile, Valencia, Málaga or Sevilla have significant airport facilities more oriented towards tourism.

Finally, this evident attempt to project themselves as global cities is complemented at a more concrete level by defining the road network that reinforces the position of these economic centres in their territory. Undoubtedly, Madrid benefits the most from a Spanish system of radial highways and high-speed trains, increasing its dominant position in the country's economy (Bel 2010; Boira i Maiques 2013). In the case of Barcelona, its potential has been hindered by a Spanish government that has favoured the primacy of the state capital. However, the Catalan city benefits from its proximity and good connections with France and the western Mediterranean, and in recent times, it takes advantage of the creation of a Mediterranean corridor that brings it closer to Valencia and the entire eastern Iberian Peninsula. Sevilla has maintained a privileged connection with Madrid for decades. For Zaragoza, the equidistance from the Spanish capital, Barcelona and the País Vasco is an undeniable advantage. Lastly, the Basque cities continue to benefit from easy access to France. In any case, the map and the dominant urban system described at the beginning of this work only reaffirm themselves based on any economic or logistical indicator used in our analysis. An analysis that has aimed to discern the determination of Spain's economic centres in the direction of consolidating a global system of cities.

References

Albadalejo García R (2014) Sectorialización del IBEX 35 y representación sobre la economía española. Trabajo Fin de Grado, Grado de Administración y Dirección de Empresas. Universidad Politécnica de Cartagena

Alberdi A (2010) Economía vasca 1980–2010: tres crisis y una gran transformación. Ekon Rev Vasca Econ 25(03):34–95

Albertos Puebla JM, Sánchez Sánchez JL (coords) (2014) Geografía de la crisis económica en España. Prensas Universitarias de Valencia, Valencia

Albertos Puebla JM, Sánchez Sánchez JL (2017) Caracterización y especialización industrial y del terciario avanzado en las regiones españolas. In: Farinós Dasì J, Olcina Cantos J (eds and coords) Geografía Regional de España. Espacio y Comunidades. Bases para una regionalización renovada del territorio español. Tirant lo Blanc, Valencia, pp 291–371

Alonso Logroño MP, Climent López EA, Escalona Orcao A (2006) Zaragoza y el sector central del eje industrial del Ebro. In: Méndez Gutiérrez del Valle R, Pascual Ruíz-Vadepeñas H (coords) Industria y Ciudad en España: nuevas realidades, nuevos retos. Thomson-Reuters Civitas, Cizur Menor (Navarra), pp 557–586

Bel G (2010) España, capital París. Origen y apoteosis del Estado radial: del Madrid sede cortesana a la "capital total". Destino, Barcelona

Boira i Maiques JV (2013) Infraestructuras y financiación en España, hacia un nuevo paradigma. In: Gómez Mendoza J, Lois R, Nel.lo O (eds) Repensar el Estado. Crisis económica, conflictos territoriales e identidades políticas en España. Universidade de Santiago de Compostela, Santiago de Compostela, pp 101–114

Capel Sáez H, Vilà Valentì J (1970) Campo y Ciudad en la Geografía española. Biblioteca Básica Salvat y Alianza Editores, Madrid

Caravaca I et al (2013) Economía creativa en la aglomeración metropolitana de Sevilla: agentes, redes locales de colaboración y principales actuaciones. Bol AGE 63:81–104

Casellas A, Dot E, Pallarès Barberá M (2010) Creación de imagen, visibilidad y turismo como estrategias de crecimiento económico de la ciudad. Finisterra 45(90):153–172

Feria Toribio JM, Albertos Puebla JM (2010) La ciudad metropolitana en España: procesos urbanos en los inicios del siglo XXI. Thomson-Reuters Civitas, Cizur Menor (Navarra)

Generalitat de Catalunya (2010) Pla Territorial Metropolità de Barcelona. Generalitat de Catalunya, Barcelona

Gutiérrez Puebla J (2014) Madrid y su entorno: estructura territorial, infraestructuras de transporte y relaciones funcionales. In: Sanz Berzal B (coord) Estructura económica de la ciudad de Madrid. Ayuntamiento de Madrid, Madrid, pp 57–86

INE (2019) Anuario Estadístico de España 2022. INE, Madrid

Jácome Estrella H, Lois González RC, López Sandoval MF, López Iglesias E (2022) Desigualdades territoriales en la inclusión financiera y económica. Universidade de Santiago de Compostela, Santiago de Compostela

Lois González RC (2019) La cambiante interpretación de la geografía española respecto a América Latina: entre el paternalismo, la hermandad y la indiferencia. In: Polo-Almeida P, Carrión A, López-Sandoval MF (coords) Debates actuales de la geografía latinoamericana: visiones desde el XVII Encuentro de Geógrafos de América Latina. Asociación Geográfica de Ecuador y Pontificia Universidad Católica de Ecuador, Quito, pp 124–132

Lois González RC, Piñeira Mantiñán MJ (2017) El modelo territorial y urbano resultante. Un sistema urbano metropolitano y policéntrico más formal que funcional activo. In: Farinós J, Olcina J (eds and coords) Geografía Regional de España. Espacio y comunidades. Tirant Humanidades y Publicaciones Universitat de Valencia, Valencia, pp 429–512

Méndez R, Abad LD, Echaves C (2015) Atlas de la Crisis. Impactos socioeconómicos y territorios vulnerables en España. Tirant lo Blanc, Valencia

Méndez Gutiérrez del Valle R (2011) Globalization, knowledge-based activities and new spatial inequalities in Madrid (Spain). Braz Geogr J Geosci Humanit Res Medium 2(2):223–237

Méndez Gutiérrez del Valle R, Sánchez Moral S (2014) Distribución espacial de las actividades económicas. In: Sanz Berzal B (coord) Estructura económica de la ciudad de Madrid. Ayuntamiento de Madrid, Madrid, pp 501–531

Molas O, Parellada M (2011) 22@ 10 anys de transformaciò económica. Rev Econ Catalunya Espec Econ Coneixement Territ 64:13–22

Nel.lo Colom O (2001) Las áreas metropolitanas. In: Gil A, Gómez Mendoza J (coords) Geografía de España. Ariel, Barcelona, pp 275–299

Pareja-Estaway M (2010) Construyendo la nueva Barcelona creativa: nuevos actores, nuevas estrategias. Finisterra Rev Portuguesa Geogr 45(90):133–152

Revista Económica de Catalunya (2011) Especial: Economia del Coneixement i Territori. Núm. 64. Colegio de Economistas de Catalunya, Barcelona

Rio-Fernandes JA, Lois-González RC, Beltrâo-Spósito ME (eds) (2022) El mundo visto de las ciudades. Tirant, Valencia

Ruíz-Urrestarazu E, Urrutia G (2008) Geografía del País Vasco. Nerea, San Sebastián

Salóm Carrasco J, Albertos Puebla JM, Fajardo Magraner F (2015) Crisis y reestructuración en el área metropolitana de Valencia. Rev Estud Andal 32:1–21

Sanz Berzal B (coord) (2014) Estructura económica de la ciudad de Madrid. Ayuntamiento de Madrid, Madrid

Torres Enjuto MC (2013) Geografía Económica del conocimiento de la aglomeración metropolitana de Bilbao. Anl Geogr Univ Complut 33(2):179–217

Part IV
Consumers and New Residents

Chapter 13
Consumer Culture and the Fashion Retail in the City of Lisbon: Avenida da Liberdade as Stage and Theater

Herculano Cachinho

Abstract The Avenida da Liberdade, in Lisbon, is quoted among the most luxurious streets in the world. This chapter examines the path traversed by this axis, from its opening in the late nineteenth century to the present day. We argue that despite the weaknesses pointed out in its conception, several factors related with centrality, symbolism, prestige and scenography allowed this boulevard to assert itself as the showcase *par excellence* of the city. Intentionally used by the nineteenth-century bourgeoisie for strolling in, to see and to be seen, in recent decades it has become the favorite stage for luxury and fashion retailing. Exploring this idea is based on a mix of qualitative and quantitative methods designed to address the different dimensions of the problem. While the reconstruction of its role as a promenade for the nineteenth-century *flâneur* relies on an active reading of scientific and literary texts, its most recent vocation as a showcase of international capital linked to luxury and fashion is founded on empirical work, which includes a survey of stores, interviews with managers of shops and field observation.

Keywords *Avenida da Liberdade* · Conspicuous consumption · Fashion retailing · Liminal space

13.1 Introduction

In a consumer society characterized by the production of spectacle (Krupar et al. 2014), the prominence of signs and images (Harvey 1989), and the economies of experience (Pine II and Gilmore 1999), fascination (Schmid 2009) and attention (Celis 2017), brands have evolved into platforms that mediate the construction of identity, shared experiences and the creation of "common social worlds" (Arvidsson 2006). Meanwhile, cities are transformed into systems of branded spaces (Klingmann 2007). Within this brandscape framework, the districts in the city associated with

H. Cachinho (✉)
IGOT, University of Lisbon, Lisbon, Portugal
e-mail: hc@edu.ulisboa.pt

R. C. Lois-González and J. A. Rio Fernandes (eds.), *Urban Change in the Iberian Peninsula*, The Urban Book Series, https://doi.org/10.1007/978-3-031-59679-7_13

luxury brands and fashion stand out in the urban setting due to their symbolic, communicative and seductive capital. There are several reasons behind this phenomenon, but the inextricable relationship between brands and the places they occupy is undeniable. The prestige of places attracts brands, and through the fascination they exert on consumers, they enhance the notoriety of places. This is evident on Rodeo Drive in Los Angeles, Fifth Avenue in New York, Avenue Montaigne in Paris, Via Monte Napoleone in Milan or Bond Street in London, to name but a few.

In Lisbon, there are two main *brandscapes* linked to luxury and fashion retailing: Chiado and Avenida da Liberdade, which perfectly illustrate the symbiotic relationship between brands and the prestige of the places. Located in the city center, both enjoy today the prestige inherited from the past. Many writers and poets have described the glamor of these city spaces in the nineteenth and twentieth centuries. Ramalho Ortigão called Chiado "the vain slope", Júlio de Castilho referred to it as "the personification of elegant Lisbon", and Norberto de Araújo tells us that "it is not just a name, it is a code of customs. It is not just a street, it is an index of Lisbon" (Janeiro 2006). Similar expressions are also used to describe Avenida da Liberdade. Moura Cabral tells us that, "According to the system of confrontation in this beautiful land (…), it is said around and about, that the Avenida is to Lisbon what the Castellana is to Madrid, the Prater is to Vienna, the Porta Venezia is to Milan, and we do not know if it is like the Bois is to Paris and Hyde Park is to London" (Cabral 1899: 77). In turn, Ramalho Ortigão (1874), Eça de Queiroz (1945), Fialho de Almeida (1893), refer to it as "the most magnificent work ever seen" or "what is best in civilization has been brought to the capital".

This paper discusses luxury and fashion retail in the city of Lisbon, based on Avenida da Liberdade. We argue that the attributes of the built environment (the architecture, location, morphology and scenography of the public space) combined with the prestige and public promenade function that the avenue performed in the past explain why it has become, in recent decades, the main showcase for international capital linked to luxury and fashion retail in the city. To establish this relationship, we have structured our presentation in three brief sections. We begin by describing the morphology of the Avenida, paying particular attention to the influence of Parisian haussmannization on the conception and design of the project. The second section describes the Avenida's functions as a public promenade and a favorite stage for the bourgeoisie. In the third section, we analyze the transformation of the boulevard into an exhibition gallery sponsored by luxury and fashion retailing. Finally, the results are discussed, and some considerations are made about the relationship between fashion retail and the social status of city spaces.

13.2 Avenida da Liberdade: A Parisian Boulevard in Lisbon

The *Avenida da Liberdade* is the most prestigious high-street in the city of Lisbon. This status goes back to the time it was first open to the public, in April 1886. Its privileged location facing the traditional downtown area—the *Baixa*—and the new business district in the so-called New Avenues (*Avenidas Novas*), together with the morphology and the symbolic power of the built environment, has made real estate prices soar in this part of the city (Fig. 13.1). High prices have therefore made it accessible only to businesses able to efficiently exploit high-income plus-values, aptly illustrated by luxury and fashion retailing.

The Avenida da Liberdade was designed to be a landmark for the city. Its opening put an end to a centuries-old cycle of linear growth in the city, which stretched along the banks of the river Tagus, and triggered a programmed expansion of the city inland, where the urbanization of Avenidas Novas was its most emblematic creation. From an urban point of view, the new avenue also meant the realization of the municipal dream of transforming the city's old Public Promenade (*Passeio Público*) into a wide boulevard that would extend the area of leisure and representation.

Despite its modernism for the time, and much having been written about its virtues and capacity for representation, the truth is that there were also many voices denouncing its negative effects on Lisbon society, describing it as a "callow channel of rustic manor houses" or a means of disseminating "bad customs acquired in the cafés and on the trottoir, encouraging ostentation and idleness" (Janeiro 2006). Ortigão (1874) voiced example of these opinions, as the passage below shows to perfection:

> The Project of making a boulevard from the promenade extending from Rossio to Campo Grande is a sadly pretentious conception, serving merely to spread the bad habits of the café and the pavement, the love of ostentation, indolence, boulevardism, the coquette, and the use of meager fashion
>
> Ramalho Ortigão (1874), As Farpas, p. 170 (Author's translation)

13.3 Avenida da Liberdade: A Public Walkway for the Bourgeoisie Exhibition

To understand *Avenida's* recent vocation as a street offering luxury and fashion retail, we need to go back in time. The journey starts in the last quarter of the nineteenth century, the time in which the *Passeio Público* was demolished to make way for *Avenida da Liberdade.* The destruction of this public garden meant the disappearance of a public space, *par excellence,* that provided a framework for social life (Silva 1994, 1997). The summer evenings with firework displays, concerts, acrobat shows, children's dances and welfare-benefit shows, not only attracted the public from the vicinity but helped to explain its nature as a place in which Lisbon society congregated

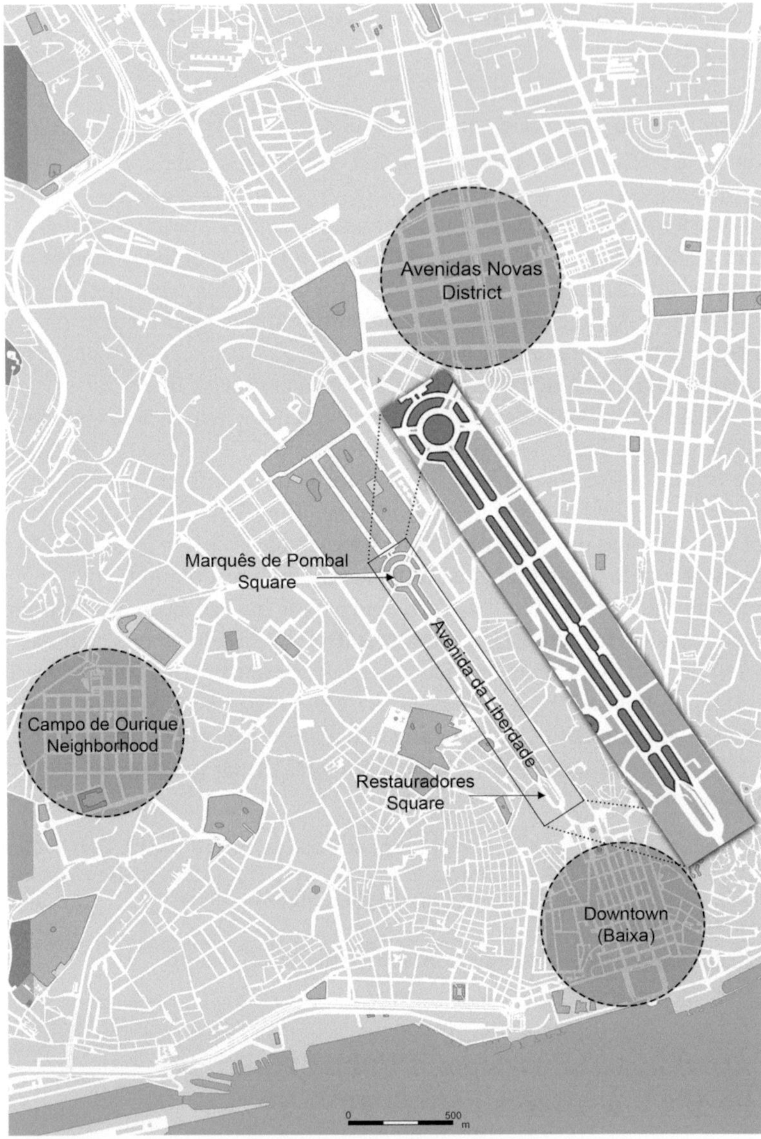

Fig. 13.1 The localization of the Avenida da Liberdade in Lisbon

and socialized. On special celebration days, it was normal for the presence of the royal family and high society, which favored the bourgeoisie association with the space, as if playing a complicated game of emulation, identification and exhibition (Le Cunff 2003); putting on a show that did not go unnoticed by those who went to the gardens with the purpose of seeing and being seen.

Therefore, due to the shortage of public places for *flânerie* and owing to the morphological and social features of the new boulevard, it was not long before it staked its claim as the fashionable new venue open to the Lisbon public, specially to the bourgeoisie who needed to show off more and gain public recognition. Indeed, the new street, that expanded the old space, attracted attention so quickly that two or three years were enough to ingrain itself into the population's habits; they now regarded it "not as an axis that expedited civilization, but rather, only as a *Passeio Público*, that was wider, prettier and more luxurious, amplifying the rituals of cosmopolitanism" (Silva 1994: 432). At the time, this fact merited the exclamation: "the best fashion in Lisbon was doing! Doing the Avenida from 6 o'clock on, just as one would do *Rua do Ouro* at 4 o'clock in the afternoon, and *Chiado* at 5 o'clock" (Pina 1886).

Several literary descriptions give us an account of the centrality of the *Avenida* with the social life of Lisbon-dwellers in its confines and its ability to change the city. Even if the opinions were far from unanimous. While some celebrated its modernity and symbolic power, others criticized the ostentation alleging that it had robbed the city of its character and identity. Cabral (1899: 80), for example, noticed that

> (…) In demolishing the *Passeio Público* and having this large street break though, life in Lisbon has been completely transformed. Dating from then, this change has indeed become visibly greater from one year to another, thus bestowing on the capital some of the most civilized airs, making her physiognomy more attractive so that the outsider may delight in contemplating her.
>
> Little by little everyone has grown used to strolling down the Avenida, doing the Avenue (…). We debated greatly about its allure in the newspapers, we spoke avidly of the uses from abroad, we fashioned the people so elegantly, inviting them to meet there every afternoon (…), that in the end, the victory was complete and today, the Avenida has given us the exact toll of our society's life, as many and as varied as its faces are.
>
> *Carlos Moura* Cabral (1899) *Lisboa em Flagrante, p. 80.* (Author's translation)

However, the Avenida's inheritance from the *Passeio Público* not only included its ornamentation but also the functions which had most projected its life: a place of entertainments, a source of information and an inter-class stage. The Avenida embodies the essence of the *Passeio Público*, showcasing an ever-evolving spectacle for all who pass through. The dynamic surroundings, changing with the time of day and seasons of the year, the constant flow of pedestrians, carriages, and the variety of goods display in the shop windows, create a vibrant atmosphere where everyone becomes both observer and participant. Everyone who walks the Avenida becomes a part of the intricate social fabric on display, offering a plethora of information and insights to those who care to look. The Avenida's effectiveness in fulfilling its role as a public space is also attributed to the captivating "social text" that unfolds before the eyes of its visitors. The interactions between people from different social backgrounds, goals and activities contribute to a rich tapestry of experiences that shape the overall ambience of the avenue. This tapestry of human interaction adds to the allure of the avenue, fostering a sense of shared experience for all those who decide to make it their meeting point and take part in the game of seeing and being seen. Ultimately, the Avenida excels in its function as a hub of information, symbols and cultural significance, enriching the daily lives of those who traverse its path. It is its intangible wealth that allows the Avenue to enhance its attractiveness as well as shape a vision of itself, stimulating the actors' powers of observation and mold their ways of daily appropriation.

Lefèbvre (1970) wonderfully depicted the potential for representation in the streets of the city, providing numerous examples. One of the most striking is the ability of women to categorize other women based on their attire, footwear, hairstyles and manner of walking. The information capital of the streets is not limited to a specific time. This capital of the streets is lost over time in cities that have the most distinctive cultural settings. This occurs because the city's public space has always served as meeting points for exchanging goods and information, besides being the space of social staging and representation (Gehl and Gemzoe 2001). Nevertheless, in the late nineteenth century, during the rise of the consumer society, the power of imagery and representational spaces in shaping the identities of social classes and the city itself gave street capital even greater significance. Indeed, Avenida da Liberdade was no exception to this trend. Its layout, grandeur, theatricality and symbolism made it a one-of-a-kind stage where the Lisbon society could exhibit themselves. This is what Cabral (1899: 84) sought to describe and as the lithograph made by João Cristino (Fig. 13.2) illustrates.

> (…) Oh! The Avenida is a grand market, a grand bazaar, the most complete and the most varied exhibition of our specimens. It is here that Lisbon may truly be assessed, its lineage, its character, its education; how it dresses, how it does its hair, how it perfumes itself, how it eats; its vivacity, its *gaucherie*, the gallantry of its coquettishness and the banality of its aspirations. It is here we may perceive who is in the limelight: who is the darling of fashion, who is the talk of the town, who is in demand; the politician who will win more public acclaim, the well-heeled who best knows how to squander his handfuls of gold; the woman who will tote up the most victims in the *salons,* in the street, at the theater, with the inconstancy of her glances and the volubility of her loves. (…)
>
> (Author's translation)

In the transition to the twentieth century, Avenida da Liberdade would play other roles, of great importance for the development of its more recent vocation as a street for luxury and a showcase for the international capital liked to fashion retail. In the absence of shopping arcades, passages, galleries and the scarcity of other spaces for consumption, leisure and dreaming that were created in Paris, London or Berlin during the nineteenth century, it was on the Avenida that the new hero of the modern city came to life: the *flâneur*. Invariably depicted as "a man of leisure, meticulously dressed in a black frock coat and top hat and carrying a walking stick or umbrella, the *flâneur* was both a familiar and yet an ambiguous figure. He was the consummate urban stroller at home on the boulevards and in the shopping arcades, restaurants, and art galleries of the city" (Gluck 2010: 272). Charles Baudelaire and Walter Benjamin had no doubt that without the aforementioned enchanted spaces, true syntheses of luxury and splendor, and without the distinctive atmosphere they created, the advent of the *flâneur* would hardly have been born. What other places in the city could serve this "gastronomy for the eyes" (Moncan 2009), the only activity, according to Balzac, to which the *flâneur* seemed to devote himself? Lisbon didn't have the charms of Paris, but it was mainly on Avenida da Liberdade, and especially on Sunday

Fig. 13.2 Panorama of *Avenida da Liberdade*, in 1905. *Source* Lithograph by João Christino, in the supplement to "*Mala da Europa*" ("*The Suitcase of Europe*") No. 488

afternoons, that the *flâneur* (and the *flâneuse*) could stroll around aimlessly and, at the same time, stroll around with the aim of showing off, of seeing and being seen. Without the enchanting spaces of Paris, it was mainly at Avenida da Liberdade, and especially on Sunday afternoons that the *flâneur* (and *flâneuse*) were able to stroll without a destination, and at the same time, dilly-dally with the purpose of showing off, seeing and being seen. This happened simply because the Avenida's atmosphere, staging and symbolic capital offered endless possibilities for exploration, a canvas for urban experiences to unfold and create meaning for this new city actor.

Nevertheless, it would be reductive to limit the Avenida's participation in the modern city's social life to the masculine gaze of the *flâneur*. Despite the descriptions of the *flâneur* quite invariably placing him among men (Wolff 1985; Clarke 2003; Mallan 2012), this did not mean women did not frequent public places with the intention of flaunting themselves, blending seduction with contemplation (Mallan 2012). One thing is certain though; the feminization of the *flâneur*'s gaze in literature seems to encompass socializing experiences offered by the large department stores in a clear attempt to bring the public space into shopping premises (Domosh 1996; Featherstone 1998). However, in the city of Lisbon, Ramalho Ortigão refers to *Chiado* as a "*ladeira vaidosa*" ("vain slope"—in the feminine) while Júlio Castilho describes it as "the personification of elegant Lisbon". As for the *Baixa*, tradition has it that at the time the capital's best shops were located downtown and everything the heart desired could be found there (Tostões 2008). However, after the Avenida had been inaugurated, bourgeois society's catwalk or *passerelle* was transferred to the new *boulevard.* Both Cabral (1899) and Brun (1910) bear witness to such a change. The expression coined at the time, "Doing the Avenue" ("*Fazer a Avenida")* aptly sums up the way in which the Parisian *Boulevard* became Lisbon society's cult site in which to pursue leisure time and seek conviviality.

13.4 Avenida da Liberdade: The Showcase of Fashion Retailing

After having been the main point of reference for Lisbon society, a place chosen by flâneurs for their Sunday leisure strolls, the Avenida succumbed to a deep sleep from which it would only wake up in the 1990s, no longer playing any active role in the city's consumer life. In the 1940s, this area was still known for its diverse shopping options, cultural attractions and entertainment venues, appealing the population for leisure activities (Janeiro 2006). However, in the following decades, the Avenida faced a decline in all aspects of city life. Its resurgence only occurred in the 1990s, driven by the emergence of luxury brands. This led to a sharp increase in real estate prices, making it an exclusive destination for high-class investors and retailers. This shift toward luxury retail transformed the Avenida into a symbol of status and distinction.

The Avenida's "new" vocation took root at a remarkable pace. By April 2014, there were 171 units operating in retail and services along the Avenida. Surprisingly, only 24% of these units had opened before 2000. Six years later, 63% of the units were in the hands of shopkeepers selling luxury goods and services or, if they did not belong to this category, at least they enjoyed a prestigious image and made part of a *shoppingscape* that could be associated with the best streets of larger cities in the world (Fig. 13.3).

Given the speed with which the Avenida has become a showcase for international capital linked to luxury and fashion retail, and one of the most prestigious boulevards in the world (PRESENCE 2005; Cushman and Wakefield 2014), the following questions may be asked: "Why did the international investors decide to open their stores along this axis?"; "What assets did the Avenida offer that made it so appealing to international capital connected with luxury and fashion?" In order to answer these questions, besides the inferences gathered from London, New York and Milan (Fernie et al. 1997, 1998; Crewe and Beaverstock 1998; Moore and Fernie 1998; Moore 2000), and the argumentation on the different facets of the luxury sector as described by Lipovetsky and Roux (2005), Kapferer (2006) or Kapferer and Bastien 2009), a survey on a sample of 120 store managers was carried out.

The results of our research clearly reveal the factors that weigh the most in the decision-making. As it may be seen in Fig. 13.4, the centrality of the *boulevard* comes at the top of the list, closely followed by its prestigious and social status of the place. The fact that shops in the same market segment are concentrated there, and the area's tourist presence are also relevant. Lastly, there is a set of reasons related with business opportunities, the safety of the place, accessibility and parking facilities or the size and quality of the stores that match their business profiles.

The importance of the symbolic dimension in the decision to set up business is also relevant of the *fast fashion industry*. Brands like Mango, Lanidor, Maximo Dutti, Fly London or Timberland belong to this second segment and the managers argue that their investments for being located on the Avenida da Liberdade have paid off. Despite the high prices of rents which stand at an average of 80€/m^2, managers

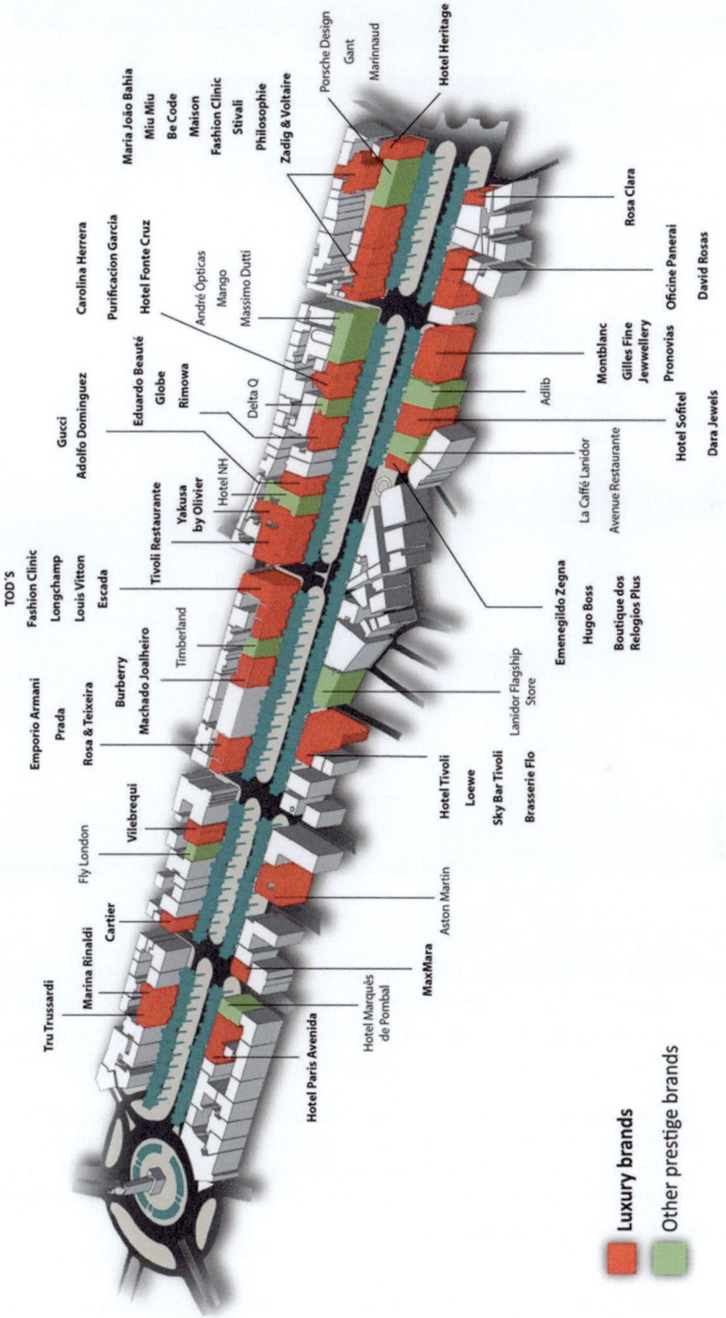

Fig. 13.3 Location of luxury and other prestigious brands along the Avenida da Liberdade, April 2014

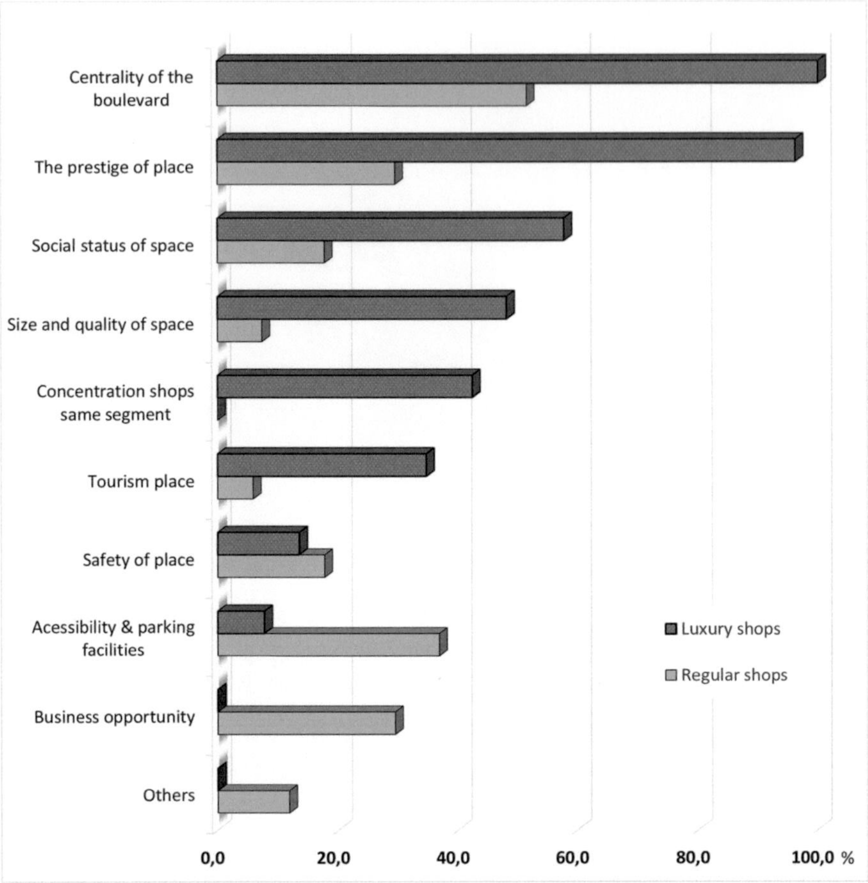

Fig. 13.4 Factors influencing the decisions of shopkeepers to set up business in the Avenida da Liberdade

recognize that there are important plus-values associated with the status and the prestige of the place that justify the presence of their flagship store that act as a driving force showcasing their brands to the rest of the country and even the rest of the world. This is very clear in the shop managers' of Lanidor and Mango:

> We must be here because the Avenida is the most famous street in the country. (…) For a prestigious Portuguese brand such as Lanidor, it's important that it is in city's prime spots. (…) The return on the location can't be measured only economically. There's the brand added value. Having a shop in the *Avenida* is a status symbol, a way of cementing the brand.
>
> *(Manager of Lanidor, April 2014)*

> Choosing *Avenida* falls within a broader strategy that has been designed for Mango. For the last half a dozen years, our top goal has been to make the brand more widely known, targeting a more demanding clientele with more refined tastes. Our most recent projects follow this line where we are working with important people such as the sisters Mónica and

> Penélope Cruz, Scarlett Johansson, Anja Rubik to promote our designer collections, or our top of the range Mango sub-designer fashions. And what better place in Portugal to show off our collections than the *Avenida*! It's not by far the shop that sells the most or that makes the biggest profit when considering our network of 60 shops spread throughout Portuguese cities; the truth is that this shop acts as a showcase for the others.
>
> *(Manager of Mango, April 2014)*

The findings of empirical research support the idea that Avenida da Liberdade serves primarily as a platform for showcasing international capital associated with luxury brands and the fashion industry. Both the physical built environment and the *Avenida's* symbolic capital contribute to this role, in much the same way as in bygone times, when it catered to the leisurely strolls of the bourgeoisie and fashionable *flâneurs*. In a society influenced by spectacle and the power of image (Debord 1967; Harvey 1989), the Avenida cannot be viewed as a mere passive space for luxury and fashion retail. Instead, it should be regarded as a dynamic arena where capital is actively created and sustained by the presence of high-end brands, luxury retailers and fast fashion outlets. More than anything else, the Avenida is a space where the offer (of goods and services) strengthens exchange values and where signs nourish the magic and seductive power that the merchandise and the shops exert upon the consumers.

The Avenida's role in creating value is evident in both supply and demand sides. As the top location in the city and the most expensive spot for property leasing in the country, it significantly impacts the discretionary value of brands. This is exemplified by the concept of the "address effect," underpinning the expression, "*tell me where you live, and I'll tell you who you are*" that Moore and Fernie (1998) transposed to fashion-design retailing. Their slogan: "*how address sells the dress*" explains how the luxury stores are all concentrated in the most expensive streets in the center of London and New York. Jansson and Power (2010) also examined the same phenomenon when discussing how the identity of Milan's shopping district, Quadrilatero d'Oro (translated as "the Golden Rectangle"), was established. The "address effect" observed on Avenida has provided entrepreneurs with the opportunity to capitalize on two key advantages. Firstly, the prestige of being located along the city's most iconic thoroughfare, and secondly, the potential for global exposure as Avenida serves as a prime showcase for the fashion and luxury industries, where only the most renowned brands are showcased.

In shaping the value of Avenida da Liberdade, consumers play a pivotal role. Numerous scholars have highlighted how products serve as vital elements in the social fabric of post-modern society, enabling individuals to express a diverse array of emotions (Firat and Dholakia 1998; Solomon 2003; Lipovetsky 2006). Some goods and services possess discretionary value, capable of conveying status and lifestyles, effectively serving as extensions of identity (Kleine et al. 1993). This function becomes particularly significant in the luxury and fashion sectors, where the survival of brands relies heavily on their symbolic capital (Elliot and Wattanasuwan 1998; Schembri et al. 2010), the allure of their stores and the prestige of their locations—all aimed at communicating brand values and identities (Moore and Doherty 2007; Klingmann 2007; Miles 2010).

Regarding the link between brand acquisitions and consumer identities, Solomon (2003: 1) succinctly describes brand social capital: "Welcome to consumerspace, where reality is branded. Where we avidly search for the products and services that will define who we are and who we want to be. Where we are what we buy – literally." While authors like Lipovetsky (2006) explore the narrative of hedonistic consumption and the diminishing significance of status, Solomon's perspective may underestimate the profound impact of luxury and fashion. In the realm of brands, individuals not only define their identities through acquired products but also through the environments where they shop. This is where Avenida da Liberdade, as one of the world's most elegant boulevards, holds undeniable symbolic value for consumers. Enthusiasts of luxury and fashion understand that this market exists on the most prestigious global stage, accessible only to those with the highest mobility and incomes.

In the "experience economy" of post-modernity (Pine and Gilmore 1999), the acquisition of luxury goods and services often holds greater significance than mere ownership. Consequently, the strategic positioning of stores within the cityscape becomes crucial for providing the desired consumer experiences. Flagship stores, acting as the calling cards of retailers (Mikunda 2006), are strategically located in exclusive areas despite the formidable financial burden of rents and operating costs (Fernie et al. 1998; Moore 2000; Moore and Doherty 2007). For the fast fashion sector, such investments in prime locations make sense as they enhance brand visibility and effectively communicate brand values, thereby attracting customers. Without these prestigious addresses, brands would struggle to convey their status and values, essential for customer appeal.

13.5 Conclusions

Viewing public spaces as arenas for display and leisurely promenades is a notion with historical roots, stretching back to at least the nineteenth century with the prominence of Parisian arcades, passages, and the enigmatic figure of the flâneur. Today, this perspective finds renewed relevance in understanding the significance of contemporary shopping environments in post-modern cities. Wrigley and Lowe (2002) provide a comprehensive overview of seminal studies in this area, highlighting the evolving role of spaces like Avenida da Liberdade in Lisbon as hubs for the (re)production of luxury and fashion capital.

Throughout its history, Avenida da Liberdade has transitioned from a mere public leisure space to a thoroughfare of style, as evidenced by accounts from store managers attesting to the embodiment of luxury and fashion. Even in its earliest days, the Avenida served as a platform for representation, where the bourgeoisie paraded and showcased their status. Today, luxury and fashion brands have seized upon the Avenida as a global showcase, turning it into a storefront open to the world. While its function in promoting merchandise remains consistent, the way the Avenida fulfills its role has undergone significant transformation.

To grasp this shift, one can draw upon Firat and Dholakia's (1998) metaphor of consumption's pervasive nature in post-modern society, likening individuals to merchandise, a concept also echoed by Amendola (2000) in describing the expansion of the show-city (*città-spettacolo*). A city where the borders between fiction and reality have faded, where city-dwellers and visitors have been converted simultaneously into actors and spectators, and the spaces of daily life have tended to become liminal. In this context, Avenida da Liberdade emerges as a stage, with luxury and fashion brands as the principal actors. Unlike the tripartite spatial organization of the traditional theater, the contemporary Avenida features only the stage, where myriad interactions unfold across economic, social and political dimensions. Here, everyone assumes the role of a player, contributing to the ongoing spectacle.

Anyone wishing to portray the *Avenida* today will have trouble in doing so without placing it *between* two worlds that have poorly-defined borders owing to many interpenetrations. From the point of view of the market, from the outset it is located between the local and global. How may the local features of the *Avenida's* individuality be identified when the *brandization* process making it an international luxury and fashion showcase, connects it with the values transmitted by global brands? It is true that the brands are in the boulevard to take advantage of the value earned by its *genius loci*, but the result of exploiting this asset cannot prevent it from being made into a two-edged knife. If on a local scale, the brands have metamorphosed the *Avenida* into an elegant *boulevard*, on an international scale the brands cannot fail to produce yet another *banalscape* common to other elegant avenues throughout the world. And what about the border between the consumers and the commodities, when the luxury and the fashion market, owing to its discretionary quality, transforms consumers simultaneously into *promoters of commodities* and the *commodities they promote*, and fashion goods into *producers* of images, signs and symbols, the real merchandise for shoppers? Paraphrasing Bauman (2007), nowadays in the *Avenida's* market theater, no line clearly separates *consumers* from what is being *consumed*, *producers* from what is being *produced* and marketing *agents* from *merchandise*. They all perform on the same stage known as the market and all of them want to be attractive and desirable; they all want to draw attention to themselves and attract demand and customers.

The transformation of the Avenue into a *brandscape* not only shaped the stage but also redefined the roles individuals play. While the affluent may choose to become actively involved, simultaneously playing the role of the scenery, the actor and the spectator, the rest of the population has the chance to take a seat in the audience and passively look on as the show unfolds, delighting in the amazing representations put on by the stores' brands and merchandises. This makeover has turned the Avenida into a liminal space, where boundaries blur, and social norms temporarily dissolve, as described by Shields (1992).

Yet, this liminality is not experienced by all. The traditional flâneur has given way to a preference for enclosed shopping malls (Cachinho 2006), while the Avenida's primary patrons, according to Global Blue Portugal, are tourists from Angola, Brazil, Russia and China, suggesting that the spectacle of consumerism extends beyond the city of Lisbon. Many examples could be quoted to illustrate the liminal nature of the Avenida in post-modern times. However, it is not in the scope of this work to develop such a challenging idea. While recognizing its high potential to explain changes regarding the fashioning of the boulevard, at the moment, we only hope to shed some light on the influences exerted by the *brandscapes* of luxury and fashion on the urban change of the contemporary metropolis.

Acknowledgements An earlier version of this paper was delivered at the 4th International Seminar "City, Urban Retail and Consumption", Università Degli Studi di Napoli L'Orientale, Dipartimento di Scienze Umane e Sociali, Italy, October 14th, 2013, where I received helpful comments. I am also grateful to my students of the course Retail and Consumption Geographies, for the valuable support provided in the fieldwork that sustains the findings of research presented in this article.

References

Almeida F (1893) Os Gatos. Círculo dos Leitores, Lisbon (edition 1992)
Amendola G (2000) La Ciudad Postmoderna. Magia y Miedo de la Metrópolis Contemporánea. Celeste Ediciones, Madrid
Arvidsson A (2006) Brands: meaning and value in media culture. Routledge, London
Bauman Z (2007) Consuming life. Polity Press, Cambridge
Brun A (1910) A Baixa às 4 da tarde. Grifo-Editores e Livreiros, Lisbon (Edition 1999)
Cabral CM (1899) Lisboa em Flagrante. M. Gomes, Lisbon
Cachinho H (2006) Consumactor: da condição do indivíduo na sociedade pós-moderna. Finisterra 41(81):33–56
Celis C (2017) The attention economy. labour, time, and power in cognitive capitalism. Rowman & Little Field International Ltd., London
Clarke DB (2003) The consumer society and the postmodern city. Routledge, London
Crewe L, Beaverstock J (1998) Fashioning the city: cultures of consumption in contemporary urban spaces. Geoforum 29(3):287–308
Cushman & Wakefield (2014) Business briefing: luxury retail in Lisbon and Porto. www.cuhmanwakerfield.pt. Accessed at 14 Feb 2014
Debord G (1967) La société du spectacle. Bouchet Chastel, Paris
Domosh M (1996) The feminized retail scape: gender ideology, and consumer culture in nineteenth century New York City. In: Wrigley N, Lowe M (eds) Retailing, consumption and capital: towards the new retail geography. Addison Wesley Longman, Harlow, pp 257–270
Elliott R, Wattanasuwan K (1998) Brands as symbolic resources for the construction of identity. Int J Advert 17:131–144
Featherstone M (1998) The flâneur, the city and virtual public life. Urban Stud 35(5–6):909–925
Fernie J, Moore CM, Lawrie A (1998) A tale of two cities: an examination of fashion designer retailing within London and New York. J Prod Brand Manage 7(5):366–378
Fernie J, Moore CM, Lawrie A, Hallsworth A (1997) The internationalization of the high fashion brand: the case of central London. J Prod Brand Manage 6(3):151–162
Firat AF, Dholakia N (1998) Consuming people: from political economy to theaters of consumption. Routledge, London
Gehl J, Gemzoe L (2001) New city spaces. The Danish Architectural Press, Copenhagen
Gluck M (2010) Flâneur. In: Ray H (ed) Encyclopedia of urban studies. Sage, Thousand Oaks, pp 272–274
Harvey D (1989) The condition of postmodernity: an enquiry into the origins of cultural change. Blackwell, Oxford
Janeiro MJ (2006) Lisboa: Histórias e Memórias. Livros Horizonte, Lisbon

Jansson J, Power D (2010) Fashioning a global city: global city brand channels in the fashion and design industries. Reg Stud 44(7):889–904
Kapferer J-N, Bastien V (2009) The luxury strategy: break the rules of marketing to build luxury brands. Kogan Page Limited, London
Kapferer J-N (2006) The two business cultures of luxury brands. In: Schroeder JE, Salzer-Morling M, Askegaard S (eds) Brand culture. Routledge, New York, pp 67–76
Kleine III RE, Kleine SS, Kernan JB (1993) Mundane consumption and the self. A social-identity perspective. J Consum Psychol 2(3):209–235
Klingmann A (2007) Brandscapes. Architecture in the experience economy. The MIT Press, Cambridge
Krupar S, Al S (2014) Notes on the society of the spectacle/brand. In: Crysler CG, Cairs S, Heynen H (eds) The Sage handbook of architectural theory. Sage, London, pp 247–273
Le Cunff F (2003) Do Passeio Público ao Parque da Liberdade. Revista Camões, 15–16, January-June, 179–186
Lefèbvre H (1970) Du Rural à l'Urbain. Anthropos, Paris
Lipovetsky G (2006) Le bonheur paradoxal: Essai sur la société d'hyperconsommation. Editions Gallimard, Paris
Lipovetsky G, Roux E (2005) O Luxo Eterno: da idade do sagrado ao tempo das marcas. Companhia das Letras, São Paulo
Mallan K (2012) Strolling through the (post)modern city: modes of being a Flâneur in picture books. The Lion and the Unicorn 36(1):56–74
Mikunda C (2006) Brand Lands, hot spots, and cool spaces: welcome to the third place and the total marketing experience. Kogan Page, London
Miles S (2010) Spaces for consumption. Pleasure and placelessness in the post-industrial city. Sage, London
Moncan P (2009) Le Livre des Passages de Paris : Aujourd'hui, Histoire, Architecture. Les Éditions du Mécène, Paris
Moore C, Doherty A (2007) The international flagship stores of luxury fashion retailers. In: Hines T, Bruce M (eds) Fashion marketing contemporary issues. Elsevier/Butterworth-Heinemann, Oxford, pp 277–296
Moore C, Fernie J (1998) How address sells the dress—an examination of fashion designer within London and New York. J Text Inst 89(3):81–95
Moore CM (2000) Streets of style: fashion designer retailing with London and New York. In: Jackson P, Lowe M, Miller DM, Mort F (eds) Commercial cultures. Economies, practices, spaces. Berg, Oxford, pp 261–278
Ortigão R (1874) As Farpas. Livraria Clássica Editora, Lisbon
Pina M (1886) Notas sobre Lisboa. A Ilustração, 20 Sept 1886
Pine BJ, Gilmore HJ (1999) The experience economy. Work is theater and every business a stage. Harvard Business School Press, Boston
PRESENCE (2005) The Quality of greeting and service in the most prestigious avenues in the World. Présence, Paris
Queiroz E (1945) Os Maias. Livraria Lello & Irmão, Porto
Schembri S, Merrilees B, Kristiansen S (2010) Brand consumption and narrative of the self. Psychol Mark 27(6):623–638
Schmid H (2009) Economy of fascination. Dubai and Las Vegas as themed urban landscapes. Gebrüder Borntraeger, Berlin
Shields R (1992) Lifestyle shopping: the subject of consumption. Routledge, London
Silva RH (1994) O Passeio Público e a Avenida da Liberdade. In: Moita I (coord) O Livro de Lisboa. Livros Horizonte, Lisbon, 425–434
Silva RH (1997) As Avenidas de Lisboa. In: UAL (ed) Lisboa Iluminista e o seu Tempo. UAL, Lisbon, pp 97–105.
Solomon MR (2003) Conquering consumerspace: marketing strategies for a branded world. AMACOM, New York

Tostões A (2008) Percursos do Urbanismo e da Arquitectura Modernos. In: Tostões A, Walter R (eds) Lisboa 1758: O Plano da Baixa Hoje. CML, Lisbon, pp 169–229
Wolff J (1985) The Invisible Flâneuse. Women Lit Mod Theory Cult Soc 2(3):37–46
Wrigley N, Lowe M (2002) Reading retail: a geographical perspective on retailing and consumption spaces. Arnold, London

Chapter 14
The Image and the Branding of the Spanish City

Luis Alfonso Escudero-Gómez

> *Mass images are relentless machines designed to foster and perpetuate stereotypes, often appearing insensitive and arrogant.*
> *Ramonet (*2000*, p. 12)*

Abstract Cities are increasingly concerned with creating an urban image that makes them more attractive to capital, mobile professionals, and tourists. City branding and city marketing have emerged as essential strategies for achieving this goal. City brands play a pivotal role in city promotion and strategic planning. As a result, the development of a public image for Spanish cities has become paramount. This chapter delves into the concept of urban image and city branding in the Spanish context, initially exploring it from a theoretical standpoint. Subsequently, it conducts an in-depth analysis of a case study: Barcelona, as a prime exemplar of city branding in Spain. The study culminates with an examination of the competitive dynamics among Spanish cities, focusing on the role of events and the tourist image. It is worth noting that many Spanish cities predominantly orient their branding efforts toward the singular objective of attracting tourists. However, these city images and urban brands have the unintended consequence of reducing Spanish cities to commodities in the highly competitive global market, often neglecting the diverse urban spaces and the well-being of their residents. It is imperative to shift the perspective and regard the city not merely as a product but as a living habitat.

Keywords Cities · City branding · City marketing · Urban Image · Brand · Spain

---

L. A. Escudero-Gómez (✉)
Department of Geography, University of the Balearic Islands, Palma, Spain
e-mail: luisalfonso.escudero@uib.cat

R. C. Lois-González and J. A. Rio Fernandes (eds.), *Urban Change in the Iberian Peninsula*, The Urban Book Series, https://doi.org/10.1007/978-3-031-59679-7_14

14.1 Introduction

Since the 1980s and 1990s, cities in Europe have been increasingly concerned with the idea of creating an urban image to make them more attractive to capital and mobile professional workers (Freestone and Gibson 2006), as well as tourists (Evans 2006). Active promotion and marketing emerged as a high-profile response to post-industrial decline (Gold and Gold 2019). As cities struggle to attain economic distinctiveness, they often choose to cultivate symbolic uniqueness, either by crafting a new image or by reshaping their existing one. Finding it difficult to achieve significant economic standing, Spanish cities opted for the development of symbolic particularities, seeking to create a new image or transform their existing one. The deindustrialisation of cities has transformed them into centres of consumption, leading them to invest in their image as a crucial factor in their success (Lyon 1996). In Spain, consequently, a public image of the city has been constructed, as reported in studies on examples such as Granada (Luque-Martínez et al. 2007), Santiago de Compostela (Escudero-Gómez 2013) and Valencia (Boira-Maiques 1992).

The link between the perceived world and the built environment is achieved through images. The human mind selects visual information, isolating, organising, interpreting and representing it. Images do not imitate, duplicate or parody; rather, they substitute reality (Moya-Pellitero 2013). They generate unsettling spectres, intentionally, as liminal objects, memories, subjects and atmospheres that stand between the visible and the invisible, the past and the present, the individual and the collective (Vanolo 2017). Boulding's seminal work (1956) showed that people's behaviour depends on the image, which he understood as all of an individuals' subjective knowledge of the world. In the case of cities, their image precedes them and mediates the relationship that exists between individuals, both residents and visitors, and the cities (Amendola 2000). Urban images are embraced by both inhabitants and users of the city. The image of a city can be described as the set of people's beliefs, ideas and impressions of it (Kotler 1997). It is a mental representation that may depend both on objective knowledge about the services and facilities provided and on more emotional and affective components (Cassia et al. 2018). In Spain, the image of the city has been a subject of study since the 1990s, focusing on various urban centres, as mentioned. However, there lacks a comparative analysis on image building in Spanish cities similar to that carried out by Hankinson (2001) for the case of 12 English localities.

This book chapter focuses on the image and branding of Spanish cities. In order to compete in the global neoliberal capitalist market, Spanish cities employ branding and marketing techniques akin to those of private companies. They craft urban images aimed at attracting economic investment and, in many cases, predominantly tourists, given their crucial role in the local economies. Simultaneously, they engage in competition with each other, often through the organization of events. In this context, Barcelona stands out as a successful model to emulate. Nevertheless, the Barcelona case is not without its share of challenges and controversies.

Following this introduction, the article's structure unfolds with two concise theoretical sections on city branding, city marketing, and the significance of branding. Subsequently, it delves into an in-depth analysis of the prominent Barcelona case. The subsequent two sections, focusing on Spanish cities, examine their competitiveness, event hosting, and the cultivation of their tourist image. The chapter concludes with a summary of findings and a list of references.

14.2 City Branding and City Marketing

A city is aware that it can create and recreate its image (Amendola 2000). The ability of a city to provide a unitary and synthetic image of itself, an ideal vision or, at least, the best possible one, is fundamental. A distinctive and positive image of the city is the first step to success in creating its brand, with city branding being the term widely used in academic publications in reference to the practices of advertising and promoting a city (Vanolo 2019). This process involves the image of cities, highlighting its positive elements. It forges place brands and urban images (Vanolo 2019). These cannot be formed freely, but must be created in detail. Today, a city's image is a variable that must be controlled and manipulated in a short time and, thus, what was once left to travellers and writers is now entrusted to marketing (Amendola 2000) and brand management for cities (Ashworth and Kavaratzis 2019). Kotler et al. (1994) explain in depth the planning required to build urban brands.

Marketing techniques, or city marketing, which refers to intense processes of place-making and collective identity formation (Vanolo 2019), are used to build a city image (Buhalis 2000; Muñiz-Martínez and Cervantes-Blanco 2010). The aim of city marketing is for cities to provide an offer that is appropriate to the demand of their various stakeholders and for the strategy to be designed in line with the identity that distinguishes it as a city, allowing it to be differentiated (Muñiz-Martínez and Cervantes-Blanco 2010). Four strategies feature in the marketing mix, namely, design, infrastructure, services and attractions (Kotler et al. 1994). In addition, cities leverage the most recent and highly evolved principles of corporate branding theory in their branding (Hankinson 2007). Such strategies seek to convince outsiders that a certain locality is an attractive place to visit, live and work (Gold and Gold 2019).

14.3 The Importance of a Brand

A brand is a promotional tool that serves to set one product apart from another. Promoting a place, such as a city, involves the conscious use of advertising and marketing to communicate selective images of specific localities or geographic areas to a target audience (Gold and Ward 1994). In the case of cities, a brand is defined as a name, term, sign, symbol and/or design, intended to identify a city and differentiate it from others (Zeybeck and Unlü 2016). It highlights the features of a city's identity and its distinctiveness (Markusen and Schork 2006). Therefore, cities build their brand and promote different strategies to this end. The names of cities become a brand and can be represented by one or more logos (Arnaud 2016). However, although the image of each city strives to be specific, brand building is, in reality, a homogeneous phenomenon that draws on similar messages, strategies and rationalities all across the world (Vanolo 2019).

Brands are relational constructs, that is, they are the result of two-way processes involving producers and consumers. In this sense, branding is clearly more than just developing a catchy slogan with a powerful logo (Vanolo 2019). Drawing on Huertas (2011), each city has its own essential components that shape its brand, as follows:

1. The elements that make up the brand identity and are associated with the city. These are communicated by the organisers through marketing:
   a. The graphic component, which entails creating a symbol and a logo.
   b. The conceptual and functional component, which consists of actual, tangible features of the city.
   c. The conceptual and emotional component, based on abstract and symbolic values to be associated with the city.
2. The elements that make up the city's brand image, how it is perceived by its stakeholders and the positioning acquired:
   a. The public perception of the graphic element, being the understanding and approval of the logo and the graphic components of the brand.
   b. The public perception of the functional elements of the city, their attractiveness, their appreciation and their disposition.
   c. The perception of the emotional values associated with the city and with which the public may identify or not.

City branding thus involves creating graphic elements, values, attractions and promises to be associated with a city, forming its identity (Botschen et al. 2017). Public institutions decide and communicate these through marketing (Blain et al. 2005). A public policy of design and management of city identification is thus developed (Fernández and Paz 2005), that is, cities are strategically planned and marketed. This is a hybrid procedural policy in which public and private sectors, economics and politics, and the market and the city are blurred and coexist (i.e. as a process) in a relational and interactive manner (Lucarelli 2018).

Marketing and branding are political objects, being clearly tied to issues of conflict, inequality, identity formation, inclusion and exclusion (Vanolo 2019). Brands institute socio-spatial division and exclusion in the city by defining the types of practices, images or attitudes that are acceptable in a particular city. In so doing, they set up cultural boundaries that serve to ward off certain people and activities (Bookman 2017). Escudero's doctoral thesis (2000) showed that in the three main cities of the province of A Coruña, namely, its homonymous capital, Ferrol and Santiago de Compostela, a public image was constructed based on its central spaces and the positive functionalities of each city, while the peripheral residential neighbourhoods and their citizens were practically ignored and given no role in this representation. In the case of A Coruña, the configuration of a new brand image based on commercial gentrification is taking place (Rodríguez-Barcón et al. 2018).

The city image clearly emphasises a specific discourse, giving prominence to certain themes and narratives and renders irrelevant other issues and spaces that disrupt the promotional discourse (Vanolo 2019). In an article on the image of Toledo (Escudero-Gómez 2018), it was argued that the image of this medium-sized Spanish inland city is stereotypical and that its brand is based on a partial view of its history and the picturesque nature of its heritage. It is this representation that attracts the vast majority of its visitors. In return, the real city is absent from both the promotional image of Toledo and visitors' perception of the locality. The stereotypical image and commercialisation of the city are the dominant force.

City marketing aims to attract economic investment, visitors and creative professionals (Evans 2006). The city's brand image must be able to forge citizens' identification with their city and, at the same time, be suitable to appear on all the products marketed with the city's name (Benach-Rovira 2000). There is a complex relationship between brands, consumption and urban life, where the brand mediates identities, lifestyles and social relations (Bookman 2017). Cities have thus emerged as consumer products, brands with identities and values (Muñiz-Martínez and Cervantes-Blanco 2010). Vanolo (2019) puts it lucidly, "[Cities] are kinds of *special* products that need to be marketed and sold with dedicated techniques in order to extract value, in a scenario of global competition among cities in order to attract global flows (investments, events, tourist, enterprises – which are basically other names for money)" (p. 4).

14.4 The Significant Case of Barcelona: Success and Falsehood

If there is a case in Spain that is evidence of the above, it is undoubtedly Barcelona. Barcelona serves as a prominent example of the application of branding principles to cities. The country's second largest city has adapted to its role as a consumer space and has been building a brand image since the 1990s (Benach-Rovira 2000). Particularly since hosting the 1992 Olympic Games, the city's image has been the object of an intense productive process (Benach-Rovira 1993). Urban marketing and international media promotion of the city (Ward 1998) have attracted investors, professionals and visitors. Borja (2004) describes this process as a success and underlines that urban design and architecture, citizen involvement and the leisure and cultural offer have made Barcelona a city of conferences, congresses and fairs, having found tourism to be a powerful and unsuspected source of industry. Smith (2005) describes this change in the city's image as the *re-imaging* of Barcelona. This transformation, grounded in urban regeneration and the commercial use of culture (Fig. 14.1)—Rius-Ulldemolins (2017) highlights that in the construction of a brand for central Barcelona, culture is

Fig. 14.1 The commercial utilization of culture in building the thriving Barcelona brand: the enormous Primavera Sound music festival and the excessive tourism at Gaudí's Park Güell (own source)

continuously used as a form of territorial differentiation and promotion—has even been termed the 'Barcelona model' (Degen and García-Cabeza 2012).

This commodification, where images of a city build a brand to sell it to investors, tourists and new inhabitants of the creative class, transforms cities into products. Cities, however, are not a product (Vanolo 2019). They are spaces where most of the world's inhabitants now live. For this reason, commercialising and putting them on sale through branding and stereotypical images is objectionable. Not everything can be marketed (Gilbert 2008). The outstanding example of Barcelona as a successful brand, as well as that of other Spanish cities such as Seville (Díaz 2013), has been criticised for its excessive commodification. Capel (2005) censures the 'Barcelona model', highlighting its limitations and shortcomings in the economic and social development of the city. This author evidences the deficits, problems and negative externalities that underlie the success of Barcelona's image. Smith (2005) points out that there has been a 're-imaging' of Barcelona that has involved the reduction of all the complexity of the city to a single image, leaving out all the density contained in spatial life. Delgado (2007) is even more critical and speaks of Barcelona as a city of lies, given that its image does not correspond to the social reality of its citizens, applying the harsh terms of fraud and misery to the 'Barcelona model'. In the same line, Palou-Rubio (2006) speaks of it as a sham city.

14.5 Competition Between Cities and Events

As the competition among cities and regions intensifies, place branding is becoming increasingly important (Asprogerakas and Mounanea 2020). In any event, cities today, as if involved in a sports tournament or an international beauty contest (Boland 2013), compete with each other and position themselves in accordance with their image. This has been measured both quantitatively (Gilboa et al. 2015) and qualitatively (Laaksonen et al. 2006). There exist league tables and brand indexes of cities (Gold and Gold 2019). In these rankings, cities are ordered from best to worst according to parameters, measurable in numbers or indicators, related to their supposed attractiveness. Local governments in multiple cities hire professionals of various types to detect potential assets with the ultimate aim being to develop strategies to climb up the rankings (Vanolo 2019). Such professional consultants specialise in conducting branding exercises to the level they feel necessary (Gold and Gold 2019). In other words, it is a question of seeking marketing actions that improve a city's positioning. Urban authorities strive to position their cities high in the hierarchy of the global urban network by making them competitive, as suggested by Vives-Miró (2011), who explains the phenomenon using the case of the city of Palma, on the Spanish island of Mallorca.

Zenker et al. (2013) refer to this competitive environment as cities metaphorically having a price tag, just like an item of clothing or another similar product on sale. Advertising and marketing are used to sell cities (Gold and Ward 1994). Municipal administrations have seen that cities can be sold, and so promote them through advertising and public relations in a game that any city can, and should, play (Hall 2000).

In pursuit of this competitiveness, one of the most common ways of positioning the city is through organising internationally recognised events, as a form of external projection and communication (Muñiz-Martínez and Cervantes-Blanco 2010). This technique of event branding is currently fashionable among city planners (Gold and Gold 2019). Since the iconic year of 1992, the fifth centenary of Christopher Columbus' first arrival in the Americas, many Spanish cities have resorted to this strategy. This has taken the form of sporting events (the 1992 Olympic Games in Barcelona being the prime example, as well as the hosting of world championships and international events, such as the 2007 America's Cup sailing competition in Valencia), international exhibitions (the Universal Exposition in Seville in 1992 and, since then, various similar events of greater or lesser importance and success, such as the Universal Forum of Cultures in Barcelona in 2004 or the International Exposition in Zaragoza in 2008) and cultural events (notably the European City of Culture in Madrid in 1993 and the European Capitals of Culture of Santiago de Compostela in 2000, Salamanca in 2002 and San Sebastián in 2015).

As an example, it is worth noting the work by Díaz (2013), which highlights the way that urban planning in Seville was put at the service of the macro event (the 1992 Universal Exposition), and that this macro project renewed the commercial image of the city, albeit at the expense of creating a mirage that is far removed from the urban reality.

14.6 The Tourist Image

Finally, the image of cities has been associated with their promotion of tourism, given that this is a key factor in the tourism development of any destination. Tourism serves as a means to promote a territory, either as a nation brand, region brand (Marine-Roig and Ferrer-Rosell 2018) or a city brand. Through their image, cities enhance their brand value and, in the face of competition between tourist destinations, the construction of a competitive image provides a fundamental difference (Anholt 2007). Destination image has the power to influence tourist behaviour as well as the success (or failure) of a destination (Tasci and Gartner 2007). Additionally, there is a close relationship between the image of a tourist destination, tourist satisfaction and loyalty to the destination (Chi and Qu 2008).

In Spain, many cities have established themselves in their role as tourism destinations, basing their image on promoting their attractions to visitors. This is the case of Toledo, one of the leading cultural tourism destinations in Spain, where a stereotypical image and a recognisable brand have been successfully consolidated in order to attract visitors, while eschewing its urban reality for a vision grounded in history, heritage and tourism (Escudero-Gómez 2018). Another example is that of Girona, where a promotional image has been created through a premeditated, specific selection of images of the city for the purpose of tourism (Galí-Espelt 2005). Other historical cities in the Iberian Peninsula, such as those with UNESCO World Heritage Sites, have engaged in such image building as a cultural tourist destination, using, for example, the possibilities of the Internet (Escudero-Gómez 2007). In this sense, city branding techniques have changed greatly due to the massive spread of new forms of digital communication and social networking technologies (Vanolo 2019). In all of these, the city is sold through a superficial sense of place (Benach-Rovira 2016), given that the perception proposed by the tourist image is picturesque, that is, necessarily fragmentary, misleading and scenographic (González-Bracco 2017).

This process transforms the city into its tourism image. The results may be successful, as shown by Marín-Cots et al. (2017) for Malaga—for example, through the iconography of the figure of Picasso (Fig. 14.2), a universally renowned painter born in Malaga, where he spent his early years, but who spent most of his adult life working in other European cities—but with the negative impacts of loss of population and social conflicts derived from the resulting overcrowding of visitors in the historical centre of the city.

Fig. 14.2 The image of Malaga is linked to the figure of Picasso (own source)

14.7 Conclusions

Spanish cities need to build positive images that create appealing and strong city brands that attract inhabitants, investors and, especially, tourists. However, these images are turning them into products in a competitive global market, in which cities seek to position themselves as well as possible. These images exclude the majority of a city's urban spaces and its citizens. Everyday life and the real economy do not conform to the adverts and are replaced by signature architecture, by the promotional and commercial use of culture (Kearns and Philo 1993) and by design and other marketing techniques. From a critical urban and geographical perspective, counter-branding practices have been advocated. These can be viewed as the demands of these neglected neighbourhoods and their inhabitants to their rights to the city. The general public—non-elites—were rarely considered a key stakeholder in the branding process (Cleave and Arku 2022).

City brands can be tools for collective resistance, supporting the search for what Massey (1994) termed a global sense of place, to give life to the spaces of a city, to make it dynamic and relevant, to emphasise its importance in the lives of the citizens who reside there.

At the same time, following Vanolo (2019), cities should be considered not as products but as habitats, given that this is their main function. Public investment in brand creation could be reverted into urban policies against inequality and for the social advancement of residents. Thereafter, having a positive image of a city is also valued, but based on an idea of identity and not as commodification, as selling a city in the market.

References

Amendola G (2000) La ciudad postmoderna. Celeste

Anholt S (2007) Competitive identity: the new brand management for nations, cities and regions. Palgrave/Macmillan

Arnaud JL (2016) Logotypes and cities representations. In: Calatrava J, Garcia F (eds) La cultura y la ciudad. Universidad de Granada, pp 23–32

Ashworth GJ, Kavaratzis M (2019) Beyond the logo: brand management for cities. J Brand Manag 16(8):520–531. https://doi.org/10.1057/palgrave.bm.2550133

Asprogerakas E, Mountanea K (2020) Spatial strategies as a place branding tool in the region of Ruhr. Place Brand Public Dipl 16(4):336–347. https://doi.org/10.1057/s41254-020-00168-1

Benach-Rovira N (1993) Producción de imagen en la Barcelona del 92. Estudios Geográficos 212:483–507

Benach-Rovira N (2000) Nuevos espacios de consumo y construcción de imagen de la ciudad en Barcelona. Revista Estudios Geográficos LX I(238):189–206

Benach-Rovira N (2016) ¿Ciudades en el mapa o en la guía turística? Venta de la ciudad y sentido de lugar. Revista CIDOB D'afers Internacionals 113:89–105

Blain C, Levy SE, Ritchie JRB (2005) Destination branding: insights and practices from destination management organizations. J Travel Res 43:328–338. https://doi.org/10.1177/0047287505274646

Boira-Maiques JV (1992) La ciudad de Valencia y su imagen pública. Universitat de València

Boland P (2013) Sexing up the city in the international beauty contest: the performative nature of spatial planning and the fictive spectacle of place branding. Town Plan Rev 84:251–274. https://doi.org/10.3828/tpr.2013.14

Bookman S (2017). Brands and the city: entanglements and implications for urban life. Routledge

Borja J (2004) Barcelona y su urbanismo. Éxitos pasados, desafíos presentes, oportunidades futuras. In: J Borja, Z Muxí (Eds) El urbanismo en el siglo XXI. Universitat Politècnica de Catalunya, pp. 171–183.

Botschen G, Promberger K, Bernhart T (2017) Brand-driven identity development of places. J Place Manag Dev 10(2):152–172. https://doi.org/10.1108/JPMD-07-2016-0051

Boulding KE (1956) The image. Knowledge in life and society. University of Michigan Press

Buhalis D (2000) Marketing the competitive destination of the future. Tour Manage 21(1):97–116. https://doi.org/10.1016/S0261-5177(99)00095-3

Capel H (2005) El modelo Barcelona: un examen crítico. Del Serbal

Cassia F, Vigolo V, Ugolini MM, Baratta R (2018) Exploring city image: residents' versus tourists' perceptions. TQM J 30(5):476–489. https://doi.org/10.1108/TQM-11-2017-0161

Chi CGQ, Qu H (2008) Examining the structural relationships of destination image, tourist satisfaction and destination loyalty: an integrated approach. Tour Manage 29(4):624–636. https://doi.org/10.1016/j.tourman.2007.06.007

Cleave E, Arku G (2022) Place branding and growth machines: implications for spatial planning and urban development. J Urban Aff 44(7):949–966. https://doi.org/10.1080/07352166.2020.1768104

Degen M, García-Cabeza M (2012) The transformation of the 'Barcelona Model': an analysis of culture, urban regeneration and governance. Int J Urban Reg Res 36(5):1022–1038. https://doi.org/10.1111/j.1468-2427.2012.01152.x
Delgado M (2007) La ciudad mentirosa. Fraude y miseria del "modelo Barcelona". Los Libros de la Catarata
Díaz I (2013) Miedo y asco en Sevilla. Especulación, espectáculo y resistencias en la ciudad post-desarrollista. In: Observatorio Metropolitano de Madrid (Ed) Paisajes devastados. Después del ciclo inmobiliario: impactos regionales y urbanos de la crisis. Traficantes de Sueños, pp 233–271
Escudero-Gómez LA (2013) La imagen urbana de Santiago de Compostela (España), un estudio de su representación pública, mediática, promocional y artística. Boletín De La Asociación De Geógrafos Españoles 62:265–294
Escudero-Gómez LA (2000) La imagen de A Coruña, Santiago de Compostela y Ferrol a través de la prensa diaria: Estudio de Geografía de la Percepción. Universidad de Santiago de Compostela
Escudero-Gómez LA (2007) Los nuevos medios de creación de imagen de las ciudades ibéricas Patrimonio de la Humanidad: páginas web en Internet. In: Actas del X Coloquio Ibérico A Geografía Ibérica no contexto europeu. Universidade de Évora
Escudero-Gómez LA (2018) The construction and perception of cultural tourism images and brands. A case study from Toledo (Spain). Lusophone J Cult Stud 5(2):61–82
Evans G (2006) Branding the city of culture-The death of city planning? In: Monclús J, Guàrdia M (eds) Culture, urbanism and planning. Routledge, eBook
Fernández G, Paz S (2005) Más allá del marketing de ciudades: hacia una política pública de diseño y gestión de los signos de identificación de ciudad. Scripta Nova. Revista electrónica de Geografía y Ciencias Sociales IX(194)
Freestone R, Gibson C (2006) The cultural dimension of urban planning strategies: an historical perspective. In: Monclús J, Guàrdia M (eds) Culture, urbanism and planning. Routledge, eBook
Galí-Espelt N (2005) La imatge turística del patrimoni monumental de Girona. Universitat de Girona
Gilbert J (2008) Against the commodification of everything. Cult Stud 22(5):551–556. https://doi.org/10.1080/09502380802245811
Gilboa S, Jaffe ED, Vianelli D, Pastore A, Herstein R (2015) A summated rating scale for measuring city image. Cities 44:50–59. https://doi.org/10.1016/j.cities.2015.01.002
Gold JR, Ward SV (eds) (1994) Place promotion. The use of publicity and marketing to sell towns and regions. Wiley
Gold JR, Gold MM (2019) Accentuating the positive: city branding, narrative and practice. Tijdschr Econ Soc Geogr 111(1):2–9. https://doi.org/10.1111/tesg.12395
González-Bracco M (2017) Guías, imágenes y suvenires: reflexiones sobre los artefactos mediadores de la práctica turística. Pasos. Revista de Turismo y Patrimonio Cultural 15(3):741–750. https://doi.org/10.25145/j.pasos.2017.15.049
Hall P (2000) La redefinición de las ciudades europeas. In: Bellet C, Llop JM (eds) Ciudades intermedias: urbanización y sostenibilidad. Milenio, pp 391–401
Hankinson G (2001) Location branding: a study of the branding practices of 12 English cities. J Brand Manag 9(2):127–142. https://doi.org/10.1057/palgrave.bm.2540060
Hankinson G (2007) The management of destination brands: five guiding principles based on recent developments in corporate branding theory. J Brand Manag 14(3):240–254. https://doi.org/10.1057/palgrave.bm.2550065
Huertas A (2011) Las claves del Citybranding. Portal de la Comunicación InCom-UAB: El portal de los estudios de comunicación, 2001–2011. https://incom.uab.cat/portalcom/wp-content/uploads/2020/01/57_esp.pdf
Kearns G, Philo C (eds) (1993) Selling places: the city as cultural capital, past and future. Pergamon
Kotler P, Haider DH, Rein I (1994) Mercadotecnia de localidades. Cómo atraer inversiones, industrias y turismo a ciudades, regiones, estados y países. Diana (Original work published 1993)
Kotler P (1997) Marketing management: analysis, planning, implementation and control, 9th edn. Prentice-Hall

Laaksonen P, Laaksonen M, Borisov P, Halkoaho J (2006) Measuring image of a city: a qualitative approach with case example. Place Branding 2(3):210–219. https://doi.org/10.1057/palgrave.pb.5990058
Lucarelli A (2018) Place branding as urban policy: the (im)political place branding. Cities 80:12–21. https://doi.org/10.1016/j.cities.2017.08.004
Luque-Martínez T, Del Barrio-García S, Ibáñez-Zapata JA, Rodríguez-Molina MA (2007) Modeling a city's image: the case of Granada. Cities 24(5):335–352. https://doi.org/10.1016/j.cities.2007.01.010
Lyon D (1996) Postmodernidad. Alianza
Marín-Cots P, Guevara-Plaza A, Navarro-Jurado E (2017) Renovación urbana y masificación turística en la ciudad antigua: pérdida de población y conflictos sociales. Ciudad y Territorio. Estudios Territoriales 193:454–468
Marine-Roig E, Ferrer-Rosell B (2018) Measuring the gap between projected and perceived destination images of Catalonia using compositional analysis. Tour Manage 68:236–249. https://doi.org/10.1016/j.tourman.2018.03.020
Markusen A, Schork G (2006) The distinctive city. Urban Stud 43(8):1301–1323. https://doi.org/10.1080/00420980600776392
Massey D (1994) Space, place and gender. University of Minnesota Press
Moya-Pellitero AM (2013) La percepción del paisaje urbano [eBook]. Biblioteca Nueva
Muñiz-Martínez N, Cervantes-Blanco M (2010) Marketing de ciudades y "place branding." Pecunia 1:123–149
Palou-Rubio S (2006) La ciudad fingida. Representaciones y memorias de la Barcelona turística. Pasos. Revista de Turismo y Patrimonio Cultural 4(1):13–28. https://doi.org/10.25145/j.pasos.2006.04.002
Ramonet I (2000) La golosina virtual. Debate
Rius-Ulldemolins J (2017) Culture and authenticity in urban regeneration processes: place branding in central Barcelona. Urban Stud 51:3026–3045. https://doi.org/10.1177/0042098013515762
Rodríguez-Barcón A, Calo E, Otero-Enríquez R (2018) Commercial gentrification in a medium-sized city: an ethnographic look at the transformation process of the historic centre of A Coruña (Spain). Urbanities-J Urban Ethnogr 8(2):87–96
Smith A (2005) Conceptualizing city image change: the 're-imaging'of Barcelona. Tour Geogr 7(4):398–423. https://doi.org/10.1080/14616680500291188
Tasci ADA, Gartner WC (2007) Destination image and its functional relationships. J Travel Res 45(4):413–425. https://doi.org/10.1177/0047287507299569
Vanolo A (2019) Cities are not products. Tijdschr Econ Soc Geogr 111(1):10–17. https://doi.org/10.1111/tesg.12385
Vanolo A (2017) City branding. The ghostly politics of representation in globalizing cities. Routledge
Vives-Miró S (2011) Producing a "successful city": neoliberal urbanism and gentrification in the tourist city—the case of Palma (Majorca). Urban Stud Res 1(2):1–13. https://doi.org/10.1155/2011/989676
Ward S (1998) Selling places. The marketing and promotion of towns and cities, 1850–2000. SPON
Zenker S, Eggers F, Farsky M (2013) Putting a price tag on cities: insights into the competitive environment of places. Cities 30:133–139. https://doi.org/10.1016/j.cities.2012.02.002
Zeybeck B, Unlü DG (2016) A view on countries' tourism logos, slogans, contents and figural characteristics within the concept of country identity. Informacijos Mosklais 76:26–40. https://doi.org/10.15388/Im.2016.76.10380

Chapter 15
The Triumph of Floating Residents in the Neoliberal City

José A. Rio Fernandes, Pedro Chamusca, and Jorge Ricardo Pinto

Abstract Cities have always been places of encounter of residents with the occasional habitant, the visitor. But, with massification of urban tourism and the increasing fluidity of short-term migration (for work, education or leisure), the floating population is increasing every year in more places, and gaining economic (and social) relevance. If cities have always been places of change, and change is permanent, it's necessary to pay attention to some of its unwelcomed effects, as it is the case those that are related with the intensity of the change and the social segregation with an exclusion of the central location for the residence of longtime residents or long-time retailers. Analyzing the general dynamic and evidence from the city of Porto (North of Portugal), we believe that a harmonious balance between the needs of traditional residents and the demands of the burgeoning tourist, student, and digital nomad is crucial for sustaining identity.

Keywords Tourism · Residence · Gentrification · Urban change

15.1 Introduction

Cities can be defined in very different ways. As places of interchange of goods and services or as a particular landscape associated with building density, compacity and some high-rise monuments or modern towers. However, the most common indicator for comparison is the number of inhabitants and no one can argue that there is no city without a certain number of residents.

J. A. Rio Fernandes (✉)
CEGOT, University of Porto, Porto, Portugal
e-mail: joseriof@letras.up.pt

P. Chamusca
CECS, University of Minho, Braga, Portugal
e-mail: pedrochamusca@ics.uminho.pt

J. R. Pinto
CEGOT, ISCET, Porto, Portugal
e-mail: jpinto@iscet.pt

R. C. Lois-González and J. A. Rio Fernandes (eds.), *Urban Change in the Iberian Peninsula*, The Urban Book Series, https://doi.org/10.1007/978-3-031-59679-7_15

But cities have always been also spaces of movement, with a special relation with their hinterland, and to those who come to shop, for services or leisure. In this direction, several authors have defined a work basin, defined in relation with daily movement and, in that direction, the notion of functional area (OECD 2020) is particularly interesting. As a consequence, cities are very incompletely identified by the total number of residents, if you take as residents only those that sleep in them most of the nights in a certain year, or who have there a permanent address.

In recent decades, with the increase of international mobility, important changes took place in the relation of cities with other places as links with other cities and distant regions became increasingly important vs contiguity (Sassen 1991). Economies became less dependent on local and regional goods, services and jobs, while urban tourism and real estate investment exploded outside the most famous destinations and a rare number of world cities (LeGates and Stout, 2016).

As a consequence, understanding urban change, from an Iberian perspective, on the increasing relevance of the non-permanent residents is necessary.

We approach the problem in three parts. On part one, we deal with the diversification and complexification of urban population in a neoliberal context, and consequent conflicts. On part two, we consider an historical perspective on the presence of city users and of the importance for some cities of non-permanent residence. On part three, we analyze the case of Porto, the largest city in the North of Portugal, as one that has suffered a very significant increase in the number of city users, where they had a very relevant association to physical, social and economic change. Finally, we leave some conclusions and suggestions for reflection and further research.

15.2 The City of Non-residents

Today, in most medium and large Iberian cities, the quantitative and qualitative relevance of floating and non-permanent residents is very important for understanding the city itself, in the context of an unprecedented diversification of the urban population. Increasingly, cities appear to be global and the traditional notion of a relatively homogeneous urban population has given way to a tapestry of individuals with varied preferences and needs, reflecting a mosaic of cultures, origins and lifestyles. Together with low-cost flights, technological advancements have played a pivotal role in reshaping the dynamics of city users. The rise of digital platforms and smart technologies has transformed the way people navigate and interact with urban environments. From ride-sharing apps to smart city initiatives, technology has enhanced the efficiency and convenience of urban living, investment and exploration.

Moreover, the advent of remote work has altered the traditional patterns of city users. The ability to work from anywhere has led to a rise in flexible living arrangements, with individuals choosing cities not only based on job opportunities but also on lifestyle preferences. This trend has implications for urban policies and planning, as cities need to adapt to the evolving needs of a workforce that values flexibility and work-life balance.

In the multifaceted tapestry, the role of international students has become increasingly significant as well for cities with universities, and more so for "university cities". Several cities have become educational hubs, attracting a diverse array of teachers, researchers, congresses with students seeking academic excellence, cultural exposure and new social experience (Teles, 2016). That injects a unique vibrancy into urban life, fostering a rich cross-cultural exchange within the city's social fabric that not only contribute to the intellectual landscape but also infuse local economies through their spending on accommodation, dining, and cultural experiences.

Nevertheless, when we analyze the diversification and complexification of city users, we need to identify "normal tourism" as the key factor. In fact, most big and middle size cities are no longer just home to residents; they are also destinations for an increasing number of short stay visitors. As a consequence, the intersection of resident and tourist experiences has become a focal point in a big number of cities in the Iberian Peninsula, as elsewhere, shaping their character and identity.

The tourism industry contributes significantly to the economic vitality of several cities, creating job opportunities and driving local businesses. However, the influx also poses challenges, such as overcrowding and strain on infrastructure, as well as increase in housing and service prices, as hotels and all type of accommodation multiplies, and restaurants and shops direct their services and goods to short time consumers. Striking a balance between catering to the needs of both residents and tourists is crucial.

Several authors call for a political shift in the approach to urban planning. But with rare and recent exceptions (Amsterdam, Berlin, Barcelona, …), European cities have tended to follow the so-called neoliberal agenda (Joy and Vogel 2021), favoring an entrepreneurial approach, where marketing and foreign investment are central. In this city, citizens are sometimes seen as "human capital", so that only the "best" are welcome (Florida 2002) and social policies just resist, sometimes with chronic underfinancing.

In this context, planning tends to have a flexible approach, in view of the need to appease to the unexcepted investment, under a strategic vision much more economically than socially oriented, and much dependent of opportunities, like events, a good place in some international ranking, or something's capital.

Also important in the neoliberal city is the scarcity of urbanized soil and public residence that promotes dispersed urbanization and car dependency, whereas public action tends to look just to fragments of the urban space, taking normally as priority the places and measures associated to attract visitors and investors.

Public policies in general and urban action gives indications that this trend is switching in view of the severe problems that we now face, several of them directly related to the overwhelming increase of floating residents and the intensity of the inequality between tourists and tourism areas v "all time residents" and normal residential spaces (Ignatiova 2023; https://www.oecd.org/cfe/urban-principles.htm; https://www.amsterdam.nl/en/policy/policy-tourism/). That is also the consequence of the severe problem of access to residence, as house and flat prices force old residents out and only the very rich or those who stay for a pair or nights or are floating around and have a very well-paid job are able to pay the speculated price of living in the central city.

A more unjust city creates a dramatic democratic problem, and tensions increase. Ultimately, cities may become no-cities as they are lived only by visitors and workers that don't sleep on it, with voters diminishing dramatically.

15.3 The Old Urban Question of the City Users in the City

Movement is everywhere. Without it, life as we know it would not exist. At its origin, humans roamed through mountains and plateaus, across vast prairies and nestled valleys, in search of food and shelter.

Even after the process of sedentarization, humans continued to move from city to city, between the peripheral areas (the Roman *suburbium*) and the core of the city, or over long intercontinental distances. Therefore, the contemporary concept of the *city user* is, in truth, an ancient reality found more or less prominently throughout the history of cities worldwide.

In classical antiquity, travel for leisure, religion, business or sports was a common practice, akin to primitive forms of contemporary tourism. This was evident in visits to ancient Egypt, where travellers arrived by boat through the Mediterranean, and complaints were heard about the destruction of heritage due to inscriptions made by visitors on the walls of temples or pedestals of monumental statues. Similarly, in the Bay of Naples, filled with urban centres like the vibrant seaside resort of *Baiae*, featuring second homes, hostels, and inns, seasonality issues similar to those seen today in the Algarve, the Balearic Islands, or any sun and sea destination were common.

During the Late Middle Ages, Europe was traversed by merchants, pilgrims, and students. The first often followed ancient Roman roads or discovered river and maritime routes connecting harbour cities like Bruges, Venice, Lisbon, or Bordeaux. They sought business, suppliers, customers, contributing to economical internationalization.

There was also extensive travel in search of knowledge for the new medieval universities, following the examples of Bologna in the eleventh century and Paris in the twelfth century. Clerics and young students circulated throughout Europe, heading to Oxford, Salamanca, Padua, or Coimbra. Later, journeys were undertaken to meet and learn from the humanist Erasmus, who welcomed intellectuals from all over Europe in his home or in the taverns of Antwerp or Rotterdam. These travellers, arriving by foot, horse, or boat, contributed to the proliferation of inns and lodgings, to a lively urban environment and to cultural diversity. Information, culture, and intellectualism were significant drivers of movement on European roads and proto-forms of tourism. One of the major events of the Renaissance took place in Frankfurt—the enormous Book Fair (still taking place nowadays). There, international booksellers and intellectuals from various countries gathered, filling the city with a floating population that typically tripled during the fair. As Feifer (1986: 71) notes, “the crowds were so dense that rubbish disposal regulations were suspended because it was often impossible to get a cart through the streets”. After intellectual work during the day, the night was filled with fireworks, alcoholic beverages, and prostitution.

This was also the era of pilgrimages, long journeys to Jerusalem and Compostela and other not so well-known places. This not only reinforced the cities of destiny but also reinforced small cities along the way, through the fields (the etymological origin of “pilgrimage”), where travellers stopped. Of all religious destinations, Rome became the reference from the fifteenth century onwards, with the definite establishment of the papacy. The city had been practically abandoned after the fall of the Roman Empire, with just over 10,000 people (from about 1 million). Over several papal generations, a program of major constructions and aligned street layouts was designed in order to create urban order and appeal to visitors. Obelisks were erected, and the gates of ancient walls were embellished to give Rome the central role in Christian pilgrimage: “mater urbium, Roma,” as Erasmus called it. The success of the intervention was evident. In 1575, the inns of Rome hosted about 400,000 pilgrims and a constant flow caused traffic problems as the streets were filled with carriages exhibited by the wealthier (Girouard 1985).

Rome was also a decisive point in the “Grand Tour”, a fundamental journey in the long staircase of knowledge for young European aristocracy, starting from the seventeenth century. With a desire to learn, armed with the knowledge acquired in their academic training and usually accompanied by a well-prepared tutor and other servants, these cultural tourists contributed to the circulation of ideas, urban and economic turmoil, and the dynamic nightlife of leisure and entertainment in the “red-light” districts, in inns, and on the streets of the many cities they visited and stayed in, in a process not very different from today “night-time zones” (Lethbridge 2022).

With the Industrial Revolution, a relevant change occurred, particularly with the advent of the train in the nineteenth century that allowed the arrival of much more visitors. Accompanying this flow, many rapidly industrializing European cities received rural workers who spent the workweek in the city and returned home on weekends, where wives and children had been exploring subsistence agriculture.

From 1851, with the International Exhibition in London, a new form of entertainment and innovation became part of the international competition among cities for visitors. Serving as a meeting place for the exchange of ideas and novelties, international exhibitions—as the one that took place in Paris (Fig. 15.1)—have been fundamental elements of attractiveness for city users up to the present day, especially in the second half of the nineteenth century and the first half of the twentieth century.

The nineteenth century also saw the emergence of a large-scale tourism movement, fuelled by increasing industrialization, improved steam transportation, and an extension of weekly rest, holidays, and vacations. (Williams 2002) Since then, in constant growth, with only a slight slowdown during the 1970s oil crisis, the September 11, 2001 events, and the dramatic interruption triggered by Covid-19, hotels, travel agencies, tourism-oriented parks, and numerous other leisure activities have multiplied. The change is especially intense since the end of the last century

Fig. 15.1 The Paris Universal Exhibition, 1900. *Source* https://www.wikiwand.com/en/Belle_%C3%89poque

(Mourão et al., 2021). And in several cases, cities have developed areas that function as a kind of playground for leisure purposes, the "Recreational Business District" (Stansfield and Rickert 1970: 215): "the seasonally oriented linear aggregation of restaurants, various specialty food stands, candy stores and a varied array of novelty and souvenir shops which cater to visitor's leisurely shopping needs".

15.4 Porto: Embracing Diversity and Growth of City Users

As several European cities, Porto has witnessed a remarkable transformation in its urban dynamics in recent years, reflecting the global trend of increased growth and diversification of city users. Traditionally known for its rich history, stunning townscape, and vibrant local atmosphere, Porto has become a magnet for a new wave of inhabitants, with a notable rise in the prominence of tourists, international students, and digital nomads.

The loss of resident population has been a reality for a long time, remaining consistent since 1981 (Fig. 15.2). Nevertheless, the dynamics of the last decade have allowed for the attenuation (initially) and reversal (in recent years) of the declining trend, with some annual estimates of population growth (Fig. 15.3). In fact, despite the natural growth consistently showing negative values (due to low birth and death rates),

Porto exhibits positive net migration balances, showing a growing trend (except for 2020, a pandemic year with various restrictions that explains the overall population decline in between 2011 and 2011). This reflects the city's capacity to attract and retain new residents, many of whom are associated with new opportunities and forms of work.

The city's allure as a tourist destination has grown significantly, drawing visitors from around the world. The influx of tourists has not only contributed to the economic prosperity of the city but has also presented challenges in managing the delicate balance between catering to visitors' needs and preserving the character of the city for its long-standing residents (Chamusca, 2023).

Between 2011 and 2022, the number of tourists experienced a growth of 139.8%, with Porto surpassing five million overnight stays annually in 2023 (Fernandes et al., 2020). This continuous growth is characterized by a predominance of foreign visitors (78.5% in 2022). Despite the decline in the years 2020 and 2021 due to the pandemic context, there is already a return to the values of 2019, with a trend towards growth (Fig. 15.4).

During this period, there has also been an increase of the number of hotel establishments, especially luxury hotels (highly concentrated in the central area), and more significantly of short-term rental units. This growth has led to an accommodation capacity surpassing the number of residents in 2019, a figure that was already exceeded in 2022, following the recovery from the effects of the pandemic (Fig. 15.5).

Another facet of Porto's evolving dynamics is the increasing presence of international students. The city's renowned universities and educational institutions have attracted a diverse cohort of students seeking academic excellence in a culturally rich environment. According to the official data, in the 2022/2023 academic year, 4500

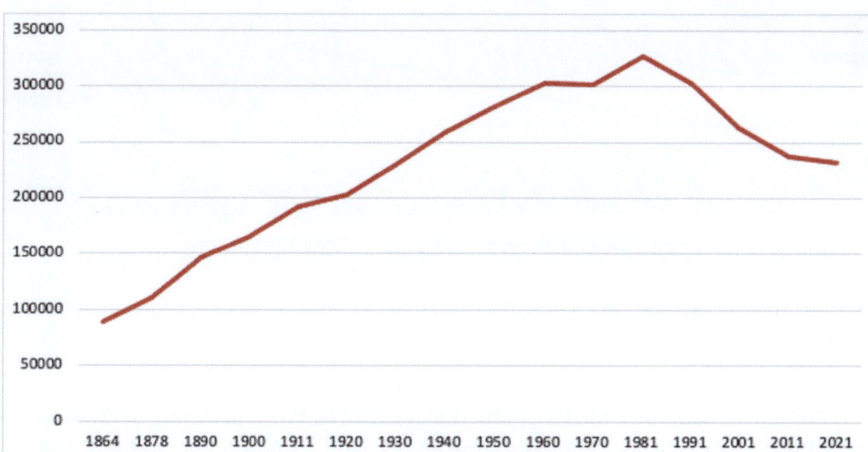

Fig. 15.2 Inhabitants in Porto (1864–2021). *Source* Own elaboration, using data from INE (2023)

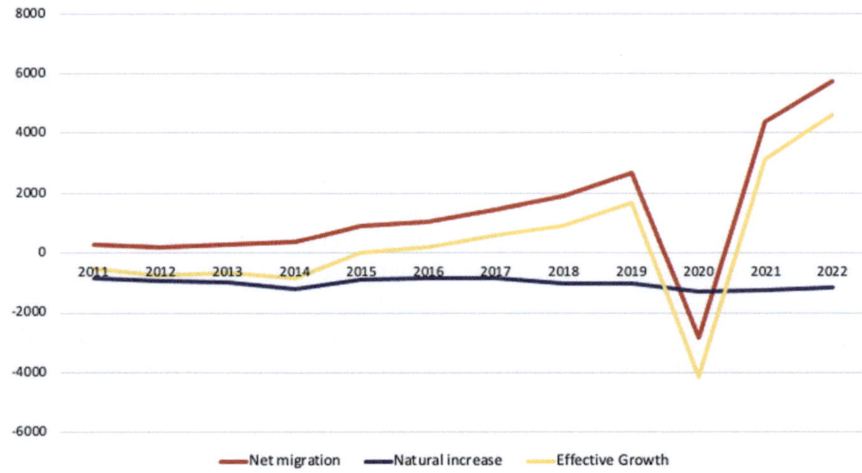

Fig. 15.3 Demographic dynamics in Porto (2011–2022). *Source* Own elaboration, using data from INE (2023)

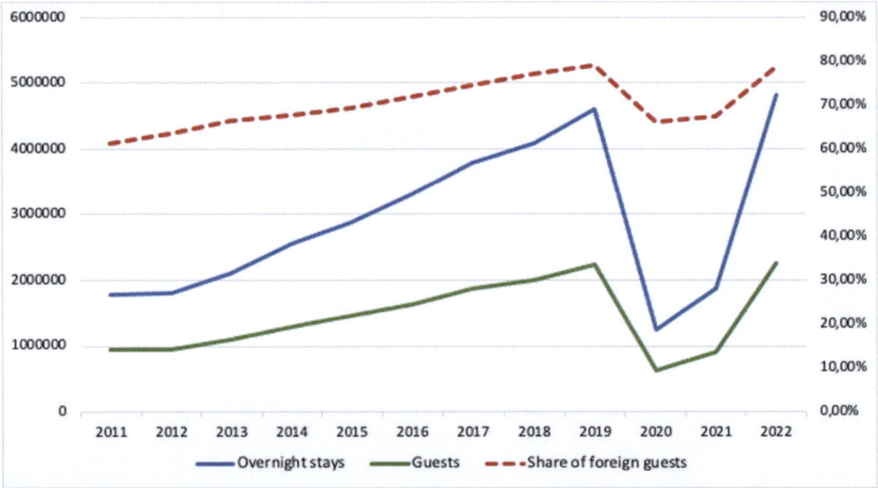

Fig. 15.4 Tourism dynamics in Porto (2011–2022). *Source* Own elaboration, using data from INE (2023)

international students attended a graduation, master or Ph.D. semester at the University of Porto (2013). These students contribute to the cosmopolitan ambiance of the city, fostering cross-cultural interactions. The integration of international students into the fabric of Porto enhances the city's reputation as an educational hub and further solidifies its standing on the international stage.

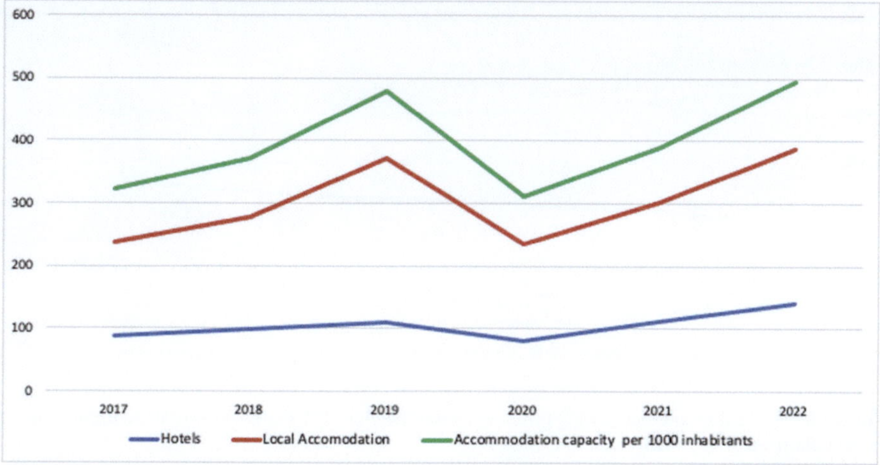

Fig. 15.5 Tourism units in Porto (2017–2022). *Source* Own elaboration, using data from INE (2023)

Moreover, there are several other students who do not reside in Porto but commute daily to pursue their education and learning, including other levels apart from university. In the academic year 2021/2022, the gross enrollment rate in secondary education in Porto was 277.3%, and in basic education it was 152.5%. This indicates that the number of students is significantly higher than the number of children and young residents in the municipality.

Simultaneously, the growth of people working in Porto and the rise of digital nomads has added a new dimension to Porto's landscape. According to the 2021 Census data, there are 203,799 people who commute daily to Porto for work or study, a slightly higher figure than the 200,352 recorded in 2011 (INE 2023) and almost the same of the "normal residents" (231,800). In addition to these movements, the city has become a haven for remote workers, drawn to its blend of modern amenities, historic charm, and a thriving community of like-minded individuals. The flexibility afforded by remote work has redefined the traditional notions of residency, with Porto emerging as an attractive destination for those seeking a balance between work and leisure in a picturesque setting.

The growth in the number of users is associated with urban rehabilitation processes and the enhancement of the city's image, helping to explain an increase in land prices also resulting from international real estate investment. The available data (INE 2023) show a 38.4% increase in the average selling price per square meter between 2019 and 2022 in the municipality of Porto, a figure that rises to 44.7% when considering only the central parish (Fig. 15.6). The rise in rental prices is even more pronounced. Between 2017 and 2022, there was an increase of 47.4% in the municipality of Porto and 51.6% in the central parish (Fig. 15.7).

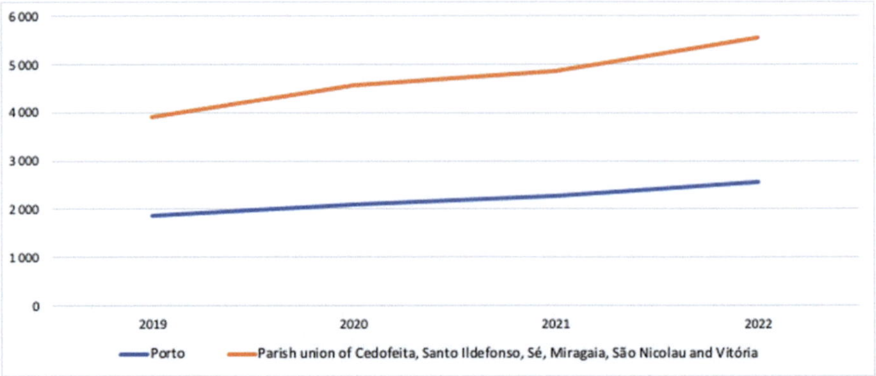

Fig. 15.6 Real estate average selling prices per square meter (2019–2022). *Source* Own elaboration, using data from INE (2023)

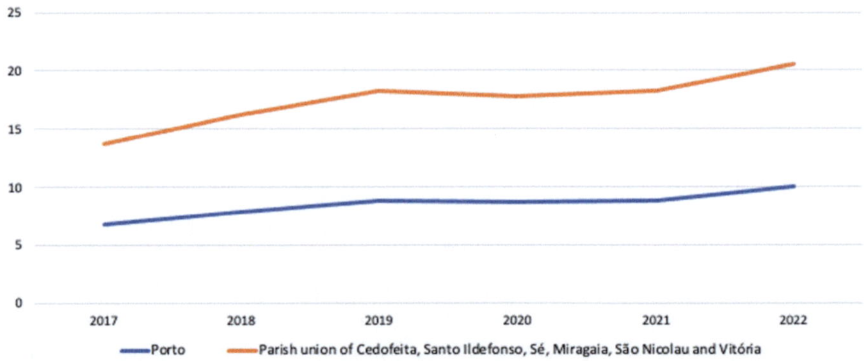

Fig. 15.7 Real estate average renting prices per square meter (2019–2022). *Source* Own elaboration, using data from INE (2023)

Overall, the evolution is carachterized by the increase and diversification of city users. In just 20 years, their number has seen a growth of over 164%, with the impact of tourism increasing significantly (Fig. 15.8).

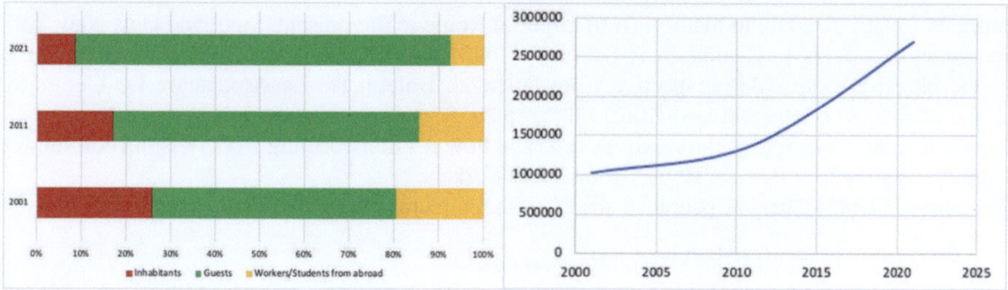

Fig. 15.8 City-users growth and diversification (twenty-first century). *Notes* 2001 data of guests is from 2002, as there is no information for 2001 and data of 2021 uses information from 2022, to avoid COVID-19 effect on the analysis. *Source* Own elaboration, using data from INE (2023)

15.5 Final Remarks

Floating residents have always been important. In fact, some cities existence and progress are very related to short time visitors, whose presence was associated, most of the time, to business, pilgrimage and education. However, on recent times, the presence on non-permanent residents has increased in a great number of cities, in numbers and diversity.

If that has to be positive, as cities have always been spaces of mixture and nodes of cosmopolitism, there are not so welcomed effects and conflicts emerge, As a consequence, while city users bring fresh energy and economic opportunities, they also underscore the importance of inclusive urban policies.

A harmonious balance between the needs of traditional residents and the demands of the burgeoning tourist, student, and digital nomad seems crucial for sustaining identity. That is certainly easier to say than to achieve. But, in this direction, we consider that initiatives that promote community engagement, preserve local heritage, and address the impact of increased tourism on housing and infrastructure are vital to ensure the continued livability and resilience of the city.

We also consider that, deep down, the final question is about equal rights: do visitors have the same rights as residents? Do they pay for them? Then, should they elect the leaders of the city they visit? We don't think so. So, we need a less consumer-oriented policy and to stop to see the city as a product.

References

Acuto M (2020) Understanding the global ecosystem of city networks. Urban Stud, 1–17

Carvalho L, Chamusca P, Fernandes J, Jorge Pinto J (2019) Gentrification in Porto: floating city users and internationally-driven urban change. Urban Geogr 40(4):565–572. https://doi.org/10.1080/02723638.2019.1585139

Chamusca P (2023) Urban planning and policy in Portugal: an overview on the role of EU funds and guidelines. Urban Res Pract 16(1):44–65. https://doi.org/10.1080/17535069.2021.1980607

Feifer M (1986) Tourism in history: from imperial Rome to the present. Stein and Day, New York

Fernandes JA, Teles F, Chamusca P, Seixas J (2020) The power of the cities and the power in the cities: a multiscale perspective from Portugal. Boletín De La Asociación De Geógrafos Españoles (87). https://doi.org/10.21138/bage.2978

Florida R (2002) The rise of the creative class: and how it's transforming work, leisure, community and everyday life. Perseus Book Group, New York

Girouard M (1985) Cities & people: a social and architectural history. Yale University Press, New Haven

https://www.amsterdam.nl/en/policy/policy-tourism/

https://www.oecd.org/cfe/urban-principles.htm

Ignatova A (2023) My neighbourhood. UN Habitat, Nairobi

INE (2023) Territorial statistics. Available at https://www.ine.pt. Accessed: 15 Jan 2024

Joy M, Vogel RK (2021) Beyond neoliberalism: a policy agenda for a progressive city. Urban Affairs Rev 57(5):1372–1409. https://doi.org/10.1177/1078087420984241

LeGates R, Stout F (eds) (2016) The city reader. Routlege, Oxford and New York

Lethbridge L (2022) Tourists: how the British went abroad to find themselves. Bloomsbury Publishing, London

Mourão LP, Ferreira L, Pinto JR (eds) (2021) Portugal e o turismo. Book Cover Editora, Porto

OECD (2020) Cities policies responses. Report, Available at: http://www.oecd.org/coronavirus/policy-responses/cities-policy-responses-fd1053ff/

Sassen S (1991) The global city. Princeton University Press, New York, London, Tokyo, Princeton

Stansfield CA Jr, Rickert JE (1970) The recreational business district. J Leis Res 2(4):213–225

Teles F (2016) Local government and the bailout: reform singularities in Portugal. Eur Urban Reg Stud 23(3):455–467

University of Porto (2013) The university in numbers. Retrieved from https://sigarra.up.pt/up/pt/. Accessed: 15 Jan 2024

Williams SW (2002) Tourism geography. Routledge, London

Chapter 16
Madrid: A City for Consumption and Investment

Carmen Mínguez

Abstract The debate on the image and urban reality of Madrid has not stopped intensifying in recent times. On the one hand, the capital of Spain projects itself through neoliberal policies and as an emblem of an entrepreneurial city. On the other hand, inequalities have deepened and a major crisis is perceived in certain public services aimed at the population as a whole. In this scenario, the city of Madrid has developed explicit policies of commercial freedom, real estate deregulation and low taxes, which have encouraged numerous consumers and collective capital funds to buy, to direct their investments to this great metropolitan centre and to speculate with processes of land revaluation. All this is analysed in the chapter. In parallel, Madrid has strengthened itself as an icon of urban tourism with differentiated characteristics. Its streets, museums, restaurants and palaces are objects of immediate consumption; a consumed city that also tries to be systematized. In short, the capital of Spain constitutes a paradigm of the city of current capitalism, hedonic, mobile and animated by quick profits. A paradigm that has been defended as a model of freedom, while its urban and metropolitan policies are criticized by important segments of the population.

Keywords Investment · Global services · Retail · Leisure · Spain

16.1 Introduction

The globalisation process is bringing about many economic and social changes which find their natural space to occur in cities, hence the consolidation of the term global city (Sassen 1991). The global city is characterised by the application of government models marked by deregulation, flexibility, competitiveness and privatisation (Díaz 2013; Sánchez et al. 2008), favouring private rights and interests above participation and the common good (Holston and Appadurai 1999). The impact of globalisation through neoliberal urban policies is manifested in an increase in inequality, in the

C. Mínguez (✉)
Universidad Complutense de Madrid, Madrid, Spain
e-mail: cminguez@ghis.ucm.es

R. C. Lois-González and J. A. Rio Fernandes (eds.), *Urban Change in the Iberian Peninsula*, The Urban Book Series, https://doi.org/10.1007/978-3-031-59679-7_16

political divide and in the creation of a city model intended for the global scale. In other words, one that is focused on international interests and the competitiveness between cities, in detriment to the local scale and the needs of the citizens (Romero et al. 2015). This is physically embodied in large urban operations or the privatisation of the space, which alters the urban dynamics with processes of gentrification and segregation and the modification of its habits and landscape (Borja 2013; Mínguez et al. 2022).

One of the most important results derived from economic globalisation is the conversion of citizens into consumers. They no longer create their identity based on territorial and shared history aspects but on multinationals and communication channels (García Canclini 2012). Therefore, the city has abandoned its Roman *civis* root, that is, a lived-in or a citizen's city and has given way to the *urbs* or business space (Cacciari and Bayrle 2010).

As a large city and the capital of the state, Madrid has not avoided the phenomenon of globalisation or its analysis. The majority of studies blame the market and capital for the processes experienced and, as a result, for the loss of power of the public institutions in the creation and consolidation of the city (De Santiago 2007; De la Fuente and Velasco 2012; Observatorio Metropolitano 2007). However, few have explored the specific actions that have been carried out within Madrid's *laissez faire* political model and how they have been embodied in the territory. In order to comprehend the recent evolution of the city, it is essential to understand the urban, fiscal and labour policies of the region and also the space granted for the actions of large corporations, whose decisions have been facilitated and assumed by the local and regional governments.

Based on evidence, this paper seeks to explain how the city of Madrid is a global city and how it cannot be understood without its region, as an important volume of the activities carried out are done so outside the administrative limits of the capital and are subject to political decisions on a regional scale. It is based on the idea that the tax benefit policies implemented in the Region of Madrid have promoted a more speculative tertiarization and have had strong spatial repercussions, creating business opportunities that have affected the configuration of area specialised in finance and global services, retail and leisure. Therefore, the political measures have converted Madrid into a space that favours consumption and investment (Mínguez 2022).

16.2 The Tertiarizations of Madrid in the Twenty-First Century

The Spanish Administration has created an environment that favours growth and investment through the private sector, relaxing the regulatory framework and reducing taxes (Blanco et al. 2017). This model incorporates the state and substitutes the Keynesian well-being system with the Schumpeterian System, characterised by being more competitive (Jessop 1993). The Administration develops new regulations that follow the logic of the market (Peck and Tickell 2002) and, on an urban scale, generate a change in the model and planning, giving rise to so-called strategic planning. This involves a transfer of the concepts and tools of business strategy planning and contemplating the city (Harvey 2001) as an entity composed of owners who are directed by municipal corporations that provide them with services, as if they were clients (Fernández Güell 2007).

Since the beginning of the 1990s,[1] the city council and regional government of Madrid have adopted a series of conservative and neoliberal policies aimed at deregulation and reducing taxes, which have moulded the capital into a city characterised by three elements: (a) significant tax benefits or a "tax haven status"[2] within the national context, (b) a growing economic activity linked to greater labour opportunities and (c) the flexibility of initiatives and of urban regulation.

The Region of Madrid is characterised by being the region with the lowest taxes in Spain. This is due, among other reasons, to the passing of a series of laws on tax measures, which have enabled personal income tax to be reduced and Inheritance and Gift Taxes to be cut by 99% and 100% respectively. In addition, rebates are also applied to wealth income tax and that of property transfer tax and stamp duties. Moreover, the Wealth Tax Law[3] was reformed in 2014, providing residents of the EU or the European Economic Area with the option of declaring their taxes based on the specific regulations of the autonomous region in which the majority of their wealth resides. According to the Inland Revenue, those taxpayers declaring assets of over 30 million euros doubled between 2011 and 2020. With respect to the distribution of the Wealth Tax by region, Catalonia is the region that collects the most, while the Region of Madrid reports an average value of wealth declared (Fig. 16.1).

This comparative tax benefit has led to a significant increase in investment, particularly in real estate, which is one of the main causes of the increase in housing prices, specifically luxury properties[4] (Fig. 16.2). This increase in investment placed Madrid in the fourth position in the world in terms of real estate appreciation in 2023, surpassed only by London, Los Angeles and Miami.[5]

The concept incorrectly referred to colloquially as "tax haven status" benefits individuals and companies who enjoy a series of competitive advantages offered by the region, such as location, legal security, the presence of other companies, institutional stability or infrastructure and services endowment. However, they also benefit from specific tax measures that make the city more attractive for investment than others. This is known as *tax dumping* which, according to the law, consists in

[1] This trend was broken in the city during the period 2015–2019 with the government of *Ahora Madrid*.

[2] A media term used to highlight the tax differences between this region and the rest, even though it does not fulfil any of the official requirements for being a tax haven, as it does have a tax burden, although lower.

[3] Law 26/2014 of 27 November, modifying Law 35/2006 of 28 November on Personal Income Tax, the revised text of the Personal Income Tax of Non-Residents, approved by Royal Legislative Decree 5/2004 of 5 March and other tax regulations.

[4] https://www.expansion.com/economia/2018/08/26/5b82cc1eca4741df378b458c.html.

[5] https://brainsre.news/venta-viviendas-lujo-espana-dispara-55-precio-aumenta-13/

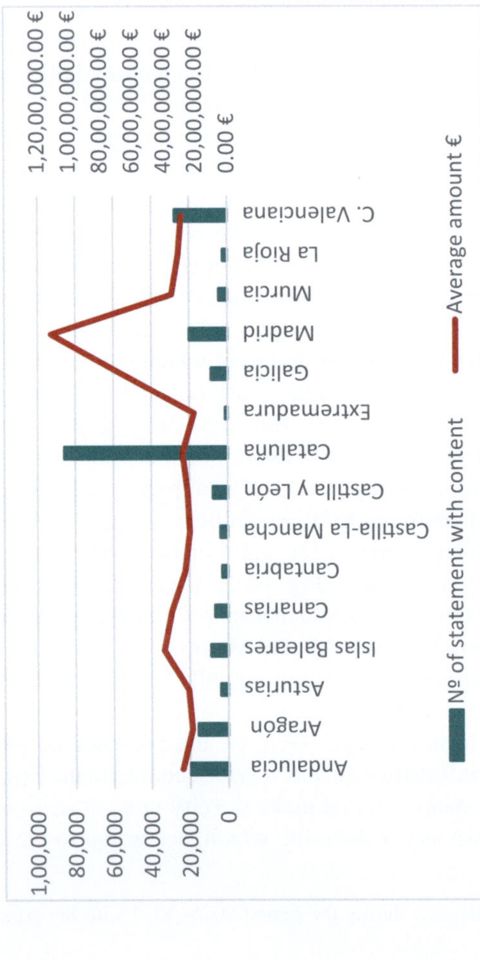

Fig. 16.1 Comparison of the number of declarants of the wealth tax and the average amount by autonomous region, year 2021

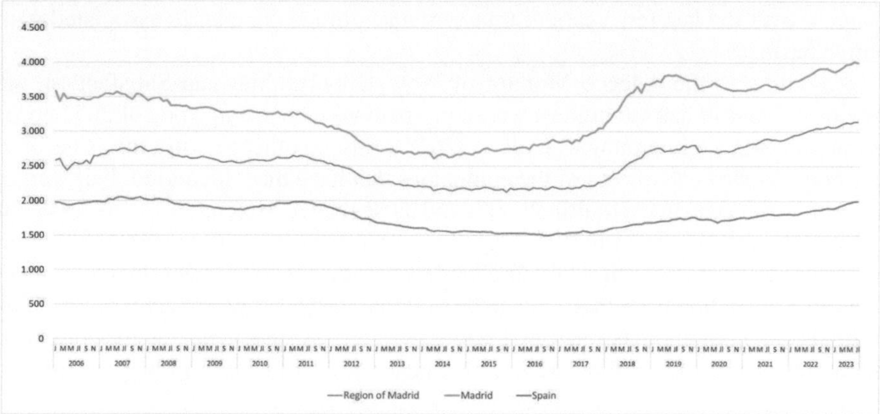

Fig. 16.2 Comparison of the evolution of land prices (€/m^2) (2007–2023)

establishing lower taxes than the rest of the regions in order to attract companies. This strategy affects the reduction of personal income tax and environmental taxes and those on assets and property. There are also mechanisms to reduce administrative obstacles and regulations,[6] and the norms are loosely enforced. The objective is to attract national and foreign investment and generate jobs. With this aim, the Regional Government of Madrid has implemented subsidy programmes for entrepreneurs and self-employed people and has created offices, such as the Centro Integral de Apoyo al Emprendedor (Centre for the Comprehensive Support of Entrepreneurship) or *Invest in Madrid.* The latter is aimed at foreign investors, similarly to the municipal office *Madrid Investment Attraction.* The presence of this type of investor has increased in the capital due to Brexit, which, according to the Foreign Investment Register, has led to Madrid receiving more than 30 financial companies previously based in London until 2021.

In light of all the above, in 2022, the European Committee of the Regions (CoR) awarded the Regional Government of Madrid with the Seal of Excellence "European Entrepreneurial Region", which is given to the regions that stand out for their entrepreneurial vision and their development of smart growth strategies, considered to be the major challenges faced by society, showing their capacity to create new opportunities and redirect the path of their economic development.

These initiatives designed to attract investment, businesses and tourists, understood as clients, directly favour the private activity, which has increased the number of jobs related to the tertiary sector, while public employment has reduced considerably. The increase has been particularly noteworthy in the hotel, information and communication sectors and in artistic, recreational and entertainment activities (Table 16.1), despite the alteration occurring due to COVID-19, which has led to a significant

[6] An idea proposed in the electoral programme of Díaz Ayuso, published on the PP website and in different spaces of the website of the Regional Government of Madrid in relation to economic growth and subsidies for companies.

increase over the last few years of transport and storage companies associated with online business.

According to the Adecco Monitor of Work Opportunities and Satisfaction, the Region of Madrid has the highest wages in Spain with an average pay of 2071 euros per month (+ 5.3% interannual) (Adecco 2022). Among other reasons, this is because it is home to the head offices of the companies that have most increased their wages, namely, large companies, multinationals and those related to technology, professional services, finance and insurance. Taking into account that a part of the population is in this situation of stability and labour abundance, together with lower unemployment rates than the national average,[7] it can be said that the spending capacity of residents in Madrid is high and has remained stable over the last few years, as the weight of the increase in the cost of living has also generated an increase in wages (Adecco 2022).

In parallel, the flexibility of the initiatives and the regulation of the city have been decisive and affect all areas of territorial and urban management. Therefore, the lack of plans, such as the Regional Plan of Territorial Strategy, the Plan for the Metropolitan Area of Madrid and the updated General Urban Plan means that the development of the Region of Madrid depends on sectoral legislation and the capacity of the municipalities (Córdoba and Morcillo 2020). Meanwhile, in the capital, the consolidated city is evolving as a result of occasional modifications to the General Urban Plan, sometimes generating legal conflicts which have been resolved in the courts. Among other aspects, this has favoured the approval of large urban projects, urban operation (PAU), renovation and refurbishment operations in the consolidated city (Brandis 2018; García 2014). However, the laxity mentioned *ut supra* also affects other activities with a spatial dimension, such as retail or property rental, and favours investment.

The region, and more specifically, the city of Madrid, are highly tertiarised, dependent on the private sector and highly polarised between activities related to the generation of services to satisfy the needs of the population and those related to economic speculation. The latter are becoming increasingly important, both in economic terms and due to the volume of commercial operations and the ways of intervening in the city. These operations are conducted to either condition spaces to improve their profitability or to abandon them and "reserve" them for future use (Sassen 2017).

16.3 Madrid as an Investment Space

In addition to clearly favouring the circulation of capital, the three aspects indicated in the previous section (tax benefits, growing economic activity and flexibility) form the basis of a city adapted to economic globalisation as they contribute three fundamental elements: an increase in investment, an increase in the population (resident and

[7] According to the EPA, the unemployment rate in the Region of Madrid in the year 2021 was 11.6%, as opposed to the national average of 14.8%.

Table 16.1 Those affiliated with the general regime and self-employed residents in the Region of Madrid, according to sector and section of activity between 2010 and 2023

	Wholesale and retail trade; repair of motor vehicles and motorcycles	Transportation and storage	Restaurant sector	Information and communications	Financial and insurance activities	Real estate activities	Professional, scientific and technical activities	Administrative activities and auxiliary services	Public administration and defense; compulsory social security	Artistic, recreational and entertainment activities
2010	200,893	67,201	91,142	90,736	62,630	10,999	123,875	137,533	70,161	20,629
2011	197,421	65,655	91,480	93,719	61,612	11,172	124,778	132,857	68,331	20,798
2012	195,572	63,442	90,934	95,050	61,439	11,096	124,762	130,602	60,525	21,065
2013	188,895	59,529	87,576	90,620	59,825	11,189	122,041	125,199	57,583	20,510
2014	182,494	61,095	86,801	89,686	58,762	11,653	124,067	122,264	55,952	20,660
2015	184,558	60,475	88,594	92,234	58,088	12,700	124,675	125,302	52,600	21,637
2016	188,003	61,252	94,477	97,562	58,619	13,805	131,421	130,137	51,635	22,832
2017	191,963	64,075	100,435	101,649	60,013	14,415	139,566	135,965	52,989	24,350
2018	195,934	67,034	104,605	104,654	61,283	15,068	142,535	142,476	53,810	25,973
2019	197,856	70,562	109,570	113,633	62,385	16,673	150,342	149,023	55,821	28,168
2020	200,664	73,579	170,790	118,599	63,490	17,574	155,217	153,529	56,535	29,465
2021	192,850	73,669	97,185	120,454	63,947	17,389	154,959	146,462	56,586	26,251
2022	198,542	75,857	106,572	125,645	63,632	18,328	160,750	153,708	57,746	29,064
2023	199,023	79,452	113,708	129,902	64,597	18,922	167,318	157,359	59,848	30,669
2010–2023	− 0.93	18.23	24.76	43.16	3.14	72.03	35.07	14.41	− 14.70	49.67

* Data 1 January each year

Source General Social Security Treasury and Director General of Economics, Statistics and Innovations of the Regional Government of Madrid Elaboration: City Council of Madrid. Sub directorate General for Statistics floating, including tourists) and liberation, that is, an increase in the public and private space and the opportunities to do business with it. In short, a sequence of actions, with a spatial element that favour the consumption in the city, creating: the city of finances and global services, the trading city and the leisure city.

16.3.1 Madrid as a Financial and Global Services Capital

Following the global trend marked by the loss of production in favour of *financialization* (Méndez 2019, 2021) in recent years, Madrid has become a benchmark city in terms of finance and global services. According to the statistics of the Bank of Spain, in 2023, 52.50% of the Spanish Monetary Financial Institutions had their head office in the capital,[8] 53.26% of the Payment Statistics Relevant Institutions,[9] 76.25% of non-monetary investment funds,[10] 72.88 of pension funds, 100% of securitization funds,[11] and 53.69% of insurance companies[12] (Fig. 16.3).

The location of the head offices of these businesses helps to define the specificity of Madrid as a services city, whose limits exceed the municipal area. The "financial cities" of Banco Santander (Boadilla del Monte and Hortaleza) BBVA (Las Tablas-Madrid) and Telefónica (Las Tablas-Madrid) have been established in the region, to which we can add the head offices of large consulting companies, such as KPMG or PWC, among others, located in Plaza Castilla. In addition, there is also the traditional CBD of Azca, where, in recent years, major building work has been carried out, such as the construction of the BBVA Tower or the Europa Tower, to attract new clients after the economic crisis of the last decade and to adapt to their current needs.

With respect to financial entities, a hub has formed around the Paseo de Castellana and is developing in the adjacent, traditionally residential neighbourhoods, such as that of Salamanca and Chamberí. Other entities are being established in northern parts of the city, close to its borders and even outside of them, with no single CBD (Fig. 16.4). Furthermore, it should be noted that these large companies still maintain the buildings in which they traditionally established their head offices, located in the Central District or close to it, and which have now been converted into representative buildings housing the cultural and social activities of the companies.

---

[8] Credit entities and financial institutions whose activity consists in receiving deposits or close substitutes for deposits from entities other than the IFM and granting loans or investing in securities on their own account.

[9] Suppliers of payment services (including issuers of electronic money) and payment systems operators.

[10] Collective investment institutions that invest in financial and non-financial assets. They do not include pension funds or monetary market funds that are included in the Monetary Financial Institutions.

[11] Companies created to isolate payment obligations of the company from those of the originator or insurance or reinsurance company and to issue securities representing debt or other debt instruments.

[12] Financial institutions primarily engaged in financial intermediation resulting from risk compensation particularly in the form of direct insurance or reinsurance.

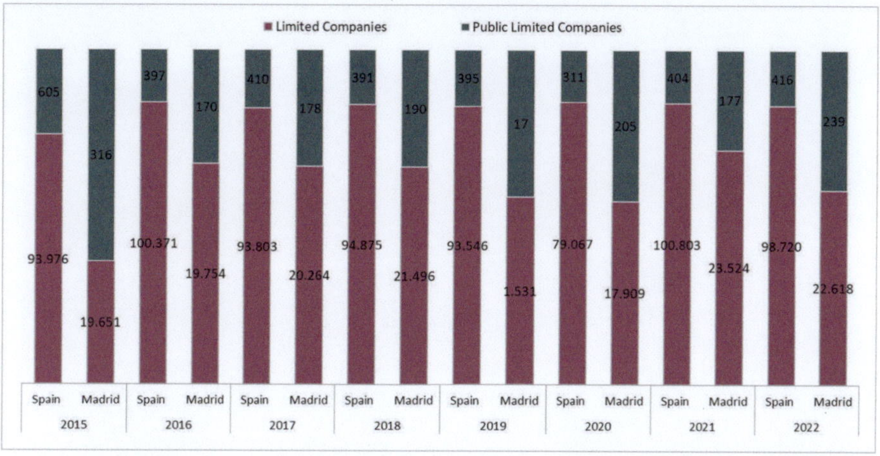

Fig. 16.3 Evolution from 2015 to 2023 of the constitution of limited companies and public limited companies in the region of Madrid. *Source* Memorias anuales de Estadística mercantil. Registradores de España 2015–2019

All of these entities generate significant economic activity, placing Madrid in 19th position in the Global Cities Report of Kearney in 2022, losing positions to Seoul, Berlin, San Francisco, Shanghai, Toronto and Sydney with respect to the years 2017 and 2018 when it held 13th place. Even so, it is still in the lead on a national level.

16.3.2 Madrid: A City of Consumption and a Consumed City

Another noteworthy occurrence over the last two decades has been the conversion of the Region of Madrid into a space for consumption. In this respect, the Law 2/2012 of 12 June on the Dynamization of Commercial Activity in the Community of Madrid has been decisive. It had the objective of activating the sector by implementing freedom to establish opening days and times. This right had already been granted to the retail outlets in the popular tourist areas. With this law, the Region of Madrid has become a large commercial hub with an impact that exceeds the regional limits.

This fact, which is also related to the increase in the population, has an important spatial dimension. On the one hand, a proliferation of shopping and leisure centres may be observed, favoured by territorial planning and its constant modifications, with a total of more than 130 in the Region,[13] including the opening in 2022 of the Caleido Tower Shopping Centre, owned by the Villa Mir group and in 2021 the Oasiz Madrid centre, owned by Phalsbourg, a group specialised in the real estate management and development. On the other hand, we can refer to the phenomenon of the so-called franchising or *boutiquisation,* consisting in the substitution of traditional retail with

[13] Source: Turismo de Madrid.

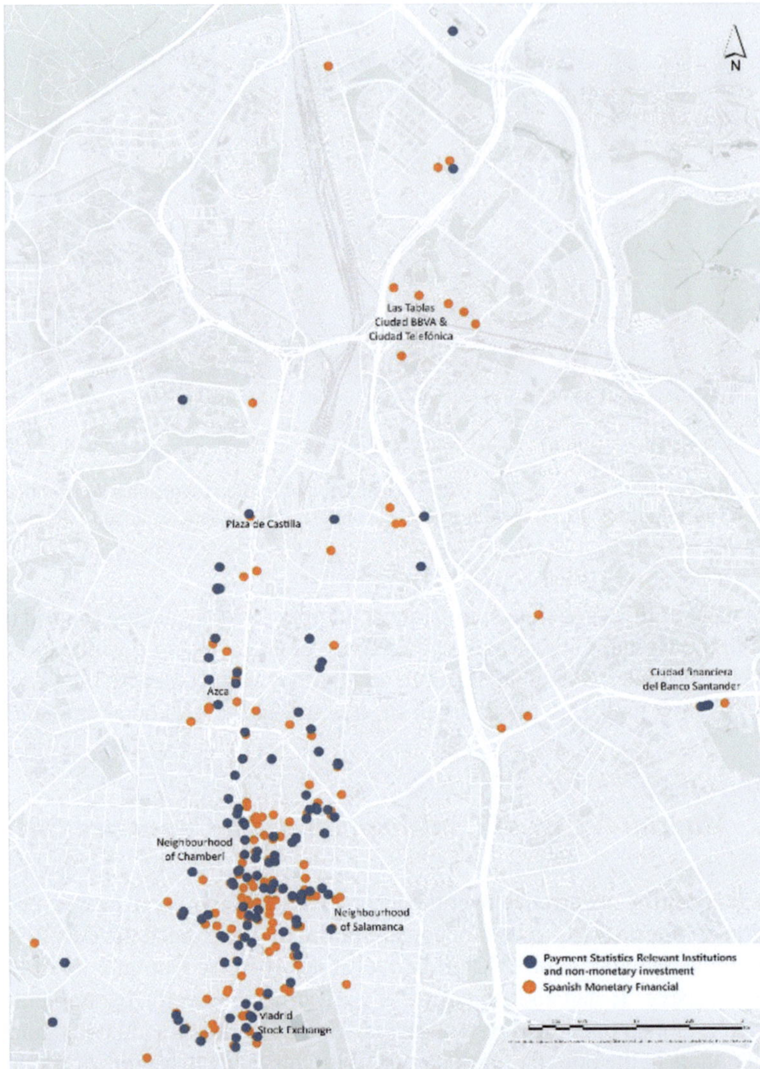

Fig. 16.4 Distribution of the financial and global companies responsible for investment Funds in Madrid. *Source* Bank of Spain. Own elaboration

large national and international businesses in the form of franchises or simply the installation of large corporations such as Inditex. This is due to the strong competition existing between businesses, together with the impossibility of small retailers to keep the same opening hours as the chains and their inability to compete due to the price of the leasing premises that increased considerably after the reform of the Law on Urban Leases, in 2015, putting an end to the old rental agreements and significantly altering retail behaviour and the landscape of retail districts.

This situation is particularly visible in shopping districts such as the Salamanca neighbourhood and the surrounding streets of Calle Preciados-Gran Vía. The former is where luxury or quality retailers can be found, with the large international companies displacing some of the small national stores. In the area around Calle Preciados and Gran Vía, there has been a proliferation of large global *low cost* retailers, which are more competitive and able to pay higher rents for large premises, former cinemas and theatres or large department stores. The volume of establishments, their size, changing supply, etc. has converted the Gran Vía into an almost themed space, whose identity is derived from the historical buildings rather than its retailers.

One of the most recent processes affecting Madrid's retail sector is the closure of stores which are sometimes replaced by restaurants and even housing. This phenomenon can be seen in the employment data and is recorded as a proven fact in the annual report published by the Mercantile Registrars Association and an increasing number of media. Therefore, according to the databank of the City Council of Madrid, in the summer of 2023, a total of 8,169 street level establishments had been transformed into housing for residents or tourists. As indicated by the Mercantile Registrars Association in its annual report of 2018, this is due to the lowering of the price of retail establishments to levels below those of 2003. This is a result of the low expectations of the promotion of businesses due to the general economic situation and, essentially, the propagation of *e-commerce.* Together with this decline in retail activity, over the last two years there has been an increase in the price of housing, which, according to the INE, rose by 10.20% in 2018. Other contributing factors are the approval in 2019 of the *Special Plan for the Regulation of the Use of Tertiary Services of Accommodation* (PEH), which obliges this type of accommodation to have a different access to that of the other residents and the high profitability that the rental of tourist accommodation has with respect to retail use (Fig. 16.165).

Fig. 16.5 Evolution of the number of premises that have been transformed into residential and tourist housing in the city of Madrid. Own elaboration

In parallel, another phenomenon has arisen with a significant spatial dimension. This is the emergence of new logistics centres[14] related to the storage and distribution of goods purchased from online retailers and the increase in the employment of unskilled workers for transport and storage (Table 16.1). Many of these logistics centres are associated with the so-called "golden logistics mile in Spain", located in the Henares corridor, covering the municipalities of Torrejón de Ardoz, San Fernando de Henares and Coslada. In the latter municipality we can find the Puerto Seco de Madrid (Dry Port of Madrid), which is an Intermodal Logistics Platform of the Port System of state interest, developed as an international goods terminal, connected with the principal Spanish ports, thanks to the rail network (Fig. 16.6). Amazon's warehouses are also in Coslada in a space commonly known as the "ground zero of e-commerce".

This complex, parallel with the A2 motorway, which joins Madrid with Barcelona, is close to the Adolfo Suarez Madrid-Barajas airport. According to Aena data, this airport moves more than 44,000 tonnes of goods each year and its air cargo centre has more than 2000 companies. However, since the pandemic, other areas of the Region of Madrid have been gaining importance, such as the southern axis, parallel with the A4 motorway, which joins Madrid with Andalusia, where logistics and retail operators such as Amazon are installed. This complex has a higher profitability, and is, therefore, attracting promoters and national and international investment funds, such as Segro, Engel & Völkers Development, Hines or DEA Capital, among others. On a smaller scale, the logistics developments close to the A3 (Madrid-Valencia) and A5 (Madrid-Extremadura-Portugal) motorways are also being extended.

The construction of logistics centres has brought about a *boom* in urban development, with a special emphasis on the SOCIMIS (Spanish Real Estate Investment Trusts), as the largest ones[15] investing in Madrid do so essentially in office buildings, shopping centres and industrial warehouses, illustrating that the business is not only associated with production, but also with land speculation.

In today's society, leisure is a fundamental part on which a significant amount of time and income is spent and is becoming commodified (Heartfield 2000). In turn, cities are undergoing a process called *ludification* (Baptista 2005), whereby the services and space dedicated to leisure activities increase and become specialised. Leisure is important because, as well as fostering the economic development of the culture industry (Álvarez and Fernández-Villarán 2012), it generates new business and employment niches (Table 16.1), following new dynamics that transform the spaces or create others ad hoc.

Over the last two decades, in the Region of Madrid, both of these phenomena have arisen. There has been a significant expansion throughout the whole region of complexes with cinemas and arcades but also theme parks, zoos or specialised

[14] Areas of Port Logistics Activities, Integrated Goods Centres, Transport and Logistics Services Centres and Corporate Logistics Support.

[15] According to the Madrid Stock Exchange, these are: Merlín Properties, Inmobiliaria colonial, Lar España, Arima, General de Galerías Comerciales and GMP Property.

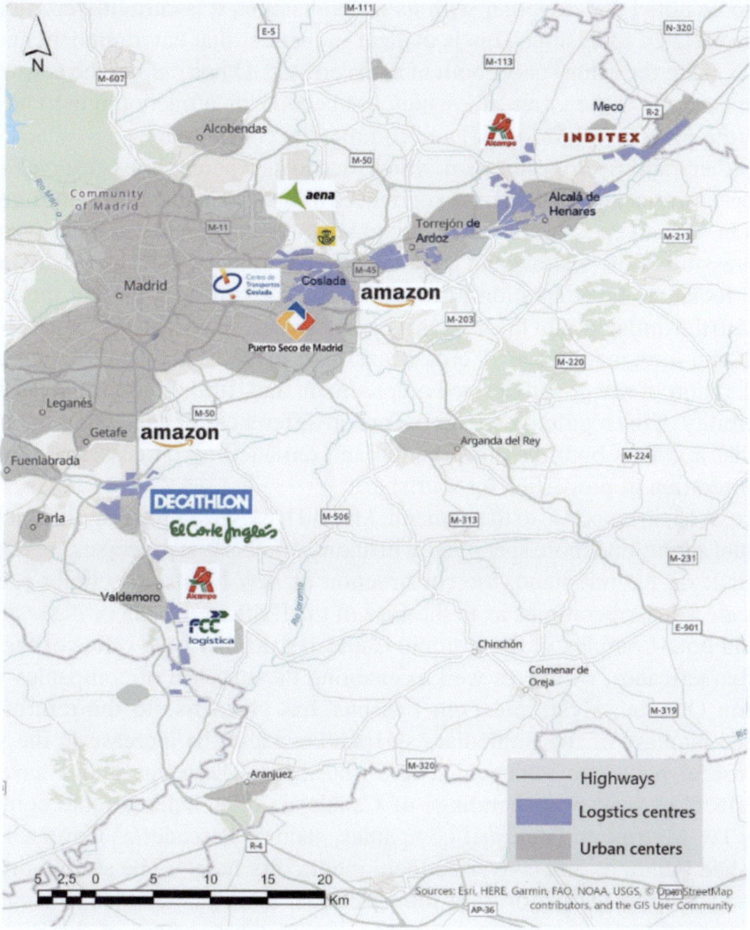

Fig. 16.6 Location of the principal logistics centres of the Region of Madrid. Own elaboration

facilities adapted to new demands (ice skating, skiing, wind tunnels, virtual reality…). At the same time, the consolidated city has also experienced a considerable transformation to adapt to these new uses. This adaptation process is closely linked to private investment (traditional restaurant businesses and nightlife and other more recent concepts, such as *escape rooms),* but also to the strategy of the local government with the conditioning of public spaces and the *museumalisation* of the city. This is evident in the major investments made for the regeneration and conditioning of large recreational areas, such as Madrid Río, and for the recovery and conversion of buildings in disuse, as a result of the productive transformations, into cultural centres, such as the Matadero and Tabacalera.

This close relationship with leisure activity can be found in two important sectors: retail and tourism. With respect to the former, to which we have already referred, it should be noted that, together with its generalisation, it is currently considered as a practice related to leisure. This is evident in the fact that entertainment and retail activities share the same space, both in areas created ad hoc, and in the consolidated city. An example is the Gran Vía, which has become an attraction due to its leisure supply (the *Broadway of Madrid*) and commercial consumption.

The other major sector is tourism, which, due to its repercussions deserves specific treatment as the city of Madrid has experienced a significant increase in the number of visitors in recent years,[16] which has brought about considerable spatial impacts due to the pressure of the visitors and the adaptation of the city. Of these many impacts, we will focus on the adaptation of tourist accommodation, given that this element pertains to the initial idea of this text as it benefits from the presence of small and large national and foreign investors (Cerdá-Mansilla et al. 2022); being closely related to the generation of direct and derived employment and, furthermore, it benefits from the flexibility of the regulations. This occurs on different scales and types of business, given that it affects both the construction and remodelling processes of hotels and the proliferation of properties for tourist use.

Since 2015, in response to the demand, Madrid has experienced significant growth in its hotel supply and more specifically in luxury hotels. Two processes are occurring in parallel: on the one hand, the construction of new hotels sometimes related to large-scale urban operations, as in the case of the Edificio España or the residential, retail and hotel complex of the Edificio Canalejas (former head office of the Banco Hispanoamericano), which, as well as enabling the entrance of companies such as Mandarin Oriental, Hyatt and Four Seasons, has given rise to the refurbishment and price increase of the immediate surroundings and the increase in the volume of the useful built area with which to do business. These operations have had a "pull effect" as in the surroundings of Canalejas, other international companies, such as JW Marriot and national companies, such as Pescaderías Coruñesas, have refurbished buildings and converted them into luxury hotels. The second process is the remodelling of the majority of the traditional luxury hotels due to a change in owner, often a foreign company, as in the case of the purchase of the Ritz by the Saudi group Olayan or that of Villamagna by the Turkish group Dogus group, among many others.

These operations illustrate the loss of Spanish ownership of the companies and the emergence of investment funds, such as Henderson Park, Parterns Groups or Blackstone and SOCIMIS that purchase hotels. Some of them belong to national chains, such as Barceló, Meliá or NH, which take advantage of the interest of international companies so as to divest in property and diversify and grow with management and rental contracts, as in the case of RIU with the retail establishments of the Edificio España.

---

[16] According to the INE, it grew by 7% between 2018 and 2019. In 2023, the pre-pandemic levels have still not been regained, although during the first semester of 2023, Spain received almost 37.5 million international tourists, that is, 23.7% more than the same period in 2022. Expenditure amounted to 46,011 euros, that is 28.3% more than in 2022.

Another phenomenon, also beginning in 2015, is that of the so-called "hotelisation of property", as a result of the emergence of the platforms initially associated with the sharing economy, of a high value at a time of economic crisis. This business opportunity has been seized by both small investors attracted to the Region of Madrid due to its tax advantages and large companies. It is considered that this has generated a rental bubble and has pushed up prices, together with a socio-economic homogenisation and, therefore a loss of diversity of the residents (Mínguez et al. 2022). This practice is also favoured by the leniency of the regulations (Martínez et al. 2020) and the lack of understanding, for years, between the local and regional governments to establish a control mechanism, which has enabled the liberation and, therefore, the increase in the space with which to do business.

16.4 What is Happening in Madrid?

The production transformations occurring in Madrid and, as a result, the social transformations are reflected in its urban space. The urban changes are not spontaneous but are a result of an urban policy strategy prevailing since the 1990s that has favoured "the urbanising expansion as an instrument to re-structure the regional production-accumulation system" (Córdoba and Morcillo 2020:80). Thanks to a permissive taxation policy and the flexibility of the regulatory framework, private companies, which have an increasing presence in the city, clearly participate in these policies. Through a multimodal network, they generate spaces of specialisation scattered throughout the region which are essential for the development of the services sector and consumption, with the objective of converting it into a global region (Córdoba and Morcillo 2020). In this respect, we can observe a productive complementarity between the city and the region, whereby the capital city surpasses its administrative boundaries. This is not only due to the availability of land necessary for certain activities or the need to provide services to the population of the region, but also because of the competitive advantages and, particularly, the political and taxation conditions of the region that attract both national and international investment.

Madrid is attracting interest for the development of certain economic activities related to retail and leisure and also to the value of the land and its possibility to generate capital gains. This phenomenon is becoming increasingly important and is associated with all production activities, as the adaptation and construction of spaces for business frequently implies urban development operations in which investment funds and SOCIMIS intervene for exclusively speculative reasons. Therefore, there is a growing presence of international capital in the Region of Madrid, showing a large capacity for intervention in the short-term design of the region, particularly the city of Madrid. Furthermore, the investment of large companies leads to spaces losing their identity and reduces the capacity of development by local businesses, prioritising service, understood as the diversity of supply and permanent availability in order to cover the needs of the demand.

Finally, it may be said that business opportunities are increasingly responsible for designing the space and, therefore, creating the city, with the support of large investments by the government, which are sometimes prioritised over the interests of the residents.

Acknowledgements This text contributes to the development of ADAPTATUR-CIUDAD—Project PID2020-114186RB-C22 financed by MCIN-AEI/https://doi.org/10.13039/501100011033.

References

Adecco (2019) VII Monitor Adecco sobre Salarios: la brecha salarial en la empresa española
Adecco (2022) Monitor Adecco de Oportunidades y Satisfacción en el Empleo
Álvarez Alday M, Fernández-Villarán Ara A (2012) Impacto económico del ocio en el siglo XXI. ARBOR Ciencia, Pensamiento y Cultura 188–754:351–363
Baptista L (2005) Territórios Lúdicos (e o que se torna lúdico um território): Ensaiando um ponto de partida. Fórum Sociológico 13–14(2):47–58
Blanco-Romero A, Blázquez-Salom M, Mínguez C (2017) Claves en la reestructuración turística de la ciudad. Naturaleza, Territorio y Ciudad en un mundo global. Actas del XXV Congreso de la Asociación de Geógrafos Españoles. AGE, Madrid, pp 1516–1524
Borja J (2013) Revolución urbana y derechos ciudadanos. Alianza, Madrid
Brandis García D (2018) Grandes proyectos urbanos y desarrollos residenciales: del urbanismo de mercado a un nuevo modelo para Madrid. Ciudad y Territorio: Estudios Territoriales 50(198):729–745
Cacciari M, Bayrle T (2010) La ciudad. Gustavo Gili, Barcelona
Cerdá-Mansilla E, Rubio N, García-Henche B, Campo S (2022) Airbnb y la turistificación de los barrios en las ciudades: un análisis de segmentación por barrios de alojamiento extrahotelero en Madrid. Investigaciones Turísticas (23):210–238. https://doi.org/10.14198/INTURI2022.23.10
Córdoba Hernández R, Morcillo Álvarez D (2020) Marco territorial de la producción de espacio en la región funcional de Madrid. Ciudades 23:71–93. https://doi.org/10.24197/ciudades.23.2020.71-93
De la Fuente Fernández R, Velasco González M (2012) La política urbana en Madrid: un relato provisional Geopolítica(s) 3(1):35–59
De Santiago Rodríguez E (2007) Madrid, 'ciudad única'. Pautas y lógicas espaciales recientes en la región madrileña: las grandes transformaciones estructurales; el despliegue del nuevo 'paradigma único' en la región urbana de Madrid. URBAN12:8–33
Díaz Orueta F (2013) Sociedad, espacio y crisis de la ciudad neoliberal. Metamorfosis urbanas. Ciudades españolas en la dinámica global. Icaria, Barcelona, pp 81–107
Fernández Güell JM (2007) 25 años de planificación estratégica de ciudades. Ciudad y Territorio. Estudios Territoriales XXIX(154):621–637
García Canclini N (2012) Consumidores y ciudadanos. Grijalbo
García Pérez E (2014) Gentrificación en Madrid: de la burbuja a la crisis. Revista De Geografía Norte Grande 58:71–91
Harvey D (2001) From managerialism to entrepreneurialism: the transformation in urban governance in late capitalism. Spaces of capital: towards a critical geography. Routledge, Nueva York
Heartfield J (2000) Great expectations: the creative industries in the new economy. Design, Londres
Holston J, Appadurai A (1999) Introduction: cities and citizenship. Cities Citizensh 1
Jessop B (1993) Towards a Schumpeterian workfare state? Preliminary remarks on post-Fordist political economy. Stud Polit Econ 40:7–39

Martínez Caldentey MA, Murray I, Blàzquez-Salom M (2020) En la ciudad de Madrid todos los caminos conducen a Airbnb. Investigaciones Turísticas 19

Méndez Gutiérrez del Valle R (2019) Ciudades en venta: estrategias financieras y nuevo ciclo inmobiliario en España. Publicacions de la Universitat de Valencia, Valencia

Méndez Gutiérrez del Valle R (2021) Financiarización urbana y burbuja del alquiler en España: tendencias y contrastes en perspectiva multiescalar. Documents D'anàlisi Geogràfica 67(3):0441–0463

Mínguez C, Blanco-Romero A, Fernández Tabales A (2022) Ciudades para vivir versus Ciudades para visitar. El turismo como factor de segregación urbana. Geografía social: permanencias, cambios y escenarios futuros. Trillo Santamaría JM, Lopez L, Lois González RC. Asociación Española de Geografía, Madrid; Santiago de Compostela: Universidade de Santiago de Compostela, Grupo de Análise Territorial (ANTE) GI1871, pp 167–177

Mínguez C (2022) Madrid ciudad de consumo y ciudad consumida. In: Lois González RC, Rio Fernandes JA, Beltrão Sposito ME (eds) El mundo visto de las ciudades. Tirant lo Blanch, Valencia, pp 455–470

Observatorio Metropolitano (2007) La explosión urbana de la conurbación madrileña, en Madrid, ¿la suma de todos? Globalización, territorio, desigualdad. Traficantes de sueños, Madrid

Peck J, Tickell A (2002) Neoliberalizing space. Antipode 34(3):380–404

Romero J, Melo C, Brandis D (2015) El giro neoliberal y las políticas para la ciudad en España. Balance a partir de los ejemplos de Madrid y Barcelona. Boletín de la Asociación de Geógrafos Españoles 69:369–386

Sánchez Moral S, Calatrava Andrés A, Melero Guilló A (2008) Las funciones comando de Madrid en la economía global: una aproximación a través del proceso de atracción de capital extranjero en Revista eure XXXIV(101):25–44

Sassen S (1991) The global city. New York

Sassen S (2017) Conferencia de inauguración del XXV Congreso de la AGE, celebrado en Madrid en octubre de 2017 (Ponencia inédita)

Part V
Mobility

Chapter 17
The End of the Car City in Portugal

Nuno Marques da Costa

Abstract Car dependency in Portugal shows rapid growth in the last 30 years. The increase on motorization rate, the expansion of motorway network and urban sprawl have contributed to the rapid increase in the importance of using the car for daily journeys, marking a new territorial organisation, less dense and disperse. At the same time, there has been a reduction in the use of public transport, which was unable to compete in terms of speed and response to new demands. In recent years we have seen a profound transformation in the organisation and supply of public transport, as well the development of initiatives for the integration of active modes in daily mobility and to get the car out from city centres. These changes give us a more optimistic outlook for reducing car dependency in Portuguese cities.

Keywords Car dependency · Portugal · Modal split · SUMP · Motorization · Public policies

17.1 Introduction

The development of the car city is focused in designing cities to accommodate the use of private vehicles, widening road networks and parking areas and paying little attention to other uses of public space, particularly for pedestrians and cyclists, and where public transport has a small share in the modal split of urban journeys.

This development is related to a strong car dependence for commuting and daily mobility needs associated to limited public transportation, urban sprawl and insufficient pedestrian and cycling infrastructures and space allocated to these uses, replaced by road and parking uses. This kind of development results in negative environmental impacts, namely air pollution, greenhouse gas emissions, and depletion of natural

N. M. da Costa (✉)
Instituto de Geografia e Ordenamento do Território/Centro de Estudos Geográficos, Universidade de Lisboa, Lisboa, Portugal
e-mail: nunocosta@edu.ulisboa.pt

R. C. Lois-González and J. A. Rio Fernandes (eds.), *Urban Change in the Iberian Peninsula*, The Urban Book Series, https://doi.org/10.1007/978-3-031-59679-7_17

resources, but also negative impacts concerning social equity, affecting disproportionately different social groups, namely those that cannot afford a car or individuals with mobility challenges.

The car dependency improved along the twentieth century, increased by car ownership and car daily use. In American cities car dependency started as early as the 1920s, where city planners and traffic engineers believed that only retrofitting traditional downtowns and neighbourhoods for cars, and removing trams to speed up car traffic, could fulfil the need for increased speed and car traffic fluidity. In fact, tram networks were mostly removed from North American cities by the 1960s (Bloom 2023).

Nevertheless, car dependency starts to become more evident after the WWII when infrastructures began to be developed centred in the car use. This idea could be seen at different scales, from the expansion of the national road network, like the Interstate Highway System in the 1950s in the United States, to zoning planning, promoting the separation of different urban functions and avoiding the mixed-use land use, and the street design, with wider roads or the parking facilities, namely on street parking, making driving easier and increment city speed. At the same time, due to the car's speed, low energy costs, and the freedom of movement, urban sprawl becomes a process that affected first the urban areas in the United States of America, and then expanded around the world, strongly related to car use, attractive land and housing prices, in low-density areas. But, also the planning regulations and options, namely the single use zoning paradigm, have contributed to this process.

As a result of this development, today car dependence individual behaviour can be divided in two groups. Those that could undertake their journeys by alternative modes, but rely on their car, as a conscious dependence, and for those that don't have a viable alternative for the use of the car, there is a structural dependence (Mattioli et al. 2016). These differences are relevant for the design and implementation of public policies. While for the former, interventions will focus on behavioural change, for the latter, interventions will have to consider the different causes of car dependence, namely absence of public transport services, economic capacities, mismatch between services schedules and personal needs, travel time differential between the use of the car and public transport, or the individual complexity of travel chains, where public transport has less capacity to respond, among others.

Three indicators, its value and the evolution trend, can be used to evaluate car dependence. These simple indicators are: motorization rate, road network extension, the share of public transport used to daily travels. The changes in these three indicators form a good approach to identify the evolution of society car dependency.

17.2 European Car Dependence

From the mid-1950s increasing traffic congestion was recognized as a major problem (Gunn 2011). The Buchanan Report, published in 1963, resulted from the evident intense use of the car in British cities, states that have intense use of the car would be a major problem that society would face in the near future. This report warns of the potential harm caused by the car use and traffic congestion and proposes actions to mitigate these problems (Buchanan 1963).

In Europe, after the United States, the car began to gain greater weight in the modal split of urban journeys (Sloman 2006). The first major urban public transport mode, the tram, started to end is operation in Paris in 1937, in line with the first great expansion of metro network, in London in 1952, which began to be replaced by trolleybuses in 1931 but soon ended their operation in 1962, or Madrid in 1972, in accordance with the expansion of metro network during the 1960s and 1970s (Marques da Costa 2011). This was a sign of transformation in European cities. The tram speed and their sensitivity to traffic congestion, competing and struggling with the car, condemned its use as a mode of urban transport in many European cities. Trams have been replaced by buses and by metro network, the latter immune to congestion, but with much high development costs.

At same time, road network, car ownership and use for daily mobility has increased. During the last 31 years, at EU-27 level, the motorway network growth was from 38 995 km in 1990 to 74,862 km in 2021, at an annual growth rate of 2.1%, and the motorization rate from 339 in 1990, to 567 cars per 1000 inhabitants in 2021, at an annual growth rate of 1.7% (European Union 2023).

On the other hand, modal shares show a significant variability from city to city (Noussan, 2019). Supported via data from the European Platform on Mobility Management for 2018, the authors, present harmonized data to 25 European cities (Table 17.1). The use of the car varies from 17% in Paris to 85% in Nicosia, and is dominant in 5 of the 25 cities. In only 10 cities, car use represents only less them a third of the daily of the trips. The use of public transport assumes more them 35% in 11 of the 25 cities, and the rest of the city's public transport represents 20% to 35%, with the exception of Nicosia, where the use of public transport is residual.

The use of bikes is relevant in cities like Copenhagen, Amsterdam, Berlin, Helsinki, or Bern, and walking is now an important mode, representing more than one fifth of the journeys in 15 of the 25 cities. The use of bike and walking are now relevant modes in urban mobility, one fifth of the journeys in 19 of the 25 cities.

Nevertheless, these figures should be read with some caution, both because of the different methods are used to collect the information and because they refer to journeys in cities rather than metropolitan areas, where the use of the car is higher and the use of active modes is less important.

17.3 Portugal Car Dependence

The use of car for daily mobility followed the European trend, nevertheless with a significant lag. In the late 1980s the use of the car for commuting accounted for just over 20%, but in 2021 that figure was 67%. Related to this evolution we must point the household's motorisation and the provision of road infrastructure.

Table 17.1 Modal split, 2018 (%)

City	Car	Public transport	Bike	Walk
Amsterdam	38	21	22	20
Athens	53	37	2	8
Belgrade	25	49	1	25
Berlin	30	26	13	30
Bem	31	31	11	26
Brussels	44	28	3	25
Budapest	34	45	2	19
Copenhagen	32	21	30	17
Dublin	57	21	7	16
Helsinki	23	34	11	32
Lisbon	48	35	1	16
Madrid	28	42	0	29
Nicosia	85	2	1	12
Oslo	36	26	5	32
Paris	17	33	3	47
Riga	45	34	2	19
Rome	65	27	1	7
Sofia	51	32	3	14
Stockholm	46	35	1	18
Tallin	25	40	4	30
Vienna	26	40	7	26
Vilnius	37	54	1	36
Warszawa	24	54	1	22
Zagreb	37	37	1	25

Source Noussan (2019)

In the 1970s the Portuguese motorization rate was as low that there were 49 cars per 1000 inhabitants, but quickly increased to 129, ten years later, and converged to the European Union average along the following years, namely after 2000s. In 2021 this rate was 544 cars per 1000 inhabitants, 96% of the European Union average (European Union 2023). Concerning motorway network, in 1970 the length was only 66 km, increased to 132 km during the 1970s and, after a slow growth in the 1990s, the network expanded to more than 3000 km in 2013.

Back to the 1940s, the Lisbon's General Urbanisation and Expansion Plan (1948) defined a zoning plan and the development of a radio-concentric street network from city centre. At the same time, the General Urbanisation Plan for the City of Porto (1940–1942), followed by the Regulatory Plan for the City of Porto (1947–1952), defined also a zoning plan and the development of a street network supported in two main axes, one North/South and another East/West. Despite the low rate of motorisation of the country, even in cities like Lisbon and Porto, those planned main axes were developed over the following decades. The Portuguese city was getting prepared to welcome the car.

Regarding the public transport services, between the end of the nineteenth century and the beginning of the twentieth century, four tram networks were installed, in Braga, Porto, Coimbra and Lisbon, replacing the preceding horsecar services. But, in the 1963 and 1979, the trams disappeared in Braga and Coimbra and during the late 1960s and 1970s the networks of Lisbon and Porto began to reduce its length, leading almost to its end (Marques da Costa 2011). Today 3 lines remain in Porto and 6 in Lisbon. The inauguration of the metro system, in December of 1959, will have contributed to the removal of services from some tram lines, but, as in the rest of the European cities, the tram showed its inability to compete for street space with the cars.

The evolution of the modal split of commuting journeys for Portugal shows the growing importance of the car for these journeys. In 1991 the car accounted only around 22% of the journeys, doubled its share in 2011 and in 2021 reached more than 67%. If we consider only the cities of Lisbon and Porto, the use of car to commuting falls to 51% for the first city and 55% for the second in 2021. The use of public transport for commuting that was of 27% in 1991 dropped to 15%, at the national level, and around 50% in 1991 for Lisbon, and 39% for Porto, fell to 28% and 22% in 2021 for Lisbon and Porto, respectively.

The evolution of the last thirty years shows that car dependency became evident in Portugal. The effects of the urban sprawl and dispersion of economic activities and the inability of public transport to deal to new travel demands, led to the decline of public transport and the increase of the car use for daily journeys.

17.4 Changing the Trend

Changing the situation of car dependency in Portugal is a challenge that public authorities have been facing. As has been said, the origin of the problem doesn't rely only on the individual behaviour, but is also structural in nature, requiring an integrated and coherent intervention, not only in the transport sector, but also in land use planning.

In recent years we have seen different interventions aimed at reducing the importance of the car in our daily lives, with the intention of reducing the negative impacts associated with its use.

These interventions can be divided into four groups: those associated with the development of planning instruments; those related to the provision of public transport; the actions in the field of mobility and public space; and incentives to decarbonise mobility.

The main planning instruments are the Sustainable Urban Mobility Plans (SUMPs). They are comprehensive sectoral plans that aim to respond to the challenges of urban mobility and, simultaneously, to environmental, social and economic challenges. They have as main objectives guaranteeing all citizens mobility options that allow access to destinations and services, improve road safety, reduce air and noise pollution, greenhouse gas (GHG) emissions and energy consumption, improving the efficiency and cost-effectiveness of transporting people and goods, and contributing to improving the attractiveness and quality of the urban environment and urban design for the benefits of citizens, the economy and society as a whole. SUMP is a result of the communication of the European Commission in 2013, where a new approach is proposed to urban mobility planning in order to break out of past approaches and develop strategies and stimulate a shift towards cleaner and more sustainable transport modes, such as walking, cycling, public transport, and new patterns for car use and ownership (European Commission 2013). In the Portuguese legal framework this instrument must be elaborated by the municipalities, municipal communities, metropolitan areas or the functional urban nuclei's, and is compulsory by the Law no 98/2021, 31st of December.

Lisbon's Sustainable Urban Mobility Plan is an example and is focused on reducing car usage and promoting sustainable transportation options. The plan includes initiatives such as expanding the public transportation network, improving cycling infrastructure, pedestrianizing certain areas, and implementing a bike-sharing program. Many other municipalities in Portugal have developed their own sustainable mobility plans to reduce car usage and promote sustainable transportation options at the local level. These plans often include measures such as enhancing public transportation, improving pedestrian and cycling infrastructure, and implementing traffic calming measures.

In the field of public transport, we can consider three types of intervention: in the legislative field, in the provision of public transport services and in supporting the reduction of emissions associated with public transport.

The first step, on the level of the legal framework, was the approval of the Legal Framework for the Public Passenger Transport Service by the Law no 52/2015, 9th of June. With this diploma, started a deep transformation concerning the planning and management of public transport service and the changing of the legal framework of the public transport service market organisation.

With this, local and regional institution became transport authorities, that are a public authority, or group of authorities, with powers and competences with regard to the organisation, operation, allocation, investment, financing and supervision of public passenger transport services, as well as the determination of public service obligations and tariffs in a given geographical area at the local, regional or national level, or any public body vested with such powers and competences by that authority. The transport authorities are responsible for defining the strategic objectives of the mobility system, planning, organising, operating, allocating, supervising, investing in, financing, publicising and developing public passenger transport services by road, inland waterway, rail and other guided systems (Lei 52/2015).

The other main legal document was the Dispatch no 1334-A/2019, that implements the Public transport fare reduction support programme. This programme allowed a very significant reduction in the cost of public transport passes that was associated with the reformulation of the fare system, namely in the metropolitan areas of Lisbon and Porto, had a significant impact in the increase of the public transport ridership, only mitigated by the crisis associated with the Covid-19 epidemic. On the other hand, the fare reform had a positive effect on commuter accessibility levels, when interpreted as the reduction of generalized costs, as a study carried out in 2023 shows (Silver et al. 2023).

In the field of public transport provision, we can point several transformations that have or will have a positive effect on public transport attractiveness. Back on 1995, a new tram (LRT) was introduced in Lisbon, and it was a first step to the development of a new network connecting the outer ring of the city, which, however, is still in the design phase. In the inner city, the metro network is in expansion, expanding the areas served. In Porto, the introduction of the metro (LRT) system became a success after its inauguration in December 2002. New extensions are in construction and in the project phase, connecting the city to the surrounding agglomerations. On the other hand, in the inner city a new BRT line was introduced. At the same time, in Coimbra a BRT system was introduced, connecting the inner city to surrounding municipalities. These are some examples of the recent transformation of the public transport supply in Portuguese urban areas.

The fleet renewal has been another of the major recent transformations as a result of the launch of public service contracts for passenger transport within the competences of the new transport authorities. More energy-efficient vehicles with lower emissions, the use of less polluting fuels or the electrification of the buses are other transformations on the field of public transport provision and could attract new users.

At the local level, several actions have been taken. The implementation of Low Emissions Zones (LEZ), the definition of Limited Traffic Zones, Low Speed Zones (30 km/h Zones), bike share systems and the extension of bike lanes and requalification of public space.

Lisbon is the only Portuguese city that had implemented a Low Emission Zone until now. As result of the air quality regulations, the municipal authority established the first zone in July 2011. Other kind of implementation was the Limited Traffic Zones, a zone where the access is only allowed in certain periods of time and for certain types of uses. They are present in many Portuguese cities, namely Chaves, Coimbra, Gaia, Guimarães, Lisbon, Monção, Porto, Santa Maria da Feira or Seixal. These measures restrict car access to certain areas of the city, particularly historic areas, preserving them for use by cyclists and pedestrians and create more livable spaces.

The transition for the use of bikes depends on the access to a bike and the availability of safe cycling infrastructure. Today a total of more of 5300 km of bike lanes are available to cycling in Portugal.[1] Many municipalities have a network and bikes parking facilities, particularly in the main urban areas. On the other hand, bike-sharing programmes are available in several Portuguese cities like Aveiro, a pioneer in making these programmes available, or Lisbon and Porto, among other examples. These programs provide residents and visitors with access to bicycles for short-term use, promoting cycling as a viable mode of transportation.

---

[1] www.ciclovias.pt.

At the same time that we observe the expansion of bike lane in the city, in more central areas, that development is associated to the requalification of public space, reshaping it for not only the use by bikers, but also for pedestrians.

In terms of decarbonization, Portugal offers various incentives to promote the adoption of electric vehicles, among public institutions, but also to private enterprises and individuals. These include tax benefits, grants for purchasing electric vehicles, and the development of a comprehensive charging infrastructure across the country.

17.5 Conclusions

Despite the initial congestion problems experienced in the 1970s in Portuguese urban areas, that only became more evident during the 1990s, when the rate of motorisation highly increased and the provision of road infrastructure had also a notable growth.

The dependence of car for daily journeys was rapid during the 1990s and bypassed half of the total of that journeys by the 2000s. Today we still have a dominance, even in the areas where the density and quality of public transport systems are higher.

In face of this situation, after the 2010s, the investment in public transportation improved and expands public transportation networks providing residents with an efficient and sustainable alternative to driving.

At the same time, we also assist to the increase of active mobility infrastructure and interventions to control the use of car in inner cities and promote more livable spaces.

But not only the investment in new infrastructures, in terms of legal and technical framework, the redesign was a major sectorial transformation, changing framework that is not suited to current conditions.

Planning mobility must be a process that must be integrated and not contradictory. Integrated at the different transport modes, including the active ones, and at land use and urban planning, and not contradictory, avoiding that planning actions on the land use and urban field have a negative impact on transport planning, and vice-versa (cf. Nunes da Silva 2004).

Let's hope that there will be an alignment of wills that will make possible to announce in the near future the end of the car city in Portugal.

References

Bloom ND (2023) The great American transit disaster. A century of austerity, auto-centric planning, and white flight. The University Chicago Press, Chicago

Buchanan C (1963) Traffic in towns. a study of the long-term problems of traffic in urban areas. H. M. Stationery Office, London

Despacho no 1234-A/2019, 4th of February. Public transport fare reduction support programme

Commission E (2013) COM (2013) together towards competitive and resource-efficient urban mobility. EC, Brussels

European Union (2023) EU transport in figures. Statistical pocketbook 2023. European Union, Luxembourg

Gunn S (2011) The Buchanan report, environment and the problem of traffic in 1960s Britain. Twentieth Century Br Hist 22(4):521–542. https://doi.org/10.1093/tcbh/hwq063

Lei no 52/2015, 9th of June. Approves the legal framework for the public passenger transport service and repeals law no. 1/2009, of 5 January, and the automobile transport regulations (Decree no. 37272, of 31 December 1948)

Marques da Costa N (2011) Mobilidade e transportes em Áreas urbanas. O caso da Área metropolitana de Lisboa. MOPT/CEG UL, Lisboa

Mattioli G, Anable J, Vrotsou K (2016) Car dependent practices: findings from a sequence pattern mining study of UK time use data. Transp Res Part A 89:56–72. https://doi.org/10.1016/j.tra.2016.04.010

Noussan M (2019) Effects of the digital transition in passenger transport—an analysis of energy consumption scenarios in Europe. Fondazione Eni Enrico Mattei, Milan

Nunes da Silva F (2004) Políticas urbanas para uma mobilidade sustentável: do diagnóstico às propostas. GeoInova 10:157–174

Silver K, Lopes A, Vale D, Marques da Costa N (2023) The inequality effects of public transport fare: the case of Lisbon's fare reform. J Transp Geogr 112. https://doi.org/10.1016/j.jtrangeo.2023.103685

Sloman L (2006) Car sick. Solutions for our car-addicted culture. Green Books, Cambridge

Chapter 18
The End of the Car City in Spain

Miguel Pazos-Otón

Abstract In Spain, the debates on the new mobility policies are concentrated in the main cities. Madrid and Barcelona, plagued by high levels of pollution, have been forced to implement measures to reduce travel by private car and to prohibit the most polluting vehicles from circulating. In Madrid, the rate of change is the subject of a bitter political debate, while the Barcelona City Council is committed to reducing car traffic in the centre of the consolidated city, with various interventions, among which the idea has become popular of the superblocks. Other cities, such as Bilbao, Seville, Zaragoza or Alicante have openly opted for the introduction of the tram and the consequent exclusion of the car from certain streets and avenues. In Valencia, the concept of new mobility acquires a much more comprehensive meaning, with the combination of various transport systems and the reinforcement of the routes reserved for the citizen who walks or goes by bicycle. Traditional urban centres that benefit from different level of comprehensive pedestrianization measures have become widespread. Even in some cases, such as Pontevedra, 70% of the urban area has been freed from cars, which, if they have to go through it, respect the priority of pedestrians. In short, the chapter narrates the different speeds that the new paradigm of mobility without a private car presents in Spain.

Keywords Mobility · Car · City · Spain

18.1 Introduction

The automobile is arguably the principal cultural symbol of contemporary life in the global north. The mass production of cars and their accessibility to the middle classes marked a significant shift not only in the daily lives of citizens but also in their lifestyles. Furthermore, the widespread adoption of motorised mobility and the ability to cover substantial distances facilitated the expansion of cities through low-density models. The concept of *suburbia* refers to this urban development that began

---

M. Pazos-Otón (✉)
Department of Geography, Universidade de Santiago de Compostela, Santiago, Spain
e-mail: miguel.pazos.oton@usc.es

R. C. Lois-González and J. A. Rio Fernandes (eds.), *Urban Change in the Iberian Peninsula*, The Urban Book Series, https://doi.org/10.1007/978-3-031-59679-7_18

expanding at those times on the outskirts with low population density. In Spanish, "suburbio" carries a negative connotation; however, in English, "suburb" embodies the American dream: living in an individual, spacious house with a garden and garage, in a residential neighbourhood on the outskirts of the city.

Although Spanish cities, like their European counterparts, have historically been characterised by compactness, the expansion of the North American model and lifestyle became more prevalent throughout the twentieth century. The "car city" began developing in the United States in the early twentieth century, allowing citizens to reside several kilometers from city centres and commute daily using highways, freeways, parkways, and other high-capacity infrastructures (Hall 2014). The transfer of this model to Europe would become widespread after World War II and reached Spain during the developing years of the 1960s, becoming distinctly evident in major cities from the 1980s onwards (Bellet and Olazábal 2019).

This transfer, often almost automatically, of the American model to Europe poses significant challenges. European cities, and Spanish ones in particular, have a very different urban structure from American cities. In Europe, the new metropolitan areas intersected by highways, ring roads, and orbitals must coexist and integrate with the traditional city. This classical city, compact, with high densities, and multifunctional (a proximity city, favourable for pedestrian movements), suddenly finds itself surrounded by a large peri-urban structure. This structure is formed by high-capacity roads, which become the new arteries of the large metropolitan entity.

In Spain, this process occurred very clearly in Madrid and Barcelona, and to a lesser extent, in other urban areas of the country, including both large and medium-sized cities, or even those of a more modest size. This is exemplified by the case of the urban area of Santiago de Compostela, consisting of a compact city with fewer than 100,000 inhabitants and an added periphery (within a radius of 10 kms) of around 70,000 inhabitants. In many instances, such as in Santiago de Compostela, urban areas did not grow solely due to a preference for residing in low-density areas. Additionally, the high price of land and housing in the central city municipalities was a key factor in explaining the increased availability of more affordable housing in neighbouring municipalities, often in the form of apartment buildings or smaller properties. Thus, we encounter an urban entity of 170,000 inhabitants, encompassed within a circle of over 300 square kilometres. Clearly, this uncontrolled territorial model implies disproportionate and unsustainable mobility, especially when the public transportation offering is weak and of low quality.

This text will focus on these new complex urban entities, consisting of two distinct parts (a compact centre and a low-density peri-urban expansion). The link between these functional spaces is precisely defined by the existence of daily urban mobility systems. These complex urban spaces grapple with longstanding issues that are challenging to resolve. In a context of disruptive mobility (Sprei 2018) due to the technological revolution and the climate emergency (which we will analyse in detail), Spanish urban spaces strive to provide a sustainable solution, albeit with varying degrees of success. Many implemented measures fail to achieve the desired outcomes due to significant constraints arising from rigid governance structures, as we will explore further.

18.2 A New Mobility is It Truly New?

We currently find ourselves in a period of change concerning the meaning of various key concepts: transportation, mobility, and accessibility. In contrast to the widespread use of the term "transportation," which is associated with something rigid, heavy, expensive, and polluting, the term "mobility" (Sheller 2011) is increasingly employed. Many local traffic police officers in cities are now referred to as "mobility agents." Numerous regional government transportation departments have transformed into "Directorate of Mobility." There is a general discourse on the need for citizens to enjoy sustainable mobility. We are facing a new paradigm: the mobility paradigm (Cresswell 2006; Sheller and Urry 2006).

However, this substitution of "transportation" with "mobility" oversimplifies the complex global issue of movements in cities and metropolitan areas. Authors such as Whitelegg argue that we should stop thinking about "mobility" and focus on "accessibility" and try to improve it (Whitelegg 2015). Whitelegg understands that citizens' quality of life does not depend on mobility, i.e., moving more and for longer periods, but on accessibility. Accessibility is determined by land uses, which must be defined through urban planning, and relies on a proper mix of residential areas with job centres, commercial sectors, public and private services, etc. The accessible city is the opposite of the *zoning* city, which encourages car use (Pozueta 2002).

Increasing our dependence on mobility is bad news, considering that the majority of today's transportation modal share depends on fossil fuels. In the Anthropocene era (Latour 2019), this has extremely negative implications for the planet's future, accelerating global warming and triggering climate change alarms.

Although a critical awareness emerged early in Spain around the 1980s, it was limited to a restricted context, such as the ecological movement. It was not until the turn of the century that society as a whole embraced the consciousness of the need to rethink global transportation policies. The commitment to decarbonise transportation and opt for more sustainable mobility was a late process in Spain compared to the rest of Western Europe.

Indeed, critical awareness of this predatory mobility model began in the 1970s, or specifically after the 1973 oil crisis. The Netherlands pioneered policies to reduce dependence on private cars, promoting bicycle use, and humanising public spaces. Today, we strive to achieve an impossible balancing act, reconciling the demand for more mobility with the need for a rapid transition from the old to the new mobility system based on clean and renewable energy.

In Spanish cities, most problems arise not from a lack of mobility but rather from its excess. The origin can be traced to the lack of urban planning in the context of an authoritarian political regime (1939–1975) and, in any case, the deficiencies in existing planning. Simultaneously, Spain's historical lag compared to other European countries led to civic movements in favour of pedestrians, bicycles, and public spaces, which started in 1973 in many European countries, particularly led by the Netherlands, taking several decades to become generalised (Hall and Hesse 2013).

Currently, within the European Union and after several decades of integration and collaboration with other urban spaces on the continent, most Spanish cities have incorporated sustainable mobility policies into their overall urban policies. However, in many cases, there is still no political consensus on the mobility model. Mobility policies, closely linked to public space policies, still depend heavily on the municipal governments' political party. In other European countries, this is not the case, as all political parties generally agree on a minimum common denominator regarding mobility sustainability. However, in Spain, the situation is still far from being so.

Perhaps the clearest and most topical example of what we are asserting is the case of "Madrid Central." This term refers to a set of planned and coherent actions implemented in the Spanish capital by Mayoress Manuela Carmena during the years 2016–19. As in other major European cities, the project involved restricting ordinary traffic in a delimited area within the city centre. The project, driven by the left-wing party *Ahora Madrid*, faced opposition from the right-wing Popular Party, which promised to eliminate Madrid Central if it came to power in the 2019 election campaign. With the change in municipal government, the new Popular Party mayor, José Luis Martínez Almeida, attempted to dismantle Madrid Central, sparking a legal battle where a significant portion of the measures taken by the previous municipal government were overturned. As of December 31, 2023, the area classified as Madrid Central covered a total of 480 hectares (Fig. 18.1).

On January 1, 2024, the City Council of Madrid, whose mayor was re-elected in the elections on May 28, 2023, announced a €200 fine for non-resident cars in Madrid which lacked the corresponding environmental badge. All of this is part of a timid shift towards implementing greener policies in the capital.

It is essential to highlight that the European Union clearly promotes the decarbonisation of transportation and advocates for softer mobilities (pedestrian, bicycle, etc.). However, in the case of Madrid, the statements of the president of the Community

Fig. 18.1 Information sign for mobility regulation in Madrid central. *Source* Miguel Pazos Otón

of Madrid (from the same political party as the mayor of the capital) have attracted attention. She stated that "traffic jams are part of Madrid's personality" and that she "doesn't know anyone who has died due to pollution." In the midst of a climate emergency and with a growing awareness among citizens, it is challenging to understand how there is no common and shared vision on mobility policies among Spanish political leaders.

Until there is consensus among all stakeholders (public and private) on the mobility policies to follow, it will not be possible to implement coherent policies. The Sustainable Urban Mobility Plans (known as PMUS), promoted by the State administration from the 2000s, always begin with a "Mobility Pact" that all stakeholders must subscribe to, reaching a consensus. It is a document on basics that must include points of agreement such as transportation decarbonisation, the promotion of collective transport, encouragement of soft mobility, an increase in pedestrian spaces, etc. The cities' mobility model must be based on what is agreed upon in these "mobility pacts." However, we are currently witnessing a moment of disruption that poses very complex challenges due to the magnitude and speed of the transformations.

18.3 Urban Mobility: Current Disruptions

The concept of disruption refers to a qualitative leap in the way things are done, leading to a different approach to problem-solving. Historically, innovations such as the printing press and the steam engine caused crucial disruptions with significant consequences in the history of society. Presently, the widespread use of the internet among the general population, smart mobile devices, and the hyper-technification of individually consumed goods are clearly causing disruption in transportation and mobility. As is often the case, changes tend to precede legal provisions and regulations, resulting in numerous issues due to the absence of clear and comprehensible standards for citizens, managers, and political leaders in cities.

Moreover, the intrusion of technology in the planning, management, dissemination, and information of mobility and transportation presents significant challenges. On the one hand, the speed at which changes occur hinders the development of a sufficiently consolidated and enduring legislative framework. The existence of different competency levels in Spain (State, autonomous regions, metropolitan areas, municipalities) further complicates the successful management of challenges arising from the relationship between technological changes and legislation. On the other hand, stemming from the rapid evolution of technology, public administrations often lack the sufficient capacity to develop their own software platforms, increasing dependence on large private companies and cloud computing services. Currently, the debate on the need to make data accessible to the entire population (open data) is highly relevant. Effective planning and management of mobility cannot exist if public administrations do not develop their own data and software, a task that seems very challenging today.

Let us, therefore, examine the main challenges that cities face today in relation to disruptive trends in mobility.

18.3.1 The Revolution in the Taxi Industry

The significant changes affecting the rental of public passenger transport vehicles with drivers, commonly known as taxis (Fig. 18.2), help us perceive the magnitude of the shifts we are referring to in cities worldwide, and more specifically, in Spanish cities.

The widespread use of smartphones, constantly connected to the internet, and the popularisation of so-called apps (short for applications) for smartphones have caused a true revolution. Applications such as Uber or Cabify facilitate the connection of any user registered on these platforms with hundreds of associated drivers. The continuous and real-time geolocation of vehicles and the user-friendly graphic interface enable on-the-go hiring of a "ride" using the internet and a simple smartphone. This is a disruptive technology because, in its early stages, it developed outside of any legislation and, therefore, outside of the tax and legal system that regulates the transportation sector in Spain.

The generalisation of the so-called "platform economy" often results in the enrichment of a series of technological companies that become mere intermediaries between the driver (service provider) and the consumer. This is the reason why there has been

Fig. 18.2 Taxis in Barcelona. *Author* Miguel Pazos Otón

strong opposition from conventional taxi drivers to the implementation of the so-called VTC (vehicle for hire) from the very beginning in Spain. The conflict has manifested over the years through protests, strikes, and significant mobilisations by the affected collective, especially in Madrid and Barcelona, demanding effective regulation of VTCs and the existence of a transport inspection that acts proactively to limit abuses. Currently, different autonomous communities have developed their legislative frameworks attempting to restrict the expansion of VTCs and promote coexistence between traditional taxis and "platform taxis."

In any case, in many medium-sized and small Spanish cities, there is an issue stemming from the insufficient availability of licenses to provide a quality service. This problem is particularly severe during night time hours, given that the profit margins for conventional taxis are very slim, if not outright non-existent.

At the core of the problem is also the significant rigidity inherent in taxi regulation, based on a guild-oriented and highly protectionist organisation. In many medium-sized and small cities, taxi services are deficient, and rather than a public service, we must refer to it as working with the so-called "captive demand." Nevertheless, citizens should be able to secure a taxi at any time due to potential medical emergencies or other urgent situations.

Municipal administrations frequently grapple with the difficulty of negotiating the issuance of new licenses with the taxi sector, balancing the pressure exerted daily by citizens and the tourism sector, which anxiously views the challenges in accessing taxi services. The fact that the provision of Vehicle for Hire (VTC) services is only permitted in large cities for intraurban mobility does not facilitate the resolution of this deadlock.

18.3.2 Electric Cars, Shared Mobility

Another major change occurring in Spanish cities relates to new ways of organising and managing individual car mobility. For many decades, the individual vehicle (referred to as "tourism"), under individual ownership, powered by fossil fuel derivatives, has been the main player in the modal distribution of Spanish cities due to its prestige, comfort, and status. In reality, it still is for many people. The so-called "car culture" is an undeniable fact in Southern European countries, while in Nordic countries, the car as an object of desire has lost almost all its appeal for decades.

We are facing a new scenario in which environmental awareness, and above all, the climate emergency, set a new agenda for the automobile. Concerns about the contribution of conventional cars to environmental pollution are becoming increasingly important. Citizen awareness of the need to reduce toxic gas emissions grows each year, and the transportation sector is seen as one of the main culprits of climate change.

Authors like Dennis and Urry (2009) talk about the "end of the car" as we know it and question whether reaching peak oil will be necessary to witness the end of the car model (individual, diesel, and owned) as we also know it (Fariña 2015). While admitting the difficulty of predicting the near future, it is true that we are witnessing changes, such as the gradual introduction of electric vehicles in cities. The autonomy of these vehicles is still limited, but many cities have already implemented measures to favour Zero Emission Vehicles (ZEV). Others directly prohibit access to certain urban areas for vehicles based on oil-dependent energies. The debate on whether electric cars are sustainable or not continues, and, along with hybrid vehicles, they are part of the set of alternatives to gasoline or diesel cars.

Another significant change we observe is the increase in the use of shared vehicles. Driven by public administrations or initiated by citizens themselves, more and more users have decided to share their vehicle or someone else's in their recurring or occasional travels. Again, we must highlight the catalytic role of various smartphone applications, among which Bla Bla Car stands out. This platform connects users who want to share rides, proving to be a powerful tool for people without a car or a driver's license, allowing citizens to organise and optimise their travels through unconventional channels.

18.3.3 Personal Mobility Vehicles

The third of the major disruptive changes to be analysed here is driven by the widespread adoption of what is already referred to as Personal Mobility Vehicles (PMV). These are compact mobility devices with electric propulsion that reinvent traditional means of transportation used for short distances, or even those considered as toys for children and young people. We are referring to devices such as rollerblades, scooters, skateboards, unicycles, or similar, which are experiencing a new golden age thanks to the introduction of small electric motors (Fig. 18.3). These devices provide users with greater mobility and minimise the effort required, especially on inclines or during long journeys.

Fig. 18.3 Shared scooter provided by the LIME platform in Madrid. *Author* Miguel Pazos Otón

Given that the size of these Personal Mobility Vehicles (VMP) is very small and they are easily transportable, they are becoming increasingly popular in Spanish cities. In addition to their low cost and the fact that no license or permit is required for their use, more and more people are using electric scooters for their daily commutes, achieving good accessibility through streets full of traffic and pedestrian spaces. These are very versatile vehicles that navigate smoothly through traffic jams, allowing for highly flexible mobility, as they can travel through practically any space. However, these mobility practices pose dangers to pedestrians, drivers, and are generally irregular, compromising the proper planning and management of urban mobility.

It is precisely this fact that is causing increasing protests in Spanish cities. The indiscriminate use of VMP on streets, roads of all kinds, and sidewalks violates the most basic traffic rules. It can be said that VMPs, in many cases, behave like cars and in others like pedestrians, adopting the role that suits them best at each moment.

As mentioned earlier, citizens and their mobile practices are ahead of legislation, which is still unable to provide an adequate response to the growing role of VMPs. Since the regulation of VMPs is a municipal competence, there is, in practice, a great tolerance for the infringement of the rules. Certainly, it is difficult for a VMP user to be aware of all the legislation in the municipalities through which they may circulate. However, this situation should lead to a change in the regulatory legislative framework, bringing it to the jurisdiction of the autonomous communities, at a minimum, or to higher levels.

18.4 The End of the Car-Centric City in Spain?

In the current context of climate alarm and energy crisis, it seems more necessary than ever to reflect on the relevance of moving towards the end of the car-centric city in Spain, as in other countries (Pazos Otón 2022). However, there are significant obstacles hindering a shift in mindset and paradigm.

In the entire situation we are outlining, it is striking that, despite mobility and transportation being one of the sectors most affected by disruption, many traditional problems persist, for which finding a solution does not seem easy. In certain cases, these are problems that technology and disruption cannot address. In others, the issue arises because too much attention and investment are directed towards secondary aspects or ones that are not relevant at all to improving mobility and urban quality of life.

This confusion in priorities is often related to policies of extravagance or mere "showcasing" and the difficulty of tackling genuine structural changes in relation to land use and mobility (Lois et al. 2013). Even after more than one-fifth of the twenty-first century has passed, integrating mobility into urban planning in Spanish cities remains challenging. It can be said that there is still a lack of acknowledgment that mobility is not only a response to the need for movement but can and should become a tool for city management, with decisive implications for residential, commercial, tourist functions, and many others (Lopez et al. 2019).

One of the main problems faced by Spanish urban and metropolitan areas is related to the governance of transport systems and, ultimately, mobility. In general, we encounter two situations. On the one hand, large cities, and metropolitan areas, which have adequate tools for infrastructure and transport system planning, and therefore for mobility management. The more complex metropolitan areas in Spain, such as Madrid, Barcelona, Bilbao, or even the so-called "Astur City," have supra-municipal transport consortia that operate as the sole transport authority and allow overcoming the limited municipal framework.

In these metropolitan areas, we observe the almost efficient management of mobility, with integrated fare policies based on zoning, intermodality bonuses, sufficient and integrated information on all means of transportation, and, ultimately, a joint operation that facilitates access to public and collective transport for the entire population. In addition, the existence of a minimum population threshold allows for greater modal diversity (buses, metro, trams, commuter trains, etc.), resulting in a satisfactory offering for the user, with the possibility to choose between several suitable travel options in many cases.

The creation of metropolitan transport consortia in large Spanish cities has been the result of the application of pragmatic policies in the face of difficulties in materialising metropolitan areas as political-administrative entities with their own competencies. The strong influence of municipal politics in Spain and the existence of significant interests linked to local politics have acted as brakes on the creation of a new level of administration between the autonomous community (or the province, in the case of Madrid) and the municipality. When we refer to pragmatism, we want to emphasise that transport and mobility are considered important enough to warrant a sectoral policy where differences and disagreements hindering a comprehensive metropolitan agreement are set aside. It is evident that a metropolitan agglomeration of more than one million inhabitants cannot function without a genuine integrated transport and mobility policy.

On the other hand, however, we encounter urban and metropolitan areas of a more modest size (medium and small) that lack adequate instruments for the management of transport and mobility. In these cases, the jurisdiction of public transport operators within the municipal term where the central city is located is usually municipal (Fig. 18.4).

However, the competencies of the transport systems integrating the rest of the urban area are often under the responsibility of the corresponding regional government, which usually operates bus lines under a concession system. The rigidity of this system and the lack of dialogue with the authorities of the central city often lead to coordination issues and uncertainty. Tariff policies, intermodality, and other advantages of territorial cooperation sometimes dissolve within attempts at specific agreements that try to alleviate specific mismatches. In the absence of transport consortia, genuine useful tools, makeshift solutions like the "Transporte Metropolitano de Galicia," voluntarily subscribed to by mayors and, therefore, uneven in its territorial implementation due often to political factors, are pursued.

Fig. 18.4 Urban bus in Santiago de Compostela. *Author* Miguel Pazos Otón

Considering urban and metropolitan areas as unitary entities is essential for transport planning and management. Developing Sustainable Urban Mobility Plans (PMUS) considering only the municipality of the central city almost always yields limited and unrepresentative results (Vega Pindado 2016). The mobility issue is not only due to movements within cities but, above all, to the inputs and outputs of the municipal framework every day, with special attention to the saturation of parking spaces in central cities. Only reducing car usage can reclaim space for pedestrians and make cities more liveable (Fig. 18.5).

The governance difficulties of transportation and mobility, in turn, explain the limitations in managing key infrastructures in the daily lives of citizens. In Spanish cities, there are four fundamental competency levels that oversee different transportation infrastructures and mobility systems. Firstly, the State (Government of Spain) manages airports, major railway stations, and ports. Similarly, it plans, constructs, and manages many of the high-capacity roads that traverse urban and metropolitan areas (Fig. 18.6).

These roads are managed based on their "general interest of the State" nature but are used daily by thousands of commuter vehicles within their urban or metropolitan area. Therefore, their management does not follow the subsidiarity principle necessary for greater effectiveness.

Secondly, the Autonomous Communities, which have their own road and, in some cases, railway networks. The regional road networks overlap with the networks of general interest of the State and emerge as a complement to that network, relieving congestion in access corridors or creating alternatives where necessary. The regional administrations also manage bus stations and some railway stations (Fig. 18.7).

Third and fourth are the Provincial Governments (provincial governments) and the City Councils, with responsibilities for the rest of the road network. They manage denser networks, with a clear dendritic and capillary character, reaching the most peripheral areas of the territory.

Fig. 18.5 Urban bus stop next to the aqueduct of Segovia. *Author* Miguel Pazos Otón

Fig. 18.6 A-8 (Cantabrian highway) passing through Vilalba (Lugo). *Author* Miguel Pazos Otón

If we consider that each of these levels of administration may have a government of a different political party or parties, we can understand the difficulty of achieving true planning and management of transportation and mobility. In reality, the State has its own planning for transportation and mobility (materialised over the years through plans like PDI, PEIT, PITVI, etc.), the Autonomous Communities have their own, and the provincial governments and municipalities also have theirs.

Fig. 18.7 Commemorative plaque for the inauguration of the intermodal station of Santiago de Compostela (May 2021). *Author* Miguel Pazos Otón

The difficulty in reaching agreements that benefit citizens in their daily lives is evident in the struggle to achieve toll exemptions on stretches of toll highways that play a significant role in metropolitan areas. One example is the lengthy process to exempt metropolitan users from tolls in the surroundings of A Coruña and Vigo. The Atlantic Highway (AP-9) plays a crucial role in structuring the Ártabro Gulf and the Rías Baixas, and metropolitan mobilities are inconceivable without it. The mix of urban and metropolitan traffic with intercity traffic also led the State to finance the expansion of the AP-9 bypass section in Santiago de Compostela due to its strategic importance. In this case, during rush hours, the traffic intensity caused by metropolitan commuters would congest the main route connecting the north and south of Galicia, disrupting mobility between the Ártabro Gulf and the Rías Baixas. Currently, the AP-9 is the only viable alternative for traveling along the Galician Atlantic Urban Axis and for many key trips in Galician metropolitan areas.

Another example illustrating the limitations of this regulatory framework is found in the new intermodal stations built and under construction due to the arrival of high-speed rail to Spanish cities. The need to reach agreements and sign agreements between different levels of administration, as well as the lack of agreement on the financial participation of each, has delayed the construction of many intermodal stations. The crisis that began in 2008 also played a role, forcing a redefinition of initial projects that relied too much on generating real estate profits derived from the sale of public land around old stations (Bellet and Gutiérrez 2011).

Lastly, the lack of coordination between administrations and electoral considerations leads to economically unviable actions with dubious profitability in the medium and long term. A very interesting case is the proliferation in the last two decades of urban trams in medium and small-sized cities, which soon proved economically unsustainable. Emblematic cases include the Parla Tram (Madrid), which has been mired in a serious crisis and economic problems since 2011, the Vélez-Málaga Tram (inaugurated in 2006 and closed in 2012), and the Jaén Tram (which only operated in a trial period in May 2011). In the latter case, the outbreak and worsening of the 2008 economic crisis prevented the inaugural trip, even though the tracks were built, and the motor units had already been acquired.

In the absence of true coordinated planning on an integrated basis, many municipalities have understood the need to "do something" to favour the reduction of car traffic and increase pedestrian space. The policies of the municipalities of Vitoria-Gasteiz and Barcelona are well-known. In the latter case, the Urban Ecology Agency, led by Salvador Rueda, experimented with the so-called "superblocks," implementing traffic calming measures and reducing regular car circulation by 300%. The model achieved successes but also had its detractors and was attempted to be exported to other Spanish cities with uneven results.

Due to its characteristics and uniqueness, one interesting case is the path taken by the city of Pontevedra in Galicia. The rise to power of the BNG in 1999 marked the beginning of an urban transformation aimed at removing cars from the centre of the compact urban fabric and returning the city to pedestrians. Undoubtedly, the modest demographic size of the city (just over 60,000 inhabitants in the urban area) was a factor that greatly contributed to the success of these policies. However, it is necessary to highlight the municipal government's commitment to a consistent, comprehensive, and determined policy for sustainable mobility. In addition to the successes of municipal policies, the City Council of Pontevedra attached great importance from the outset to the dissemination, awareness, and education of young people in terms of road safety and sustainable mobility. As a result of these efforts, the "Metrominuto" stands out, an infographic based on metro network graphs adapted to demonstrate the feasibility of pedestrian mobility in the city. The "Metrominuto" model has been adapted by many other cities in Spain, Europe, and the world for its simplicity and excellent visual communication.

18.5 Conclusions

Spanish cities are currently undergoing a period of change. They have inherited a generally chaotic situation due to the lack of urban and metropolitan planning and integration for much of the twentieth century. It could be said that it was not until the late twentieth century that the need to change the prevailing mobility model in Spanish cities began to be understood, while in other European countries in our surroundings, the movement started in the early 1970s.

However, despite this awareness, there is still a lack of logical interrelationships between transportation and mobility planning and general planning. For most cities, it is challenging to assume that both planning processes must be carried out simultaneously, and that one conditions the other. Only the case of Vitoria-Gasteiz can be cited as an example of good practices. In this city, the urban expansion to the north (Abetxuko) was accompanied by the introduction of a tram line connecting it to the city centre, integrated into the general planning.

Although the term "mobility" has become a fashionable term and has been incorporated into almost all discourses and political agendas, many of the policies practiced are still, in reality, transportation policies. There continues to be an obsession with the construction of large road infrastructures and little commitment to the railway mode (with the exception of the country's largest cities). It is noteworthy that Spain, being the second country in the world by the number of high-speed railway kilometres, has one of the worst percentages in Europe regarding the railway's share compared to the road in freight transportation. Likewise, there is significant room for improvement in suburban and regional rail networks.

The lack of urban and metropolitan planning and management structures acts as a brake on achieving an alternative model to the current one, based on cars, especially in medium and small cities.

Therefore, we have a dual situation. In large urban agglomerations, there is an acceptable modal diversity, with sufficient planning and management structures and fare policies that encourage citizens to leave their cars in the garage. On the contrary, in smaller cities, the lack of a good public transportation offer leads to a higher use of private cars. This car usage does not reach disproportionate levels because, in general, distances are much shorter, facilitating walking in many cases. But in general, externalities (noise, pollution, lack of public space, etc.) can be as significant or even greater in medium and small cities as in large ones.

The gradual awareness of citizens and authorities about the need to change the model, mainly due to the climate emergency and energy crisis, can be an important factor for change. However, it is necessary to decisively promote a model change by increasing pedestrian routes, promoting the integration of bicycles in cities, and encouraging alternatives to private cars. Spanish cities can draw inspiration from successful models in European cities that have been able to reverse the situation for many decades.

References

Bellet C, Gutiérrez A (2011) Ciudad y ferrocarril en la España del siglo XXI. La integración de la alta velocidad ferroviaria en el medio urbano. BAGE: Boletín de la Asociación de Geógrafos Españoles (55):251–279. ISSN 0212-9426, ISSN-e 2605-3322

Bellet C, Olazabal E (2019) De la ciudad compacta a la ciudad extensa.: procesos de urbanización recientes en áreas urbanas españolas articuladas por ciudades medias. Anales de geografía de la Universidad Complutense 39(1):149–175. ISSN 0211-9803

Cresswell T (2006) On the move. Routledge, London

Dennis K, Urry J (2009) After the car. Wiley

Fariña Tojo J (2015) Cambiar el modelo urbano. Ciudades: Revista del Instituto Universitario de Urbanística de la Universidad de Valladolid, ISSN 1133-6579, No. 18, (Ejemplar dedicado a: La urbanística contra-reformista: administradores, promotores, financieros, propietarios de suelo, políticos, ideólogos, profesionales y francotiradores, contra la ciudad), pp 69–79

Hall P, Hesse M (eds) (2013) Cities, regions and flows. Routledge, Abington
Hall P (2014) Cities of tomorrow. An intellectual history of urban design and urban planning since 1880. Wiley–Blackwell
Latour B (2019) ¿Dónde aterrizar? Taurus, Madrid
Lois González RC, Pazos Otón M, Wolff JP (2013) Le tramway entre politique de transport et outil de réhabilitation urbanistique dans quelques pays européens: Allemagne, Espagne, France et Suisse. Annales de géographie (694):619–643. ISSN 0003–4010
Lopez L, Pazos Otón M, Piñeiro Antelo Á (2019) ¿Existe overtourism en Santiago de Compostela? Contribuciones para un debate ya iniciado. BAGE: Boletín de la Asociación de Geógrafos Españoles (83). ISSN 0212–9426, ISSN-e 2605–3322
Pazos Otón M (2022) Perpetuum mobile. Movilidad en tiempos pandémicos. Tirant lo Blanch, Valencia
Pozueta Echavarri J (2002) Movilidad y planeamiento sostenible: Hacia una consideración inteligente del transporte y la movilidad en el planeamiento y en el diseño urbano. Cuadernos de investigación urbanística (30):1–109. 2000. ISSN 1886–6654
Sheller M, Urry J (2006) The new mobilities paradigm. Environ Plann (A). SAGE J 38(2):207–226
Sheller M (2011) Mobility. Sociopedia. https://doi.org/10.1177/205684601163
Sprei F (2018) Disrupting mobility. Energy Res Soc Sci 37:238–242
Vega Pindado P (2016) Una década de planes de movilidad urbana sostenible en España 2004–2014. Anales de Geografía de la Universidad Complutense 36(20)
Whitelegg J (2015) Mobility. Straw Barnes Press, London

Chapter 19
Active Mobility in Portugal: The Case of Lisbon

Miguel Padeiro

Abstract This chapter explores the role of active mobility, particularly non-motorised transportation such as walking and cycling, in urban policies. Active mobility is recognised for its potential to enhance physical health, reduce psychological stress, improve air quality, and mitigate the burden of chronic diseases. Additionally, it aligns with decarbonisation and energy independence objectives by reducing greenhouse gas emissions and dependency on fossil fuels. The case of Lisbon is examined, with a specific focus on cycling, within the context of its urban challenges. The chapter analyses the promotion of active mobility in Lisbon, highlighting initiatives such as cycle lane construction and the Gira bike-sharing system. While significant progress has been made, questions regarding equity in access and distribution of cycling infrastructure persist. Gender disparities in bike-sharing usage and geographical inequalities in infrastructure quality are explored, along with their implications for socioeconomic disparities. There is a need for comprehensive urban planning and policy measures to address these equity concerns and promote cycling as a viable mode of urban transportation.

Keywords Active mobility · Cycling infrastructure · Urban challenges · Equity in transportation · Sustainable urban planning

19.1 Introduction

Active mobility, understood here as non-motorised transportation (primarily walking and cycling), although this definition is not universally agreed upon (Moura and Kalakou 2021; Zhou et al. 2020), is now widely recognised as a predominant instrument of urban policies. Two categories of associated benefits are highlighted by an increasing number of scientific studies, official reports, and strategic planning documents. On the one hand, active mobility is generally associated with an increase

M. Padeiro (✉)
CEGOT—University of Coimbra, Coimbra, Portugal
e-mail: jmnp@uc.pt

R. C. Lois-González and J. A. Rio Fernandes (eds.), *Urban Change in the Iberian Peninsula*, The Urban Book Series, https://doi.org/10.1007/978-3-031-59679-7_19

in physical activity (Pan et al. 2021; Pucher et al. 2010), a reduction in psychological stress (Lorenz 2018; Ma et al. 2021; Synek and Koenigstorfer 2019), and an improvement in air quality (Johansson et al. 2017). Public health benefits stem from these three advantages: a reduction in comorbidities such as obesity (Brown et al. 2016), cardiovascular diseases (Alessio et al. 2021; Boone-Heinonen et al. 2009), seeking other health improvements, along with a reduction in costs related to disease management (Jarrett et al. 2012). These advantages appear to be more significant than the potential drawbacks linked to using a bicycle, such as exposure to pollution and accident risks (De Hartog et al. 2010; Mueller et al. 2018; Tainio et al. 2016). Where infrastructures reduce these direct exposures, the gains are even more evident (Kondo et al. 2018; Monsere et al. 2012; Parker et al. 2013) and also act as a leverage effect capable of ensuring an increase in modal shift (Buehler and Dill 2016; Dill and Carr 2003; Hong et al. 2020; Hull and O'Holleran 2014; Wardman et al. 2007; Xing et al. 2010). On the other hand, active mobility is also seen as a means to achieve decarbonisation goals, especially through the reduction of greenhouse gas emissions (Bucher et al. 2019; McQueen et al. 2020; Zhang and Mi 2018), and of energy dependence (Bucher et al. 2019; Zhang and Mi 2018).

These reasons explain why an increasing number of cities have developed increasingly ambitious policies promoting active mobility over the past decades. Here, we will focus on the case of Lisbon, and more specifically on the use of bicycles, although it must be acknowledged that the case of pedestrian mobility would deserve further exploration, both from the perspective of the efforts made by the municipality and the obstacles still present.

19.2 The Context

Spanning 100 square kilometers, the capital city of Portugal, Lisbon, was home to 546,000 residents as per the most recent population census in 2021 (INE 2022). This figure shows a slight decrease from the 565,000 inhabitants recorded in 2001 and a minor drop from the 548,000 residents in 2011. Although the population in the city centre of Lisbon is on the decline, the demographic significance of its surrounding metropolitan area (LMA—Lisbon Metropolitan Area) is on the rise. The LMA's population expanded from 2.66 million in 2001 to 2.82 million in 2011 and further increased to 2.87 million by 2021. Currently, the LMA accounts for 28% of Portugal's total population.

Lisbon presents several fundamental characteristics stemming from various historical, institutional, and geographical considerations. Firstly, Lisbon has an irregular topography and is perceived as a hilly city, although 54% of the streets are considered flat (< 3% inclination) while 75% are below a 5% grade, suitable for cycling. Secondly, the growth of Portuguese cities, including Lisbon, began later than in most European countries. This delay in both urban and political development has significant implications, leading to the creation of an expansive and largely unregulated urban landscape. Since the 1950s, rapid demographic growth has led to the emergence of disjointed urban areas, forming a patchwork of functionally isolated territories. This resulted in a weak urban fabric and disorganisation, challenging the LMA to refuge functional connections and spatial continuity. Lisbon faces considerable hurdles in addressing urban sprawl and reliance on cars, which have profoundly impacted its urban planning and transportation systems. The dominance of automobiles is a major hurdle for urban development, particularly given the limited and underdeveloped public transport alternatives.

Portugal, with a surge in highway construction since the late 1980s, ranks fourth in the OECD in terms of highway length per capita and second in the European Union, behind Luxemburg. The LMA is particularly notable for having the highest highway density in Europe at the NUTS II level, with over 220 km of lanes per 1000 km^2. This focus on highway infrastructure, often at the expense of substantial investments in public transport networks, has paved the way for new urban developments characterised by increased suburbanisation. This trend has brought with it challenges related to housing accessibility and population mobility. Moreover, urban development in response to highway expansion has often neglected the integration of transport infrastructure and public amenities, leading to increased mobility demands for residents. This has resulted in an overreliance on personal vehicles, contributing to longer travel distances and the transformation of urban areas into quasi-highway landscapes.

These patterns of urban sprawl and car dependency have further repercussions, such as traffic congestion and air pollution in the Lisbon Metropolitan Area. With around 89% of commuters using private vehicles and an average car occupancy rate of only 1.2 passengers per vehicle, the city confronts the challenges of a transportation system heavily tilted towards individual mobility.

Finally, Lisbon contends with a notably low cycling modal share, presenting a urban mobility scenario dominated by private car use and limited adoption of active transportation modes. The prevailing mode of commuting is private car usage, comprising 34% of total trips, overshadowing public transport at 36%, and active modes, primarily walking and cycling, at nearly 18%. Cycling's contribution stands at a mere 0.19%, significantly below the national average. The city's cycling modal share has shown marginal improvement over the years but remains modest. In 2011, Lisbon reported a cycling modal share of 0.2%, substantially lower than the EU average of 8%. The challenging backdrop, characterised by hills, tram rails, and various pavement types, contributes to the reluctance of residents to choose cycling. Despite 54% of streets being almost flat and 75% having a grade below 5%, conducive for cycling, the absence of dedicated cycling infrastructure and concerns about safety further impede the adoption of bicycles for daily commuting. A noteworthy shift has been observed in recent years, with the cycling modal share reaching 0.6% in 2017, as per the National Mobility Survey. Lisbon faces challenges in enhancing its cycling modal share, rooted in topographical complexities, inadequate cycling infrastructure, and safety concerns. While there has been a visible increase in cycling, particularly attributed to municipal investments, the city is yet to reach cycling modal shares comparable to European averages, demanding sustained efforts to promote cycling as a viable and attractive mode of urban transportation.

19.3 The Increasing Promotion of Active Mobility

Consequently, over the past decade the city of Lisbon has developed a number of measures to promote the use of bicycle (Lisbon City Council 2020), including a set of initiatives and campaigns to promote the use of bicycles in the city, and many operations reconfiguring public space in favour of pedestrian and cycling modes. Two policies can be highlighted here: the construction of cycle lanes, and the creation of a public bike sharing system called GIRA.

The city experienced significant growth in cycle lane construction since 2008, with a notable increase in segregated cycling infrastructure from 2007 to 2015, spanning 60 km and incorporating 300 parking facilities. Bicycle promotion campaigns were initiated during this period. Subsequently, municipal enhancements from 2017 onwards saw the development of an extensive, connected, and commuter-focused bicycle network, expanding to 210 km of dedicated cycling infrastructure. By 2022, the city boasted 175 km of cycle lanes, including 152 km of visually or physically segregated lanes, constituting a noteworthy portion of the overall 1662 km road network (Fig. 19.1).

Fig. 19.1 Cycling infrastructures. *Source* Lisbon city council open data

In 2017, the Gira bike-sharing system was introduced, featuring 131 docks (as of April 2022) and over 1000 bicycles, including 1400 bicycles, of which 70% are e-bikes. Complemented by promotional campaigns and initiatives like free passes during the system's initial phase, these measures led to a significant surge in bicycle usage, marked by a 3.5-fold increase noted by Félix et al. (2020) following the cycling network expansion. The first year of the bike-sharing system saw high demand, averaging about 7 trips per bicycle per day and surpassing one million trips (Félix 2020). Despite these gains, the continued growth of bicycle use relies on ongoing enhancements to cycle infrastructures, as highlighted by Félix et al. (2019) and Moura et al. (2017). The system's affordability, with an annual fee of 25 euros compared to the monthly public transport pass of 30 euros, contributed to its popularity, attracting around 18,500 subscribers. City initiatives aim to extend the cycling infrastructure to 200 km by 2021, as depicted in the expansion plan illustrated in Fig. 19.1.

Consequently, there has been a notable rise in the population of cyclists. Analysing data derived from the 2011 and 2021 Population Census, it becomes evident that in Lisbon, the proportion of commuters who use bicycles as their mode of transport has increased significantly. Specifically, the modal share of bicycles in commuting has experienced a remarkable growth, escalating from a mere 0.2% in 2011 to a substantial 1.3% in 2021. This astonishing development signifies a staggering 539% increase in the number of trips undertaken for commuting purposes by bicycle within this decade.

19.4 Equity at Stake

The significant progress recorded in recent years is not without raising several questions regarding the equity of its distribution, as demonstrated by several recent studies (Cunha and Silva 2023; Félix et al. 2020, 2019; Padeiro 2022; Taborda et al. 2023). On the one hand, a significant gender gap persists both in the use of private bicycles and in the use of the public bike-sharing system: 70 to 75% of GIRA system users are male (Taborda et al. 2023). This over-representation has slightly decreased to reach 67% according to Taborda, but it persists. On the other hand, not all geographical areas of Lisbon are served by cycling infrastructures, nor with the same quality. Regarding private bike-sharing systems, there is no conclusive study, but several controversies have arisen in the press due to the impossibility for users of private bike-sharing systems to access certain parts of the city such as Alto da Ajuda, Lumiar, Marvila, Chelas neighbourhoods. As for the public system, significant disparities emerge between the most and least deprived block groups concerning the presence of bike lanes, bike-sharing docking stations, and the overall quality and accessibility of associated infrastructures. Notably, these discrepancies consistently favour less deprived areas. The allocation of funds towards expanding cycling and bike-sharing networks may not have mitigated existing inequalities in active mobility access; in fact, it is likely reinforcing these disparities. Analysis of population density and average slopes indicates that access inequalities are intertwined with deeper socio-spatial disparities ingrained in the city's territorial and physical layout. The cycling and bike-sharing networks in Lisbon tend to bypass hillier and densely populated neighbourhoods, disproportionately affecting low-income residents. Consequently, the cycling network and local bicycle-sharing systems exhibit notable inequities in terms of coverage, distance, and connectivity, particularly disadvantaging the most deprived areas.

Various factors contribute to these inequalities. Firstly, public authorities may be hesitant due to concerns about lower bicycle utilisation and financial sustainability in the most deprived neighbourhoods, often characterised by limited amenities and higher unemployment rates. Although low-income areas tend to use bicycles less, targeted campaigns and affordability measures have proven successful in increasing usage among these groups in certain contexts. Secondly, infrastructure planning may prioritise technical aspects and costs over social considerations, leading to the concentration of bike lanes in areas with easier or less expensive construction. In Lisbon, this approach may explain the avoidance of steep, narrow streets in the historic centre. The lack of integration of social equity considerations into infrastructure policy is noted, with arguments focusing more on local quality of life and air quality improvement. Thirdly, a higher demand for cycling infrastructure may exist among younger and more affluent populations, as supported by the study's models and consistent with other research. Lastly, cycling infrastructures can contribute to the visual quality and local character of neighbourhoods, symbolising livability and sustainability. However, this may inadvertently reinforce exclusionary dynamics, particularly in the context of financialisation, gentrification, and touristification. Expanding cycling networks to low-income areas could potentially trigger gentrification processes, warranting careful monitoring and attention from public authorities.

The previous Mayor of Lisbon had set a goal to expand the cycling network to 200 km by the conclusion of 2021. This objective had been consistently communicated in the Portuguese media starting from 2016. However, a new mayor, who has not articulated a definitive stance on cycle lane policies, was elected in September 2021. Consequently, there has been no advancement in the development of the cycling infrastructure since the change in leadership.

19.5 Conclusion

Active mobility, encompassing non-motorised transportation modes like walking and cycling, has gained recognition as a vital component of urban policies. This recognition is fuelled by its potential to improve physical health, reduce stress, enhance air quality, and contribute to decarbonisation efforts. The case of Lisbon illustrates both the progress and challenges associated with promoting active mobility, particularly cycling, in a complex urban context. While the city has expanded its cycling infrastructure and introduced bike-sharing systems, questions of equity in access and distribution persist. Gender disparities and geographical inequalities in infrastructure quality raise concerns about the equitable distribution of the benefits of active mobility. To ensure that active mobility contributes to a more sustainable and equitable urban future, comprehensive urban planning measures and policy interventions are needed to address these disparities and promote cycling as a viable and attractive mode of transportation.

References

Alessio HM, Reiman T, Kemper B, von Carlowitz W, Bailer AJ, Timmerman KL (2021) Metabolic and cardiovascular responses to a simulated commute on an e-bike. Transl J Am Coll Sports Med 6(2):e000155

Boone-Heinonen J, Jacobs DR Jr, Sidney S, Sternfeld B, Lewis CE, Gordon-Larsen P (2009) A walk (or cycle) to the park: active transit to neighborhood amenities, the CARDIA study. Am J Prev Med 37(4):285–292. https://doi.org/10.1016/j.amepre.2009.06.006

Brown BB, Tharp D, Tribby CP, Smith KR, Miller HJ, Werner CM (2016, 1 Sept) Changes in bicycling over time associated with a new bike lane: relations with kilocalories energy expenditure and body mass index. J Transp Health 3(3):357–365. https://doi.org/10.1016/j.jth.2016.04.001

Bucher D, Buffat R, Froemelt A, Raubal M (2019, 1 Oct) Energy and greenhouse gas emission reduction potentials resulting from different commuter electric bicycle adoption scenarios in Switzerland. Renew Sustain Energy Rev 114:109298. https://doi.org/10.1016/j.rser.2019.109298

Buehler R, Dill J (2016) Bikeway networks: a review of effects on cycling. Transp Rev 36(1):9–27

Cunha I, Silva C (2023, 1 March) Assessing the equity impact of cycling infrastructure allocation: implications for planning practice. Transp Policy 133:15–26. https://doi.org/10.1016/j.tranpol.2022.12.021

De Hartog JJ, Boogaard H, Nijland H, Hoek G (2010) Do the health benefits of cycling outweigh the risks? Environ Health Perspect 118(8):1109–1116

Dill J, Carr T (2003) Bicycle commuting and facilities in major US cities: if you build them, commuters will use them. Transp Res Rec 1828(1):116–123

Félix R, Moura F, Clifton KJ (2019, 1 Dec) Maturing urban cycling: comparing barriers and motivators to bicycle of cyclists and non-cyclists in Lisbon, Portugal. J Transp Health 15:100628. https://doi.org/10.1016/j.jth.2019.100628

Félix R, Cambra P, Moura F (2020, 1 June) Build it and give em bikes, and they will come: the effects of cycling infrastructure and bike-sharing system in Lisbon. Case Stud Transp Policy 8(2):672–682. https://doi.org/10.1016/j.cstp.2020.03.002

Hong J, McArthur D, Raturi V (2020) Did safe cycling infrastructure still matter during a COVID-19 lockdown? Sustainability 12(20):8672

Hull A, O'Holleran C (2014, 1 Jan) Bicycle infrastructure: can good design encourage cycling? Urban Plann Transp Res 2(1):369–406. https://doi.org/10.1080/21650020.2014.955210

INE (2022) Censos 2021. Instituto Nacional de Estatística. Retrieved 15–06–2022 from https://www.ine.pt/scripts/db_censos_2021.html

Jarrett J, Woodcock J, Griffiths UK, Chalabi Z, Edwards P, Roberts I, Haines A (2012) Effect of increasing active travel in urban England and wales on costs to the national health service. The Lancet 379(9832):2198–2205

Johansson C, Lövenheim B, Schantz P, Wahlgren L, Almström P, Markstedt A, Strömgren M, Forsberg B, Sommar JN (2017, 15 April) Impacts on air pollution and health by changing commuting from car to bicycle. Sci Total Environ 584–585:55–63. https://doi.org/10.1016/j.scitotenv.2017.01.145

Kondo MC, Morrison C, Guerra E, Kaufman EJ, Wiebe DJ (2018, 1 March) Where do bike lanes work best? A Bayesian spatial model of bicycle lanes and bicycle crashes. Saf Sci 103:225–233. https://doi.org/10.1016/j.ssci.2017.12.002

Lisbon City Council (2020) Move LISBOA Visão Estratégica para a Mobilidade 2030

Lorenz O (2018) Does commuting matter to subjective well-being? J Transp Geogr 66:180–199

Ma L, Ye R, Wang H (2021, 1 April) Exploring the causal effects of bicycling for transportation on mental health. Transp Res Part D: Transp Environ 93:102773. https://doi.org/10.1016/j.trd.2021.102773

McQueen M, MacArthur J, Cherry C (2020, 1 Oct) The e-bike potential: estimating regional e-bike impacts on greenhouse gas emissions. Transp Res Part D: Transp Environ 87:102482. https://doi.org/10.1016/j.trd.2020.102482

Monsere CM, McNeil N, Dill J (2012, 1 Jan) Multiuser perspectives on separated, on-street bicycle infrastructure. Transp Res Rec 2314(1):22–30. https://doi.org/10.3141/2314-04

Moura F, da Silva JM, Santos LP (2017) Growing from incipient to potentially large cycle networks: screening the road network of the consolidated urban area of Lisbon. Europ J Transp Infrastruct Res 17(1):170–190

Moura F, Kalakou S (2021) Active modes and sustainability. In: Leal Filho W, Azul AM, Brandli L, Lange Salvia A, Wall T (eds) Industry, innovation and infrastructure. Springer International Publishing, pp 10–26. https://doi.org/10.1007/978-3-319-95873-6_7

Mueller N, Rojas-Rueda D, Salmon M, Martinez D, Ambros A, Brand C, de Nazelle A, Dons E, Gaupp-Berghausen M, Gerike R, Götschi T, Iacorossi F, Int Panis L, Kahlmeier S, Raser E, Nieuwenhuijsen M (2018, 1 April) Health impact assessment of cycling network expansions in European cities. Prev Med 109:62–70. https://doi.org/10.1016/j.ypmed.2017.12.011

Padeiro M (2022) Cycling infrastructures and equity: an examination of bike lanes and bike sharing system in Lisbon, Portugal. Cities Health 1–15. https://doi.org/10.1080/23748834.2022.2084589

Pan X, Zhao L, Luo J, Li Y, Zhang L, Wu T, Smith M, Dai S, Jia P (2021) Access to bike lanes and childhood obesity: a systematic review and meta-analysis. Obes Rev 22(S1):e13042. https://doi.org/10.1111/obr.13042

Parker KM, Rice J, Gustat J, Ruley J, Spriggs A, Johnson C (2013) Effect of bike lane infrastructure improvements on ridership in one new Orleans neighborhood. Ann Behav Med 45(suppl_1):S101–S107. https://doi.org/10.1007/s12160-012-9440-z

Pucher J, Buehler R, Bassett DR, Dannenberg AL (2010) Walking and cycling to health: a comparative analysis of city, state, and international data. Am J Public Health 100(10):1986–1992. https://www.ncbi.nlm.nih.gov/pmc/articles/PMC2937005/pdf/1986.pdf

Synek S, Koenigstorfer J (2019) Health effects from bicycle commuting to work: insights from participants of the German company-bicycle leasing program. J Transp Health 15:100619

Taborda S, Zuzarte M, Magalhães L, Carvalho G, Banza M (2023, 1 Jan) Addressing the gender gap in cycling: Lisbon's journey for a more inclusive cycling ecosystem. Transp Res Procedia 72:2401–2408. https://doi.org/10.1016/j.trpro.2023.11.735

Tainio M, de Nazelle AJ, Götschi T, Kahlmeier S, Rojas-Rueda D, Nieuwenhuijsen MJ, de Sá TH, Kelly P, Woodcock J (2016, 1 June) Can air pollution negate the health benefits of cycling and walking? Prev Med 87:233–236. https://doi.org/10.1016/j.ypmed.2016.02.002

Wardman M, Tight M, Page M (2007) Factors influencing the propensity to cycle to work. Transp Res Part a: Policy Pract 41(4):339–350

Xing Y, Handy SL, Mokhtarian PL (2010) Factors associated with proportions and miles of bicycling for transportation and recreation in six small US cities. Transp Res Part d: Transp Environ 15(2):73–81

Zhang Y, Mi Z (2018, 15 June) Environmental benefits of bike sharing: a big data-based analysis. Appl Energy 220:296–301. https://doi.org/10.1016/j.apenergy.2018.03.101

Zhou Q, Che M, Koh PP, Wong YD (2020, 1 March) Effects of improvements in non-motorised transport facilities on active mobility demand in a residential township. J Transp Health 16:100835. https://doi.org/10.1016/j.jth.2020.100835

Chapter 20
New Forms of Mobility: The Case of Madrid

Daniela Arias-Molinares and Juan Carlos García-Palomares

Abstract The evolution of Madrid during the last decades has implied a densification of the central city and a notable growth of its peripheries. The densification of the capital municipality does not imply a significant increase in its number of inhabitants, but it does imply an increase in its activities, endowments and functions as a central place, be they commercial, service or command of highly relevant economic activities. This causes an accelerated trend towards an increase in traffic and mobility demand. On the other hand, the development of the peripheries, which end up accounting for a population greater than that of the city, introduces the need to provide new connection networks, more public transport vehicles and efficient road lines. Mobility becomes metropolitan, with intense periphery-centre flows, but also centre-periphery and between the different localities located outside the capital node. This complex panorama is analysed and interpreted in the chapter. At the same time, the text raises the necessary intermodality, complementarity of modes of transport and reduction of emissions that should lead to achieving a sustainability objective. Mobility has gone from being citizen to metropolitan, it is becoming enormously complex and intense, and, in this context, the objective of reducing the resulting environmental impacts becomes a fundamental commitment that the urban space as a whole must assume for the coming decades.

Keywords Shared mobility · Micromobility · Mobility as a Service (MaaS) · Madrid (Spain)

D. Arias-Molinares · J. C. García-Palomares (✉)
Research Group On Transport, Infrastructure and Territory (tGIS), Geography Department, Universidad Complutense de Madrid, Madrid, Spain
e-mail: jcgarcia@ucm.es

D. Arias-Molinares
e-mail: daniar02@ucm.es

R. C. Lois-González and J. A. Rio Fernandes (eds.), *Urban Change in the Iberian Peninsula*, The Urban Book Series, https://doi.org/10.1007/978-3-031-59679-7_20

20.1 Introduction: The Burst of New Forms of Mobility

We live in a digital era. People access information, buy things or travel to places only with the use of their smartphones. This new era is rapidly changing people's lifestyles and their mobility behaviour. Furthermore, with the current climate emergency, achieving sustainable mobility is at the heart of many global and regional public policies, like the 2030 Agenda for Sustainable Development (United Nations 2015), the 17 Sustainable Development Goals set by the United Nations (United Nations 1987) or the more recent Sustainable and Smart Mobility Strategy (European Commission 2020). The task is clear: a transition from a car-oriented model into a proximity one, in which humans and the environment are placed at the core. Two factors are key to boosting new forms of shared mobility. On the one hand, lifestyle changes, as well as the new demands of the younger generations (millennials and Generation Z), who base their consumption on proximity and "access to", instead of ownership. On the other hand, continuous advances in technology, the development and high penetration of smartphones, connectivity speed (5G networks), mobile payments, GPS systems, as well as the Internet of Things are facilitating a transformation that is disruptive rather than dynamic, owing to the speed of these innovations. Due to the importance of technology in the development of these mobility models, people have started to refer to them with the concept of Smart Mobility.

Big Spanish cities have strongly embraced these mobility models, especially in their urban centres. Looking backwards to the first decade of the twenty-first century, our cities have seen the consolidation of shared mobility services with car-sharing and bike-sharing systems. In its early stages, bike-sharing systems operated essentially, with the use of transport/intelligent cards and as station-based models (also called docked services), meaning that the bike is picked-up from and returned to a fixed station. The next decade, especially from 2017 on, has been marked by the burst of new forms of mobility, even coining the term micromobility. In this second decade, two main innovations have taken place: (1) most of the services are dockless (also called free-floating), meaning that the vehicle is picked up and returned to any location within a certain geographic area (commonly called geofence); and (2) most of them are rented using only a smartphone (no need for transport/intelligent cards). Lastly, the beginning of the third decade of the century is bringing new challenges: the more complex the mobility ecosystem becomes, the more necessary it is to integrate all the mobility offer in a single place, reducing the hassle for citizens. This idea is known as Mobility as a Service (hereafter, MaaS), and it is part of a new mobility paradigm.

This new mobility paradigm has important challenges ahead. Due to their nascent stage of development, on many occasions the new services are not being fully integrated into the transport and mobility planning agendas. Urban and transportation planning must consider all available mobility services. A quick search in Google Play yields the inevitable: with more than 60.000 travel apps and shared mobility services available, these new forms of mobility are here to stay (Li and Voege 2017). The combination of public transport and shared mobility is key for MaaS development. The sooner we plan for them, the better the mobility ecosystem will work.

In the broad Spanish panorama, big cities have embraced these new forms of mobility especially in their city centres. Cities such as Barcelona, Valencia, Bilbao or Seville have shared mobility systems, initially public bicycle schemes, and after, these have expanded to include cars, motorcycles or electric scooters, or private bicycles distributed around the city without being anchored to stations. However, it is undoubtedly in Madrid where these modes of transport have been introduced more intensely, even converting the city into an authentic European "living lab" of shared mobility. In this chapter we will describe and discuss, through the case of Madrid, how are shared mobility and micromobility services evolving and their role in a future MaaS offering.

20.2 Shared Mobility and Micromobility Development

Shared mobility has been defined as a short-term access to shared vehicles, according to user needs and convenience, rather than requiring vehicle ownership. In its most recent Smart Mobility Strategy, the European Commission has highlighted the importance of shared mobility services as they have demonstrated to be crucial for intermodality and the public transport system's resilience (especially after the COVID-19 pandemic outbreak) (European Commission 2020). In this context, shared mobility is being one important urban transport revolution and even the automobile sector is starting to envision itself as a shared mobility provider, instead of just car manufacturers (Fulton 2018).

Except for car-sharing, all other vehicles are for individual use and have low speeds (most up to 30 km/h). In this sense, the term micromobility was coined to refer to these small, lightweight and low-speed shared modes as mopeds, bicycles and scooters (Shaheen and Cohen 2019). Micromobility offers the same benefits of shared mobility in general, such as the reduction of noise and air pollution and the promotion of intermodality with public transport. But in addition, it offers greater flexibility by avoiding congestion, reducing parking space needed and in the case of mopeds, they provide greater speed, the possibility of traveling medium distances and a very attractive way to overcome the slopes in cities that are not flat (Aguilera-García et al. 2020). All these advantages are well known by users, and to a large extent, they are the cause of micromobility's popularity in the recent years. Additionally, these new forms of movement require a much lower vehicle production compared to what will be needed if each vehicle was dedicated to a single person (who would probably keep it parked most of the time).

The most recent European Shared Mobility Index (Fluctuo 2022), which monitors more than 30 European cities and 408 services, indicates that in third quarter of 2022 (June to September) there were 100 trips made in shared modes, resulting in a 20% growth in ridership compared to the same quarter of 2021. With a global European fleet calculated in more than 500 thousand vehicles, most of these trips are made in shared scooters (43%), followed by station-based bike-sharing (32%). Madrid is one of the leading cities with a shared mobility index of 20, as Paris and Milan and just behind the city of Berlin.

Micromobility companies basically manage a fleet of vehicles and offer them to their subscribers as a rental service for a given time (hours, minutes, etc.). They generally work through applications on smartphones, which has become widespread among a large part of the population. Users usually access an interface with a city map in their applications and locate the closest vehicle to their location. Once the vehicle has been identified, they unlock it from the app and begin the journey. User data and bank information are saved when registering for the service in order to make payment for the use of their fleets. Once the user reaches their destination, they park and relock the vehicle from the app. As we can see, one of the most important reasons that has made it possible for these services to be extended in cities is that just by having a mobile, you can make a vehicle reservation, start and finish the trip (Monzón et al. 2019).

Micromobility services allow greater flexibility and the possibility of door-to-door trips. However, a critical issue from the operational point of view is the need to redistribute vehicles since there is often an imbalance between the origins and destinations of trips according to hours of the day. Thus, for example, in early morning hours, most trips originate from residential zones, while they end at employment areas and the contrary tends to occur in the afternoon (commuting patterns). This causes vehicles to be concentrated in some neighbourhoods and scarce in others, forcing operators to redistribute vehicles in vans, with the consequent increase of operation costs, congestion, and pollution. In addition, dockless services require more operational logistics for maintenance and recharging tasks. In regard to the spatial distribution of these services, they are often offered mainly in the city centres, where operators find more profitability, as land uses are diverse, and a high concentration of public transport and activities offered. However, the concentration of shared services in the city centre may be competing with public transport and active mobility and hence, a more homogeneous distribution is desired reaching also peripheric areas, usually the most car-dependent ones, and where these services could be a solution to the last-mile trips.

Moreover, unlike the first public bike-sharing systems that were mostly financed or promoted by public organisations, the new shared mobility services are mostly incentivised by private initiatives. Large private companies are behind many of these initiatives (e.g., BMW with Car2Go, Renault with Zity or Kia with Wible). These services are usually implemented with hardly any limitations and at a relatively lower costs, since most of them are dockless and do not require any physical infrastructure. However, the lack of regulation has also caused many conflicts, especially when these vehicles are parked invading sidewalks. Since their irruption worldwide in the summer of 2017 and 2018, cities are now just beginning to react to this phenomenon with the introduction of regulations (see for example, Comunidad de Madrid 2019). In any case, the fact is that micromobility services such as BiciMAD in Madrid or Bicing in Barcelona are becoming important forms of mobility in their cities. The success of these systems is helping to promote a more sustainable modal shift and increase policies oriented to extend the needed infrastructure to make them safer and more attractive. With the burst of shared mobility and more recently, of dockless micromobility services, there is a growing need to have all the mobility offer centralised in a single place, which has enabled the concept of Mobility as a Service (MaaS).

20.3 The Umbrella Concept: Mobility as a Service (MaaS)

With many shared mobility and micromobility options available, citizens are finding it difficult to navigate through the plethora of applications. MaaS is born from the idea of servicing mobility by integrating the existing traditional public transport systems with the new emergent shared mobility services. The result of this combination is precisely the "bet" on MaaS, as it could cover all transport needs required to substitute a private automobile. Instead of owning individual vehicles, MaaS customers will join an offer that includes different transport options.

This concept arose in 2006 when the entrepreneur Sampo Hietanen proposed, what he called mobility packages, and idea oriented to bundle our mobility services in a similar way as a mobile phone contract. Hietanen described Maas as a distribution model that delivers users' transport needs through one single interface of a service provider, combining different transport modes to offer tailored mobility packages (Hietanen 2014) (Fig. 20.1). MaaS is referenced as an intelligent, multimodal and completely user-oriented mobility distribution and management system, in which a provider integrates multiple transport operators (public and private) providing easy access to them, through a digital platform and allowing them to plan, pay and access their mobility services quickly and conveniently (Mulley, 2017). In general, the central idea of MaaS is to concentrate the entire mobility offer in a single application that allows users to navigate through all the options and to move within or outside their city, making it possible for them to obtain information, plan routes and make a payment for the services consumed, all quickly and conveniently.

The first element of MaaS is related to having a unique app and the needed functionalities. Having a unique channel makes it easier for users, avoiding the hassle of checking multiple mobility and travel applications, information sources, journey planners and ticketing platforms. Besides having a single app, users have to be able to plan, pay and access their mobility services through this unique channel. The planning stage of the trip involves having real-time information on as many services as possible, offering an intermodal journey planner, as well as booking functions. Real-time information needs to be dynamic, meaning that it changes or adapts throughout the day according to incidences/anomalies, and for all mobility services included. Moreover, the MaaS journey planner should be intermodal, because most of the existing journey planners do not offer information for intermodal trips, meaning that they do not combine more than one transport mode -with the exception of walking, which is usually the access and egress mode)- and only include some of the transport modes available in an area. Another important aspect that should be considered in

Fig. 20.1 Main elements of MaaS. *Source* (Arias-Molinares et al. 2022b)

the MaaS app is payment integration, which basically means allowing consumers to pay through the app. The access stage of the trip has proven to be important, offering ticketing functions for users to validate and access mobility services (Li and Voege 2017). Nowadays, the latter is generally done by carrying an intelligent transport card, which at most, integrates public transport, but not shared-mobility services. Being able to pay, access and travel through all modes available with nothing more than their smartphones is the desired in a MaaS app.

The second element of MaaS involves offering multiple and diverse mobility services. Many apps only integrate the public transport systems, but no shared-mobility services. An ideal MaaS will try to include as many mobility services as possible, in order to cover most of the transport offer in the city. It is recommended to have as varied transport modes as possible: scooters, bicycles, mopeds, automobiles, buses, tram, subways, rail, etc. Nevertheless, studies have shown that the vital service to be included is the public transport offer, as it is the core service that users will look for when joining a Maas platform (Strömberg et al. 2018).

Finally, the third core element in MaaS is mobility packages. Mobility packages are bundles of various transport modes that include a certain amount of km/minutes/points that can be utilised in exchange for a monthly payment. Some argue this is the real "new" aspect offered by MaaS, since information and payment integration already existed before MaaS became a topic (Lyons 2020). With mobility packages, users join a monthly subscription (just like a phone plan) to cover all their travel needs and they receive a unique invoice at the end of the week/month/year with all the mobility expenses.

When reviewing MaaS pilots worldwide, Arias-Molinares and García-Palomares (2020) have observed that the majority of apps are just mere information aggregators or customised journey planners as they still lack some of the most important features like being able to do everything in a single app or offering mobility packages. One of the main challenges for the development of MaaS may be oriented to finding a governance framework to enable the concept, as technological solutions seem to be in place. And as governance issues are always linked to a particular context, in the next section we will try to land and understand how these concepts are evolving in Madrid.

20.4 The Case of Madrid: Breeding Ground for Shared Mobility, Micromobility and MaaS

Madrid is the capital of Spain and the country's most populated urban area. It holds a population of 3.3 million (municipality) and reaches 6.8 million at the regional level. The city also concentrates jobs and a great supply of social, educational and cultural facilities, offering various entertainment events and public activities. It has a robust public transport system network that spreads through metro, suburban rail and city and inter-city buses. Madrid is one of Europe's cities with the highest level of public-transport use, as well as non-motorised transport, according to the (European Commission 2014). The last Household Mobility Survey of Madrid Region showed that there are 15.8 million displacements occurring on a working day out of which 3.8 million employ public transport (24%), showing the important of the public transport system in the city (Comunidad de Madrid 2018). The modal share in the Municipality of Madrid is distributed in 33% of trips with private car, 32% in public transport, 32% walking and 3% in other modes, while the city centre shows a more sustainable pattern with a distribution of 40% of trips walking and 35% in public transport. When zooming into the share of the category "other modes", we find that motorcycles are the most important in both the Municipality of Madrid and the city centre, followed by taxi and ride-hailing services, bicycles and other modes.

Madrid's shared mobility ecosystem has experienced rapid growth and is commonly recognized as one of Europe's most significant shared mobility testing grounds. Since 2010, numerous companies have entered the shared mobility sector, particularly since the summer of 2018 when many micromobility companies began operations (see Fig. 20.3). Unlike the public bicycle sharing service (BiciMAD), which is promoted by the local government, almost all new shared mobility models are associated with private companies. From a business perspective, shared mobility models in many Spanish cities are still in the 'deployment phase' of the service. As of the beginning of 2020, the oldest shared mobility companies had only been operational for a couple of years, and most had yet to turn a profit. Moreover, this is particularly true for major cities, where numerous operators emerged within a short period, offering their services almost simultaneously. In 2019, Madrid alone had over twenty shared scooter companies. Many companies had also shut down, and there was a high level of financial speculation. Mergers and acquisitions between companies became common, resulting in the formation of larger companies aiming to establish themselves in an emerging, highly competitive sector where profits were expected in the medium or long term. Many startups, associated with smartphone app development, either closed or were acquired by larger firms. The sector's capital requirements and the need to restructure their business models toward the future of shared mobility and electric mobility have driven investments from large companies such as Acciona, Bosch, Seat, and even Uber itself into some of the major motorbike and scooter sharing companies currently operating in Madrid. In 2019, due to this emerging ecosystem, the city had a fleet of approximately 30,000 vehicles and nearly 30 different operators (see Fig. 20.2).

However, after the COVID-19 pandemic outbreak, and also as a consequence of several new regulations that are taking place, Madrid has experienced an intense restructuring process in the sector. This process has caused there to be less but more consolidated operators (Fig. 20.3). Despite the wide availability of shared mobility services, their usage is still limited. Micromobility services were more pandemic-resilient, as their usage dropped only 10% in comparison with the 80% downfall reported by public transport during the second half of 2020 (Arias-Molinares

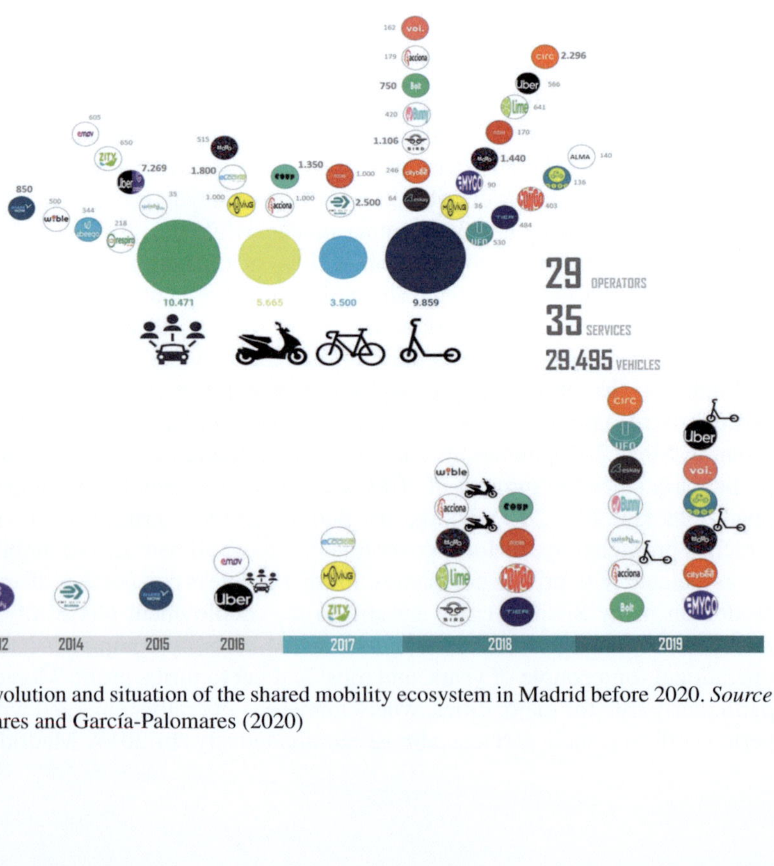

Fig. 20.2 Evolution and situation of the shared mobility ecosystem in Madrid before 2020. *Source* Arias-Molinares and García-Palomares (2020)

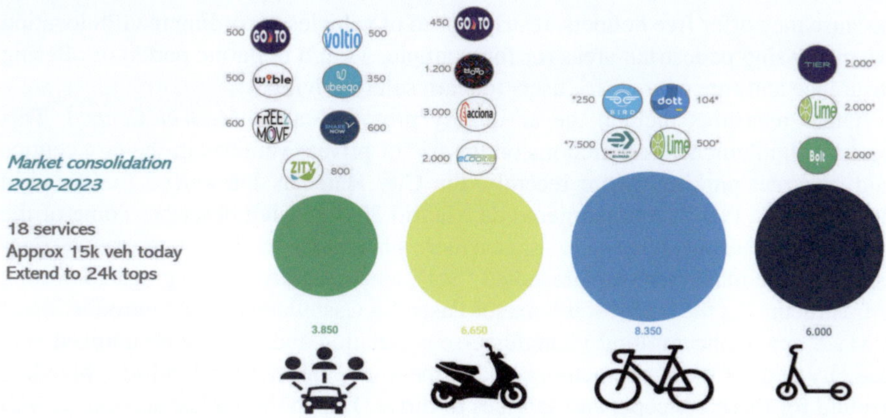

Fig. 20.3 Madrid shared mobility ecosystem today (2023). *Source* Prepared by the authors

et al. 2022a). However, the subsequent recovery of public transportation has been significant, returning to its pre-pandemic share in the modal distribution.

With regards to shared mobility preferences in Madrid, some recent studies have found that approximately 12% of BiciMAD's users (Madrid's public bike-share system) are also subscribed to other dockless services like shared mopeds and scooters (Arias-Molinares et al. 2021). This study also reported that even though BiciMAD is being mostly used by people who walked or used public transport (unsustainable shift), it is also substituting 10% of car trips. When asked what they will use if BiciMAD stopped being operative, many of the respondents have also chosen dockless shared modes (mopeds/scooters), from which we can infer that the bikeshare system is making users keener to try other new forms of mobility.

The City of Madrid is starting to take actions regarding micromobility regulations. The new City Ordinance is an effort to exhort BiciMAD and micromobility subscribers to circulate on the road at no more than 30 km/h (Comunidad de Madrid 2019). It states that micromobility users may use bike infrastructure (non or segregated ones). Additionally, it restricts the use of micromobility in bus lanes and pedestrian space. An exception is made in the cycling infrastructure inside parks where users can reach up to 5 km/h circulating using BiciMAD and other dockless micromobility services. It also recommends the city and operators to work together to provide adequate public spaces for parking. Some companies have started to introduce the concept of *virtual stations*, which are designated areas for parking micromobility vehicles. These designated areas are located usually in sidewalks when they are wide and offer adequate circulation for pedestrian and micromobility users and/or on the road substituting previous parking spots. The services' apps inform subscribers that they should pick up the vehicle at these designated areas. In the case of parking after completing the ride, the app instructs customers on where to leave the vehicle, and they are required to take a picture of it properly parked; otherwise, the app will continue to charge them for the current ride. Some private initiatives are also relevant, because they offer free helmets, restrict speed of vehicles according to GPS location when crossing pedestrian areas (or for example, 5 km/h in public parks) or offering insurance and free courses for users to learn safety driving.

Other measures include the air-quality protocol called *Madrid Central*. This protocol implemented restrictions on the use of private vehicles in the city centres and on-street parking. More recently, the City Hall has introduced the "Madrid 360 Strategic Plan" (Ayuntamiento de Madrid 2019) which describes some of the measures to be considered to promote a more sustainable mobility as the pedestrianisation of the main city's square called "Sol", a new 6.5 km long segregated cycling infrastructure in the north–south axe of Paseo La Castellana and the introduction of a MaaS app. Concentrating young/diverse population and jobs, being a mixed land use city and all these characteristics and measures are making Madrid a breeding ground for these concepts and services to thrive. Hence, in the last section we will discuss how all these factors may contribute to create a future for a possible MaaS offering in the city.

20.5 The Near Future: Where Does Madrid Go from Here?

Madrid is an example of the growth of shared mobility, micromobility and MaaS. As many studies agree on, shared mobility is here to stay and hence, planners should start considering these new forms of mobility in the modelling and planning processes. Madrid seems to have all the ingredients in place to cook a possible future MaaS offering (robust public transport, shared mobility and micromobility offer, MaaS developments and the concentration of highly diverse and young population). For now, we can highlight two main MaaS pilots that are being developed. The first one, coming from the public sector, the Municipal Transport Company of the city of Madrid (EMT) has developed the test prototype of an application called "Madrid Mobility 360". This application offers real-time information on different services (intercity trains, subway, urban bus and shared services) and the bus occupancy both when calculating a route and when waiting at a stop. In addition, payments can be made through the application to purchase single bus tickets and use the services of BiciMAD. Nevertheless, for now, the app is just an information aggregator for the case of dockless services, with developments needed in the payment integration. The second pilot has been incentivised by the company Renfe, which operates the primary rail network of Spain. They have developed a MaaS prototype called "DOCO". The train operator is envisioned itself as a mobility service provider in the future, expanding its revenues by involving the emergent concept in their core essential activities. The app was tested during months with around 500 people and is now available to download from Google Play. These two main initiatives are differentiated by two main aspects: their reach and their focus. The Madrid 360 app is a local initiative (Madrid's municipality) pushed by the City Hall and it is currently more oriented to add or integrate as many mobility operators as possible. On the other hand, DOCO has national coverage as it is managed by a national rail company involving light and interurban trains as well as interurban buses.

From a positive perspective, we could highlight that having two very different initiatives can result on interesting findings and comparisons. However, all these initiatives, at the present, are mainly in their early stages and have a partial level of integration. Additionally, there are scarce regulation frameworks for MaaS to work, and collaborations between public transport and shared micromobility operators are just starting to take place (especially promoted by the City Hall with the Madrid 360 app). However, important advances are observed from micromobility operators. On the first hand, their willingness to share data (which is one of the most important elements to enable MaaS) as operators want to stablish collaboration agreements with academic institutions. On the other hand, they believe that becoming part of the MaaS app of Madrid could increase their exposure to citizens that are not yet clients, which could result in an increased market share. To that end, many of these micromobility operators are willing to make discounts and arrange convenient prices of their services in order to gain more exposure through the MaaS app. Advances have been also announced from the Consorcio Regional de Transportes, Madrid's public transport authority, regarding the development of payment and ticketing ICT infrastructure that could allow users to buy tickets with only the use of their smartphone and a single app, which is a basic requirement to integrate the public transport systems as the backbone of MaaS.

In conclusion, governance appears to be the main challenge in the new mobility paradigm. There is low collaboration between MaaS possible stakeholders. Some of this lack of collaboration may be due to COVID-19 pandemic impacts, as public transport and micromobility operators are more focused on retaining or recovering their market share and ridership levels than discussing the concept of MaaS. In the near future, cities need to push the MaaS agenda and the promotion of micromobility services in order to create the best environments of collaboration. We are immersed in an initial moment of change in the urban mobility model. However, the emergence of the electric vehicle, shared mobility models or mobility services are beginning to transform the ways in which citizens move around.

Acknowledgements Grant NEWGEOMOB (PID2020-116656RB-I00) funded by MCIN/AEI/ https://doi.org/10.13039/501100011033.

References

Aguilera-García Á, Gomez J, Sobrino N (2020) Exploring the adoption of moped scooter-sharing systems in Spanish urban areas. Cities 96(July 2019):102424. https://doi.org/10.1016/j.cities.2019.102424

Arias-Molinares D, García-Palomares JC (2020) Shared mobility development as key for prompting mobility as a service (MaaS) in urban areas: the case of Madrid. Case Stud Transp Policy 8(3):846–859. https://doi.org/10.1016/j.cstp.2020.05.017

Arias-Molinares D, Julio R, García-palomares JC, Gutiérrez J (2021) Exploring micromobility services: characteristics of station-based bike-sharing users and their relationship with dockless services. J Urban Mobility 1(November):100010. https://doi.org/10.1016/j.urbmob.2021.100010

Arias-Molinares D, García-Palomares JC, Gutiérrez J (2022a) Micromobility services before and after a global pandemic: impact on spatio-temporal travel patterns. Int J Sustain Transp 1–16. https://doi.org/10.1080/15568318.2022.2147282

Arias-Molinares D, García-Palomares JC, Gutiérrez J (2022b) On the path to mobility as a service: a MaaS-checklist for assessing existing MaaS-like schemes. Transp Lett 1–10. https://doi.org/10.1080/19427867.2022.2038987

Ayuntamiento de Madrid (2019) Avance de la Estrategia de Sosteniblidad Ambiental. https://www.madrid360.es/

Comunidad de Madrid (2018) Documento síntesis: encuesta domiciliaria de movilidad de la Comunidad de Madrid 2018. https://www.crtm.es/conocenos/planificacion-estudios-y-proyectos/encuesta-domiciliaria/edm2018.aspx

Comunidad de Madrid (2019) Proyecto de Ordenanza de Movilidad Sostenible. In Proyecto de Ordenanza. https://doi.org/10.15446/dfj.n15.50535

European Commission (2014) Special Eurobarometer 422a. Quality of transport. https://doi.org/10.2832/783021

European Commission (2020) Sustainable and smart mobility strategy—putting European transport on track for the future. In: SWD (2020) 331 final

Fluctuo (2022) European shared mobility index report: Q3

Fulton LM (2018) Three revolutions in urban passenger travel. Joule 2(4):575–578. https://doi.org/10.1016/j.joule.2018.03.005

Hietanen S (2014) Mobility as a service—the new transport model? Eurotransport 12(2):2–4

Li Y, Voege T (2017) Mobility as a Service (MaaS): challenges of implementation and policy required. J Transp Technol 07(02):95–106. https://doi.org/10.4236/jtts.2017.72007

Lyons G (2020) Walking as a service—does it have legs ? s e. Transp Res Part A 137(March):271–284. https://doi.org/10.1016/j.tra.2020.05.015

Monzón A, Cascajo R, Romero C, Calzado R, López C (2019) Observatorio de la Movilidad Metropolitana. Informe OMM-2017. 638-19-072-5

Mulley C (2017) Mobility as a Services (MaaS) – does it have critical mass? Transp Rev 37(3):247–251. https://doi.org/10.1080/01441647.2017.1280932

Shaheen S, Cohen A (2019) Shared micromoblity policy toolkit: docked and dockless bike and scooter sharing. https://doi.org/10.7922/G2TH8JW7

Strömberg H, Karlsson M, Sochor J (2018) Inviting travelers to the smorgasbord of sustainable urban transport: evidence from a MaaS field trial. Transportation 45(6):1655–1670. https://doi.org/10.1007/s11116-018-9946-8

United Nations (1987) Our common future: report of the World commission on environment and development

United Nations (2015) Transforming our world: the 2030 agenda for sustainable development. https://sdgs.un.org/2030agenda

Part VI
Governance

Chapter 21
Urban Governance and Public Policies in Portugal

Pedro Chamusca

Abstract This paper delves into the evolving landscape of urban governance and public policies in Portugal, examining their role in shaping cities from 2000 to 2030. This study scrutinizes the transformative shift towards inclusive governance paradigms. Emphasizing stakeholder engagement, participatory frameworks, and sustainable planning, it analyzes Portugal's efforts to integrate diverse actors into decision-making processes impacting urban spaces. The research underscores the significance of effective governance and policies in steering the trajectory of future urban development. Notable initiatives, such as participatory budgeting, collaborative planning platforms, and grassroots associations, exemplify Portugal's commitment to inclusive governance models. The outlined policy recommendations prioritize participatory frameworks, collaborative partnerships, technology integration, and sustainable planning as pivotal strategies. This exploration concludes by highlighting the promising trajectory towards resilient, liveable, and responsive cities in Portugal, while also suggesting potential areas for further research in the field of urban governance and development.

Keywords Urban transformation · Stakeholder engagement · Participatory frameworks · Sustainable urban planning

21.1 Introduction

Urban centers stand as the bustling crucibles of our societal evolution, where the intricate interplay between governance structures and public policies molds the very fabric of these cities (Sassen 1991; Ascher 1998). At the heart of this synergy lies urban governance, the framework that orchestrates decision-making, resource allocation, and community engagement within city boundaries. Urban governance encapsulates the structures, processes, and institutions through which cities are managed

P. Chamusca (✉)
Communication and Society Research Centre (CECS), University of Minho, Braga, Portugal
e-mail: pedrochamusca@ics.uminho.pt

R. C. Lois-González and J. A. Rio Fernandes (eds.), *Urban Change in the Iberian Peninsula*, The Urban Book Series, https://doi.org/10.1007/978-3-031-59679-7_21

and governed (Hall and Pain 2006). It involves the collective decision-making, policies, and strategies formulated by various stakeholders, including government bodies, civil society organizations, private sector entities, and local communities, to address the complex challenges and opportunities within urban areas (Kim 2020; Chamusca et al. 2022). Parallelly, public policies serve as the guiding principles shaping urban landscapes, delineating the course of development, and addressing the multifaceted challenges inherent in modern urbanity (Chamusca 2023).

In this discourse, we embark on an exploration of these intertwined realms, delving into their pivotal roles in shaping cities while navigating the spatial and societal challenges of our times. The significance of urban governance and public policies in the contemporary era cannot be overstated. As cities worldwide grapple with burgeoning populations, strained resources, environmental concerns, and societal disparities, effective governance structures and robust policies emerge as linchpins for sustainable, equitable urban development. The Portuguese urban landscape, in particular, has undergone a notable evolution from the turn of the millennium until the projected horizon of 2030. Within this timeframe, Portugal's cities have experienced a kaleidoscope of changes, witnessing demographic shifts, economic transformations, infrastructural advancements, and social dynamics that have reshaped their identities (Fernandes et al. 2020).

Against this backdrop, this paper sets out to illuminate the intricate tapestry of urban governance and public policies in Portugal from 2000 to 2030. It endeavors to dissect the evolution of governance structures, analyze the efficacy of policies in tackling contemporary urban challenges, and shed light on the dynamic interactions among stakeholders shaping Portugal's urban destiny. Moreover, this study aims to not only dissect the current state of affairs but also prognosticate potential trajectories, offering insights into policy formulations that can steer Portuguese cities toward sustainable, inclusive, and resilient futures.

By navigating through historical shifts, contemporary challenges, and future trajectories, this paper aspires to contribute to the broader discourse on urban change within the Iberian Peninsula. Through a nuanced exploration of urban governance and public policies in Portugal, this work seeks to offer a comprehensive understanding of the forces driving urban evolution while paving the way for informed, forward-thinking approaches in urban governance and policymaking.

21.2 Research Outline and Theoretical Framework

21.2.1 Methods

This research is anchored in a comprehensive process involving bibliographic review and the construction of a theoretical framework, drawing upon the analysis of both national and international studies pertaining to urban governance and public policies.

The methodology encompasses a detailed examination and synthesis of scholarly works, research papers, and authoritative publications addressing the nuances of urban governance, thereby establishing a robust theoretical underpinning.

Complementing this theoretical foundation, the study undertakes an in-depth analysis and discussion of a spectrum of documents—ranging from laws and regulations to programs and initiatives—implemented in Portugal within the context of urban development. This analysis aims to assess the impact of these documents concerning their efficacy in addressing concrete challenges faced, the enhancement of governance structures, and the effectiveness of public policies in fostering sustainable urban development.

By scrutinizing and synthesizing the insights gleaned from academic literature alongside the examination of policy documents and initiatives within Portugal, this research strives to offer a comprehensive evaluation of the country's urban governance landscape. It seeks to identify patterns, gaps, and successes in policy implementation, highlighting areas where urban governance and policy effectiveness can be bolstered for more sustainable and responsive urban development strategies.

21.2.2 Theoretical Framework

Urban governance, a dynamic and complex construct, transcends the traditional confines of government structures. It embodies a paradigm of collaborative and inclusive decision-making involving diverse stakeholders—governmental bodies, civil society, private sectors, and community representatives. This approach acknowledges the multifaceted nature of urban challenges, emphasizing the need for coordinated efforts and participatory mechanisms to address these complexities effectively. Urban governance integrates elements of leadership, participatory processes, transparency, accountability, and efficiency in managing urban affairs (Brenner 2004a, b; Merrifield 2013; Teles 2016).

In parallel, public policies represent deliberate courses of action formulated and implemented by governmental or public entities to address societal issues and steer societal development. They manifest as the outcome of a systematic process involving problem identification, analysis, and the design of strategies or programs aimed at influencing behaviors, redistributing resources, and shaping societal outcomes (Bieber 2018; Sennet 2016). These policies span diverse domains, encompassing economic, social, environmental, and governance spheres, serving as instrumental tools in societal development.

The theoretical underpinning of this research contends that the efficacy of public policies is inherently linked to the quality of urban governance structures. Governance models that prioritize transparency, effective participation mechanisms, and accountability mechanisms tend to catalyze more efficient policy formulation and implementation processes. Transparent governance fosters trust among stakeholders and facilitates informed decision-making, thereby enhancing the legitimacy and effectiveness of policies. Moreover, effective participation mechanisms that engage diverse stakeholders in policy formulation ensure a broader representation of societal needs, promoting policies that are more responsive and tailored to the intricate urban fabric (Chamusca 2012; McCann 2015; Innerarity 2020).

Furthermore, the accountability inherent in robust governance models enhances the monitoring and evaluation of policy outcomes, ensuring their alignment with societal goals. This interconnected relationship between governance structures and policy efficacy elucidates the pivotal role of governance frameworks in shaping sustainable urban development trajectories. It underscores how a governance model underpinned by transparency, participation, and accountability mechanisms acts as a catalyst, fostering the design and implementation of more equitable, competitive, and sustainable policies to navigate the intricate dynamics of urban spaces (Chamusca 2023).

21.3 Public Policies, Governance and Urban Development in Portugal

21.3.1 The Governing Evolution: Looking for Decentralization

Portugal's urban governance has traversed a rich historical tapestry, reflecting the nation's evolution across epochs. From medieval municipal councils fostering local autonomy to the centralized governance of the absolute monarchy, each era left its imprint on the governance structures shaping Portuguese cities (Fernandes and Seixas 2018; Teles 2018). However, the most profound shifts in governance mechanisms emerged during the twentieth and twenty-first centuries.

The last decades witnessed a paradigmatic shift in Portugal's urban governance landscape. Embracing democratic values after the democratic revolution in 1974, the Constitution of 1976 reinforced the role of municipalities in urban governance. It marked a departure from centralized authority, granting greater autonomy to local governments and emphasizing their responsibility in urban planning and development. This marked the beginning of a trajectory toward decentralized governance, with municipalities assuming a pivotal role in shaping the future of Portuguese cities (Chamusca 2012).

Recent decades have seen concerted efforts toward decentralization and empowering local authorities. Policies such as the New Municipal Finances and the Local Administration Reform Law aimed to bolster municipalities' capacities, enabling them to effectively govern and manage their urban spaces. This shift towards empowering local governance has been crucial in fostering responsive and participatory decision-making, aligning policies more closely with the needs and aspirations of diverse urban communities (Lourenço et al. 2023).

However, amidst these advancements, challenges persist. Fiscal constraints, spatial planning intricacies, and the need for enhanced coordination between different levels of government continue to pose hurdles. The trajectory of Portugal's urban governance in the twenty-first century is deeply intertwined with the quest for sustainable urban development, digitalization for efficient governance, and inclusive decision-making. The ongoing endeavor involves fostering collaborative frameworks that prioritize sustainable urbanization, leverage technological innovations, and ensure citizen engagement to shape resilient and livable cities for the future.

The evolution of urban governance in Portugal reflects an ever-evolving journey, where historical legacies intersect with contemporary aspirations. It's a journey characterized by shifts toward greater decentralization, participatory governance, and a concerted drive to tackle the complexities of modern urban challenges.

21.3.2 Contemporary Urban Challenges

Portugal's urban landscape confronts a mosaic of contemporary challenges, reshaping the fabric of its cities. Foremost among these challenges is the persistent pressure of urbanization (Fig. 21.1). The influx of population into cities strains infrastructure, housing availability, and public services (Carvalho et al. 2019; OECD 2020). This burgeoning concentration within urban spaces, particularly in major cities like Lisbon and Porto, amplifies housing affordability concerns, creating an increasingly stratified socio-economic landscape. High property prices (Fig. 21.2) and rental costs compromise accessibility, posing a threat to social inclusion and fostering spatial segregation within these vibrant urban hubs (Acuto 2020).

Environmental sustainability is another critical challenge. Portugal faces the brunt of climate change, with urban areas experiencing the adverse effects—heat island effects, air pollution, and heightened vulnerability to disasters. These environmental stressors demand urgent measures for sustainable urban design, resilient infrastructure, and green initiatives to fortify cities against the impacts of climate change (Parnell 2016; Goh 2020).

Mobility and transportation present their set of hurdles. Congestion plagues urban centers while insufficient public transport infrastructure hinders efficient mobility. The inadequate connectivity between urban cores and suburban regions compounds the issue, affecting accessibility and exacerbating environmental concerns due to increased reliance on private vehicles.

Socio-economic disparities form another layer of complexity. Marginalized communities within urban spaces encounter challenges accessing fundamental services like education, healthcare, and employment opportunities (Nello 2015; Cave et al. 2020). These disparities perpetuate social exclusion, hindering cohesive community development and perpetuating a cycle of inequality (Piketty 2014; Harvey 2010).

These challenges collectively reshape the dynamics of urban governance and policymaking in Portugal. The need to address multifaceted, interlinked urban issues demands adaptive governance structures capable of accommodating complex

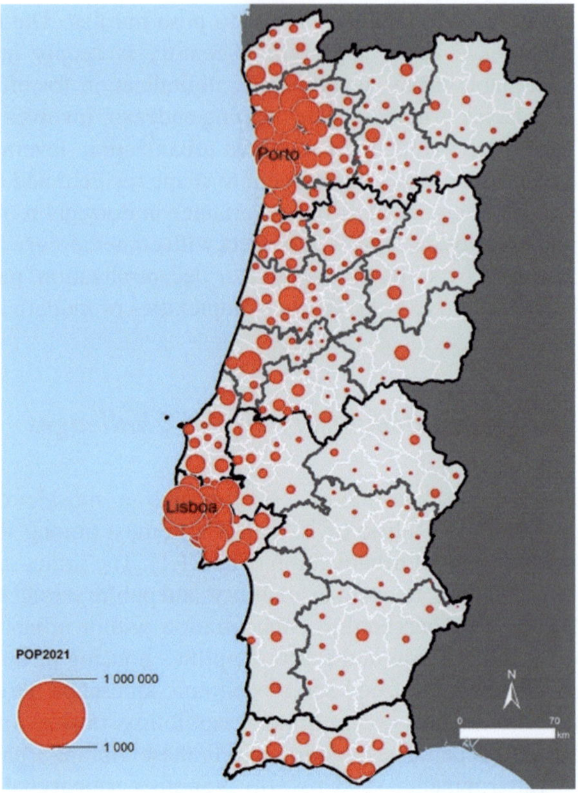

Fig. 21.1 Inhabitants (2021). *Source* INE (2023)

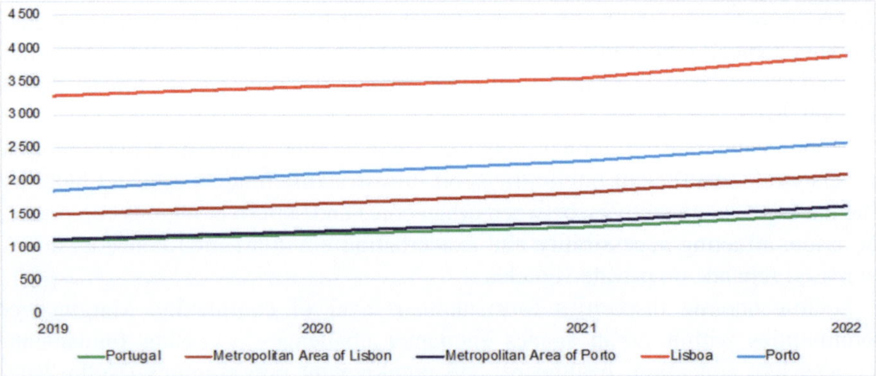

Fig. 21.2 Average real estate selling prices (€/sqm). *Source* INE (2023)

challenges. Policymaking must transcend traditional silos, fostering collaboration among diverse stakeholders. Innovative solutions and long-term strategies become imperative to navigate this intricate web of challenges effectively.

Moreover, the complexities of these issues underscore the necessity for data-driven decision-making, citizen engagement, and a holistic approach to urban governance (Subirats 2016; Innerarity 2020). Crafting policies that address the urban challenges in an integrated manner becomes pivotal. Creating resilient, sustainable, and inclusive cities requires a concerted effort, strategic planning, and dynamic governance mechanisms capable of responding to the evolving needs of Portugal's urban landscape. Thus, the interplay between these contemporary challenges and urban governance is symbiotic, emphasizing the pivotal role of policy formulation in sculpting the future trajectory of Portuguese cities.

21.3.3 Public Policies for Urban Development

The recent evolution of urban development in Portugal has been underpinned by transformative legislative changes and initiatives aiming to reinforce governance principles, bolster participation, and facilitate multiscalar approaches in urban governance. This transformation unfolds within the framework of a hierarchical governance structure encompassing four levels: central, regional, intraregional, and local. However, this structure exhibits a dual centralism, heavily emphasizing national government and municipalities as the primary administrative and political entities, sidelining regional autonomy except for the autonomous regions of Madeira and Azores.

In response to this centralism, the national government initiated decentralization efforts and participatory processes to strengthen governance principles. Notable initiatives include the establishment of Intermunicipal Communities and Metropolitan Areas, facilitating intraregional planning, and coordinating municipal investments of shared interest. Additionally, Law No. 50/2018 expanded the scope of decentralized competencies, covering areas like education, health, culture, and transportation, empowering municipalities to delegate authority to parish structures through inter-administrative contracts.

Further changes were catalyzed by urban development strategies, notably POLIS XXI, which advocated participatory models and regeneration strategies based on partnerships and networking across various sectors. This philosophy resonated through the Sustainable Cities 2020 Strategy, aligning with EU-supported Strategic Urban Development Plans and Urban Regeneration Action Plans within Portugal 2020. These legal frameworks reinforced participatory logics in planning and decision-making processes, complemented by funding mechanisms like Jessica and IFFRU 2020 (Decree-Law 104/2004; Decree-Law No. 312/2007; Decree-Law No. 307/2009; Law No. 5-A/2002; Law No. 45/2008; Law No. 75/2013; Law No. 50/2018).

However, the practical application of these initiatives revealed complexities and limitations. Urban regeneration policies adopted a multidimensional approach intersecting social, economic, and environmental dimensions. Yet, the implementation faced challenges in integrating these diverse objectives, often privileging physical rehabilitation over social and economic perspectives. The evolution of policies demonstrated variations in emphasis between governments, with right-wing initiatives favoring private initiatives, while left-wing policies prioritized social aspects and participation.

The evolution of governance and participatory approaches witnessed a shift from Polis XXI's comprehensive urban rehabilitation to a more market-oriented focus for economic competitiveness and innovation. Initiatives like the Partnerships for Urban Regeneration (PUR) initially sought to foster citizen participation but faced hurdles in realizing effective governance and participation in decision-making processes.

Despite significant strides in policy instruments and strategic plans advocating governance principles and citizen involvement, effective civic engagement remains a challenge. Municipalities retain exclusive competence in territorial management, often collaborating with large private groups in public–private partnerships, limiting involvement from civil society and associative sectors. While the intent to engage citizens persists, there's an evident gap between the strategic objectives of urban rehabilitation support programs and the practical realization of effective civic participation in decision-making processes.

Nine constitutional governments were in office, led by the Socialist Party—PS (Progressive Alliance of Socialists and Democrats in the European Parliament) or the Social Democratic Party—PSD (European People's Party—Christian Democrats) in a government coalition with the christian democrats of CDS-PP: PS (1999–2002); PSD, with CDS-PP (2002–2004 and 2004–2005); PS (2005–2009 and 2009–2011); PSD, with CDS-PPP (2011–2015 and 2015); PS with the support of left wing parties (2015–2019); PS (2019–2024). In essence, while Portugal's legislative landscape and policy frameworks emphasize governance principles and citizen involvement, translating these intentions into effective participatory decision-making processes remains a considerable challenge within the territorial management and planning domain, especially in urban areas (Table 21.1).

21.3.4 Stakeholder Engagement and Participation

The landscape of urban governance in Portugal has embarked on a profound transformation, signaling a departure from historical top-down methodologies towards a more inclusive and participatory approach. This shift signifies a pivotal recognition of the indispensable role that stakeholder engagement plays in not just shaping cities but also in ensuring their vibrancy and sustainability.

In recent times, Portugal has consciously embraced a more holistic framework, acknowledging the multifaceted contributions that diverse stakeholders offer in the

Table 21.1 Major urban policies in Portugal (21st century)

Date	Urban policy instrument	Type	Major goals (simplified)	Governance impacts	Government in office
2004–2018	Urban Rehabilitation Companies	Territorial strategy Operational	To promote the urban rehabilitation of historic areas and critical urban areas	Public–private partnerships are fostered Strategic planning and community empowerment and participation are understood as nuclear to urban areas management	PSD
2007–2013	POLIS XXI City Policy	Territorial strategy	To qualify and integrate different spaces in inclusive, coherent, sustainable and well informed urban operations with the participation of citizens To strengthen and diversify the human, institutional, cultural and economic capital of each city To qualify and intensify the integration of the city in the surrounding region To innovate for qualification	Local/intraregional cooperation, partnerships, area-based initiatives and multi-level urban actions are promoted Governance principles are introduced to urban policies and strategies	PS

(continued)

Table 21.1 (continued)

Date	Urban policy instrument	Type	Major goals (simplified)	Governance impacts	Government in office
2007–2010	Partnerships for Urban Rehabilitation	Legal/ Financial	To support actions for the revitalization of critical neighborhoods, abandoned urban areas, historic centers and river front, considering environmental, physical, economic and social dimensions To develop new forms of urban management, strengthening citizen participation and cooperation	Local partnerships between public entities, private sector and civil society is promoted Governance principles are introduced to urban policies and strategies	PS
2009	Legal Regime of Urban Rehabilitation	Legal	To qualify and revitalize cities, in particular their most degraded areas, and qualify the housing stock, seeking a more harmonious and sustainable urban life for all, ensuring decent housing	Area based initiatives and strategic planning within communities is fostered	PS
2009–2021	JESSICA Fund Portugal	Financial	To support private rehabilitation operations of higher risk	Public–private partnerships are fostered, with direct funding to private rehabilitation of buildings	PS

(continued)

Table 21.1 (continued)

Date	Urban policy instrument	Type	Major goals (simplified)	Governance impacts	Government in office
2014–2020	Strategic Plans for Urban Development	Legal/ Financial	To promote low-carbon strategies To develop measures that improve urban environment, recover and decontaminate abandoned industrial areas To support the physical, economic and social regeneration of disadvantaged communities in urban and rural areas	Public-centered strategies for urban development Private/civil society participation is valued Governance models are promoted	PSD
2015	Sustainable Cities 2020 Strategy	Territorial strategy	To qualify in an integrated way, promoting the functional, cultural, social and economic development of diverse problematic urban areas, including historical centers	Area-based initiatives and community empowerment and participation are promoted Governance principles are relevant for urban policies and strategies success	PSD

(continued)

Table 21.1 (continued)

Date	Urban policy instrument	Type	Major goals (simplified)	Governance impacts	Government in office
2018–2021	Financial Instrument for Urban Rehabilitation and Revitalization	Financial	To provide easy access to financing for investment promoters in the area of urban rehabilitation To promote the integral rehabilitation of buildings aged 30 years or older, including social housing and industrial units	Public–private partnerships are fostered, with direct funding to private rehabilitation of buildings	PS
2021–2026	Digital commercial neighborhoods (Recovery and Resilience Plan)	Financial	Promote the digitalization of the economy, either through technological adoption by economic operators and the digitalization of their business models, or through raising awareness and training workers and entrepreneurs	Public–private partnerships are fostered, with direct funding to digital transition and the capacitation of retailers and entrepreneurs	PS

(continued)

Table 21.1 (continued)

Date	Urban policy instrument	Type	Major goals (simplified)	Governance impacts	Government in office
2023–2030	Portugal 2030	Territorial Strategy	To support development strategies at local level, promoting social and territorial cohesion, and supporting sustainable urban development, based on the concept of interconnection of networks, centered on people's needs. Urban rehabilitation and regeneration, focusing on the challenges associated with attracting population to urban centers, the provision of inclusive and accessible spaces that enhance intra- and inter-urban mobility (including the NEB concept), and energy efficiency objectives	To be implemented and evaluated	PS

Source Fernandes et al. (2020) (updated)

urban governance sphere. This recognition extends beyond traditional actors like local authorities and incorporates a broad spectrum of voices and expertise. From residents who intimately understand their neighborhoods to civil society organizations advocating for specific causes, from the private sector offering innovative solutions to academia contributing research and knowledge—these stakeholders collectively form the tapestry of engagement, each thread essential in weaving a cohesive vision for urban development (Nye 2011; Seixa and Albet 2012; Teles 2018).

The evolution from rigid, top-down governance to more inclusive models reflects a profound paradigmatic shift in how decisions are made regarding urban spaces. It's a departure from unilateral directives to a symphony of voices where decisions are not imposed but collectively crafted (LeGates and Stout 2016). This departure is not merely a change in methodology; it embodies a fundamental shift in governance philosophy—one that acknowledges the inherent value of diverse perspectives and the power of collaborative decision-making.

These shifts are evident in various initiatives across Portugal. In fact, several municipalities in Portugal have implemented participatory budgeting programs, allowing citizens to directly influence the allocation of a portion of the municipal budget. Cities like Porto and Lisbon have successfully engaged residents in deciding on community projects, infrastructure improvements, and social initiatives, fostering a sense of ownership and active citizenship (Fernandes and Seixas 2018; Chamusca 2023).

Digital platforms have been instrumental in democratizing urban planning, providing avenues for citizens to contribute ideas and critiques. Initiatives such as digital platforms for collaborative urban planning have emerged, enabling citizens to contribute ideas, provide feedback, and engage in dialogue regarding urban development projects. These platforms, exemplified by initiatives like the 'Planos Locais de Habitação' (Local Housing Plans), facilitate transparent decision-making by integrating citizen input into planning processes.

Grassroots projects and associations have also empowered communities to actively engage in neighborhood revitalization, acknowledging the intrinsic importance of local knowledge and involvement in shaping their living spaces. Initiatives like 'Bairros Saudáveis' (Healthy Neighborhoods) and 'Viver a Mouraria' in Lisbon are notable examples. These grassroots initiatives empower residents to actively participate in neighborhood revitalization efforts, addressing social, cultural, and environmental aspects of urban life.

Finally, urban regeneration projects, such as those carried out in the context of the POLIS XXI initiative, have sought to involve multiple stakeholders in revitalizing urban areas. While challenges exist in achieving perfect integration of diverse interests, these initiatives have aimed to engage stakeholders ranging from local authorities to private investors in shaping urban renewal strategies.

These examples underscore Portugal's commitment to fostering inclusive and participatory governance models, enabling diverse stakeholders to contribute their expertise, perspectives, and aspirations towards co-creating more livable and resilient urban environments.

The broader recognition of stakeholder engagement as a cornerstone of urban governance signifies a maturing understanding that sustainable, vibrant cities are not merely built through infrastructure but through inclusive, collaborative, and transparent decision-making processes. This evolving approach seeks not just to address immediate urban challenges but to lay the groundwork for resilient and adaptable cities that embody the aspirations and values of their inhabitants. However, challenges persist, including ensuring equitable representation of marginalized groups, navigating power dynamics, and translating participatory input into meaningful policy outcomes.

21.4 Policy Recommendations for Future Urban Development

The landscape of urban governance in Portugal has experienced a notable evolution, marked by a growing acknowledgment of stakeholder engagement's critical role in shaping sustainable and vibrant cities. This recognition reflects a broader global shift towards inclusive and participatory governance models, where decision-making processes involve diverse actors. Portugal, in recent years, has been actively exploring multifaceted stakeholder engagement strategies to harness the expertise and perspectives of residents, civil society organizations, the private sector, academia, and local authorities in shaping urban spaces. Thus, based on the previous discussion and on the international experience, a set of recommendation can be proposed, considering:

- Enhancing and institutionalizing participatory mechanisms across governance levels, namely by encouraging municipalities to adopt participatory budgeting, digital engagement platforms, and regular community forums would ensure diverse voices actively contribute to urban policies.
- Fostering collaborative efforts among stakeholders will be pivotal, which requires the design of frameworks that incentivize partnerships for urban development projects to leverage collective expertise and shared responsibility.
- Investment in programs aimed at enhancing the capacity of local governments and stakeholders in participatory governance is imperative. Supporting knowledge-sharing platforms, training, and best practices exchange will ensure effective implementation and sustained engagement.
- Developing user-friendly digital platforms that offer real-time information and facilitate transparent communication between authorities and citizens will foster greater accountability.
- Prioritizing policies that ensure inclusivity and equity in urban development is paramount, as creating targeted programs to address social disparities, promote affordable housing, and prioritize the needs of marginalized communities will prevent exclusion in the development process.
- Embracing sustainable urban planning approaches, prioritizing environmental conservation, green infrastructure, and climate resilience through principles of circular economy and renewable energy integration will ensure long-term urban sustainability.
- Implementing robust monitoring and evaluation mechanisms to assess the impact of policies and projects, as embracing adaptive governance models will allow flexibility and iterative adjustments based on feedback and evolving urban dynamics.

- Developing comprehensive national and regional policy frameworks aligned with European Union strategies will provide clear guidelines and incentives for participatory governance, sustainable development, and innovation in urban planning.
- Encouraging local initiatives and grassroots-driven projects is important and must be aligned with supporting grassroots organizations will enable active community participation in shaping neighborhoods and fostering a sense of ownership.
- Cultivating a culture of continuous engagement and communication among stakeholders is crucial, as open dialogue, information sharing, and feedback loops will ensure ongoing collaboration and adaptive decision-making.

21.5 Conclusion

The exploration of urban governance trends in Portugal unveils a transformative shift towards inclusive governance paradigms. Stakeholder engagement, participatory frameworks, and sustainable planning emerge as critical determinants shaping the trajectory of the nation's cities. The integration of collaborative decision-making involving diverse stakeholders signifies a pivotal transition towards more responsive and resilient urban development strategies.

Recent initiatives in Portugal, including participatory budgeting, collaborative planning platforms, and grassroots associations, exemplify the country's commitment to inclusive governance models. These strategies, woven into the fabric of urban governance, instill a sense of ownership among citizens and stakeholders, resulting in sustainable, vibrant, and equitable urban spaces.

The significance of effective governance and well-crafted policies in steering Portugal towards future urban development goals cannot be overstated. The outlined policy recommendations—centered on participatory frameworks, collaborative partnerships, technology integration, and sustainable planning—underscore the necessity of proactive and inclusive approaches. Implementation of these recommendations stands as a linchpin in ensuring resilient, liveable, and responsive cities amid evolving challenges.

Portugal's trajectory towards participatory governance and sustainable urban development presents avenues for deeper exploration. Investigating the enduring impacts of participatory initiatives on urban resilience, conducting comparative analyses of governance models across municipalities, and assessing the scalability of successful participatory programs warrant deeper investigation. Furthermore, research exploring the effectiveness of technology-driven transparency in urban decision-making and examining the social implications of inclusive governance strategies on community cohesion would augment our comprehension.

In conclusion, Portugal's urban governance evolution signifies a promising pathway towards inclusive and sustainable cities. Embracing effective governance frameworks, underpinned by robust policies, is instrumental in fostering cities characterized by collaboration, inclusivity, and adaptability. This trajectory sets the stage for a resilient urban future in Portugal, promising continued progress and innovation in urban development.

References

Acuto M (2020) Understanding the global ecosystem of city networks. In: Urban studies, pp 1–17

Ascher F (1998) Métapolis, acerca do futuro da cidade. Celta Editora, Oeiras

Bieber F (2018) Is nationalism on the rise? Assessing global trends. Ethnopolitics 17,5:519–540. https://doi.org/10.1080/17449057.2018.1532633

Brenner N (2004) New state spaces. Urban governance and the rescaling of statehood. Oxford University Press, Oxford

Brenner N (2004b) New state spaces. Oxford University Press, Oxford

Cave B, Kim J, Viliani F, Harris P (2020) Applying an equity lens to urban policy measures for COVID-19 in four cities. Cities Health. https://doi.org/10.1080/23748834.2020.1792070

Carvalho L, Chamusca P, Fernandes J, Jorge Pinto J (2019) Gentrification in Porto: floating city users and internationally-driven urban change. Urban Geogr 40(4):565–572. https://doi.org/10.1080/02723638.2019.1585139

Chamusca P (2012) Governança e regeneração urbana: entre a teoria e algumas práticas. FLUP. Porto. 400p. Tese de doutoramento

Chamusca, P.; Marques, J. L.; Moreno Pires, S. & Teles, F. (2022) Territorial cohesion: discussing the mismatch between conceptual definitions and the understanding of local and intra-regional public decision-makers. In: Territory, politics, governance. https://doi.org/10.1080/21622671.2022.2044899

Chamusca P (2023) Urban planning and policy in Portugal: an overview on the role of EU funds and guidelines. Urban Res Pract 16(1):44–65. https://doi.org/10.1080/17535069.2021.1980607

Decree-Law 104/2004. Available at https://data.dre.pt/eli/dec-lei/104/2004/05/07/p/dre/pt/html. Accessed: 05/01/2024

Decree-Law No. 312/2007. Available at https://data.dre.pt/eli/dec-lei/312/2007/09/17/p/dre/pt/html. Accessed: 05/01/2024

Decree-Law No. 307/2009. Available at https://data.dre.pt/eli/dec-lei/307/2009/10/23/p/dre/pt/html. Accessed: 05/01/2024

Fernandes JA, Seixas J (2018) Cities and Urbanisation in Democratic Portugal in Méditerranée, Revue géographique des pays méditerranées, número 127 https://journals.openedition.org/mediterranee/10698?lang=fr

Fernandes JA, Teles F, Chamusca P, Seixas J (2020) The power of the cities and the power in the cities: a multiscale perspective from Portugal. Boletín De La Asociación De Geógrafos Españoles (87). https://doi.org/10.21138/bage.2978

Goh K (2020) Flows in formation: the global-urban networks of climate change adaptation. Urban Stud 57(11):2222–2240. https://doi.org/10.1177/0042098018807306

Hall P, Pain K (eds) (2006) Polycentric metropolis: learning from mega-city regions in Europe. Earthscan, London, Sterling, VA

Harvey D (2010) The enigma of capital and the crisis of capitalism. Profile Books, London

INE (2023) Territorial statistics. Available at https://www.ine.pt. Accessed: 05/01/2024

Innerarity D (2020) Una teoría de la democracia compleja. Gobernar en el siglo XXI. Galaxia Gutenberg, Barcelona

Kim S (2020) Inter-municipal relations in city-region governance. Cities (104). https://doi.org/10.1016/j.cities.2020.102771

Law No. 5-A/2002 of 11/01. Available at https://data.dre.pt/eli/lei/5-a/2002/01/11/p/dre/pt/html. Accessed: 05/01/2024
Law No. 45/2008 of 27/08. Available at https://data.dre.pt/eli/lei/45/2008/08/27/p/dre/pt/html. Accessed: 05/01/2024
Law No. 75/2013 of 12/09. Available at https://data.dre.pt/eli/lei/75/2013/09/12/p/dre/pt/html. Accessed: 05/01/2024
Law No. 50/2018 of 16/08. Available at https://data.dre.pt/eli/lei/50/2018/08/16/p/dre/pt/html. Accessed: 05/01/2024
LeGates R, Stout F (eds) (2016) The city reader. Routlege, Oxford and New York
Lourenço R, Ponte C, Chamusca P, Lameiras M (2023) Atores, conceitos e debates: o que está em causa em discursos mediáticos sobre regionalização. In: Carmo A, Rego C, Branco M, Baltazar M (eds) (Re)visitar a Regionalização 25 anos depois. Edições Almedina, S.A., pp 143–171
McCann P (2015) The Regional and Urban Policy of the European Union: cohesion, results-orientation and smart specialisation. New Horizons in Regional Science series. Edward Elgar Pub ISBN-10: 1783479507. 304 p
Merrifield A (2013) The urban question under planetary urbanization. Int J Urban Reg Res 37(3):909–922
Nello O (2015) La ciudad en movimiento. Días & Pons, Madrid
Nye J (2011) The future of power. Public Affairs, New York
OECD (2020) Cities policies responses. Report, Available at: http://www.oecd.org/coronavirus/policy-responses/cities-policy-responses-fd1053ff/
Parnell S (2016) Defining a global urban development agenda. World Dev 78:529–540
Piketty T (2014) O capital no século XXI. Temas e Debates, Lisbon
Sassen S (1991) The global city. Princeton University Press, New York, London, Tokyo, Princeton
Seixas J, Albet A (2012) Urban Governance in Southern Europe. Ashgate, Londres
Sennett R (2016) Building and dwelling: ethics for the city. Allen Lane, London
Subirats J (2016) El poder de lo próximo: las virtudes del municipalismo. Los libros de la catarata, Madrid
Teles F (2016) Local government and the bailout: reform singularities in Portugal. Eur Urban Reg Stud 23(3):455–467
Teles F (2018) Animal farm: the new political leaders in mass democracies. In: Tony Middlebrooks (ed.) Symposium. J Leadersh Stud 12(4):69–70

Chapter 22
Governance and Urbanism: The Search for a Better City Challenges and Recent Advances in Spain

Joaquín Farinós Dasí

Abstract The chapter intends to review the major stages of urban local government in Spain since the mid-twentieth century. In the first place, the city apparently regulated by General Municipal Planning Plans, which had to specify which new sectors were reserved for urbanization, to channel the growth of the nucleus, as well as which were left aside in the following years, has been surpassed. The government based on plans solved the housing problem in the main cities, even if it was at the cost of a certain precariousness of urban planning in public spaces or in endowments for the community. In addition, it reinforced for a long time the vertical and compact city model, positive in some aspects, but which has been associated with chronic congestion problems. With the economic and real estate crisis that began in 2008, largely the result of a speculative bubble, the rate of expansion of cities and urban spaces has slowed drastically. The General Plans lose many of their functions and, on occasions, are blocked by judicial resolutions that prevent their execution. For these reasons, and secondly, a new stage of urban governance has opened, which calls for participation and sustainability. This model is still in the process of being defined, although the Urban Agendas or sustainable urban development strategies, supported by the European Union, can show us some aspects of the future of real city governance in the coming period.

Keywords Urban planning and management · New urbanism · Local urban agendas · Metropolitan governance · Spain

This chapter has been prepared on the basis of other previous works by the author, updated and adapted to the objective of this publication, in the framework of the Research Project GOBEFTER-III entitled 'Proposal for institutional and community design for comprehensive territorial planning in the transition towards a sustainable economy', PID2021-128356NB-I00, financed by MCIN/AEI/10.13039/501100011033 and by FEDER A way of making Europe.

J. F. Dasí (✉)
IIDL and Department of Geography at the University of Valencia, València, Spain
e-mail: joaquin.farinos@uv.es

R. C. Lois-González and J. A. Rio Fernandes (eds.), *Urban Change in the Iberian Peninsula*, The Urban Book Series, https://doi.org/10.1007/978-3-031-59679-7_22

The analysis of urban governance in the Spanish case is approached from the recognition and diagnosis of its different dimensions: citizen participation (more present in the second section), intersectoral and interdepartmental coordination (in the third one), territorial cooperation and multilevel coordination and cooperation relationships (mainly in the fourth one).

If one understands governance as innovation/evolution in terms of government-planning-management of the city, we must refer to the evolution of urban policy (in the first section of this chapter). As well as its performance and to what extent it has allowed or it can allow certain levels of improvement to overcome its limitations (usually forced, as explained in the second section). Largely because of the change in context conditions (as discussed in the third one). Advances in terms of more sustainable and 'do-ocratic' urban planning (vid. Lloret and Farinós 2018) are mostly the result of the crisis and citizen demands than political will and leadership. The fourth and last section focuses on the specific issue of metropolitan governance.

Cities must face new challenges and needs, which has led them to formulate new Urban Agendas. As a result of international influences (Habitat III, the European Urban Agenda, and the Integrated Sustainable Urban Development Strategies—ISUDS), a more comprehensive vision of the city is making its way in Spain, with greater prominence on the social dimension. However, if this new urbanism is possible, it has not yet been fully generalized. A deep-rooted urban tradition, highly regulated and based on the generation of capital gains through changes in land use rights, continues to be the main way of managing the city (in an uncoordinated, sectoral, and project-based manner, not as a real system; besides more focused on the physical than on the social). Few of the Spanish cities get away from this, and, if they do, they only can do it in an ephemeral way, to stagger when there is any political cycle change. Real estate and building businesses continue to be the great object of desire, and they are seen, together with tourism, as the most desirable engines for economic recovery.

Despite political discourse easily talks about a new and updated model (of sustainability, of a climate emergency to face the appropriate strategies, of the 'New Green Deal', of sustainable mobility, of adaptive urbanism, of integrated sustainable urban development…), the truth is that progresses have not been so many nor so easy. The great influence of tradition (politics, administration, spatial planning and management) continues to be decisive, coming with innovations usually from outside. In other words: so much 'path dependency' and very little 'path creation'.

22.1 The Bursting of the Real Estate Bubble and the Subsequent Economic Crisis as the Main Ally for Good Urbanism and Sustainability

Since 2007 (the year in which the housing bubble burst and the economic and financial crisis started), nothing seems to have contributed as much to control growth and urban sprawl and to the sustainability of the territorial model that such a crisis. In the Spanish case, its Land Law of 1956 made the management of capital gains associated with the reclassification of land uses (from undevelopable to developable land) the way to produce a city. Despite the questionable distribution of these gains in favor mostly of land owners instead administrations responsible for such reclassification (which benefit from the associated fees and taxes for this change of land uses, as well as the reserve of 10% of the developable area). The biggest nonsense will take place with the approval of Law 6/1998, of April 13, on Land Regime and Land Valuations (also impugned in court and partly annulled), later topped with the Royal Decree-Law 4/2000, of June 23, on Measures Urgent Liberalization in the Real Estate and Transportation Sector. From then on, the burden of proof is reversed, making it necessary since then to justify the decision to classify land as undevelopable (that is, justify the reasons for its protection), being developable otherwise. That largely contrasts with the pre-existent rule and tradition in which that was proposed by the classification was the new developable land. The impacts of such a new legal framework on the territory and its sustainability were devastating.

It was the reaction (in the opposite direction, after the change of government in 1996: from progressists–PSOE-to conservatives –PP-) to the initiative of the Consolidated Text of the land law of 1992, afterward annulled by the Constitutional Court through its well-known STC 61/1997. Among other effects, it caused each Spanish Autonomous Region–CCAA-(which have this exclusive competence in its hands since then) to have its urban law. For this purpose, they followed three models (Farinós et al. 2005): the CCAAs that would continue with the aforementioned Land Law of 1956 style (Andalusia, Catalonia or the Basque Country); those looking for a progressive installation of neoliberal criteria tending to land deregulation and commodification (with Madrid at the forefront); in the very specific case of the Valencian Autonomous Region through the adaptation to this same goal (what it was not difficult) of some previous figures such as those of the Land Developer Agents. All this led to a spatial development model based on building and real estate speculation, as well as to a progressive dependence on local financing of urban sprawl.

A new stage began with the new Land Law of 2007, again challenged in the TC, giving rise after its ruling to a new text, the Royal Legislative Decree 2/2008, of June 20, which approves the text recast of the Land Law. This one, together with the subsequent Law 8/2013, of June 26, on urban rehabilitation, regeneration, and renewal, gave rise to the currently in force Royal Legislative Decree 7/2015, of October 30, by which the revised text of the Land and Urban Rehabilitation Law is approved (at this moment under revision; see https://www.congreso.es/public_oficiales/L15/CONG/BOCG/A/BOCG-15-A-19-1.PDF). This put an end to the prodigious decade of urban growth in Spain (Burriel 2008) and to the so-called casino capitalism (Romero 2005), with associated cases of political corruption that have already been tried today (others are still pending in the courts). From then on, the option for a new urbanism opened up. This was more focused on the rehabilitation, regeneration, and renewal of the consolidated city and on the control of urban growth (the EU has in its roadmap to reduce to 0% net urban growth by 2050).[1]

22.2 City: From Speculative Product to Recovered Living Space; the New Urban Agendas. Territorial and Urban Conflicts as a Trigger for Government Changes

The goals of: integrated sustainable urban development (and its Strategies –ISUDS-, in which 5% of the European Regional Development Fund –ERDF- of each country was invested in the period 2014–2020)[2]; a healthy and inclusive city; reduction of obliged mobility and to make it intermodal and low carbon; have the necessary open and community spaces integrated into a Green Infrastructure at several scales (from the neighborhood to regional); all of them are part of the new Urban Agendas (UN Habitat III, European Urban Agenda, Spanish Urban Agenda, Autonomous Urban Agendas, and Municipal Urban Agendas). With this, it goes beyond the traditional morphological urbanism and focuses on its most functional dimension and the improvement of the quality of life of citizens (now more involved through the corresponding public participation plans).[3]

Cities are privileged spaces for change and transformation, where social innovation, new forms of relationship and coexistence, and new forms of citizenship happen. New Urban Agenda processes at the local and supranational level stem from the conviction that cities are places where the place where sustainable development challenge is won the most. However, the institutional statute of cities does not always adjust to being able to give an adequate response to increasingly complex challenges. There is a certain mismatch between the capacities they have and the importance of the challenges they face (González Medina and De Gregorio Hurtado 2019). For

---

[1] European Commission (2016): "Science for Environment Policy. Future Brief: No net land take by 2050?". https://www.eea.europa.eu/data-and-maps/indicators/land-recycling-and-densification/no-net-land-take-by (acceded on 12.05.2023).

[2] In 2016, these ISUDS (Integrated Sustainable Urban Development Strategies), continuation of the Urban Initiative, began to be developed in 83 Spanish cities, for an amount of €710 million. They were followed by another 40 in the second call, for an approximate amount of €300 million. Finally, in the third call, €357 million (De Gregorio Hurtado 2016).

[3] An outstanding example of this is the detailed review of the current General Plan for Urban Planning of the city of Valencia (from 1988), giving rise to the "Special Plan of guidelines for the improvement of the urban quality of the neighbourhoods of the city of Valencia" (approved on 09.07.2020). Its main criteria are: ensuring adequate access to basic services at the neighbourhood level (with a vision of the whole system) and the concept of micro-scale Green Infrastructure as key element for spatial articulation (vid. Gregori et al. 2019).

this reason is proposed the transition from a bureaucratic logic (based on hierarchy and competition principles), to new practices facilitating stakeholder mobilization and interrelation, the optimization of resources, and the solution of problems in a collaborative way (urban and metropolitan governance).

The Minister of Public Works presented the Spanish Urban Agenda at the Council of Ministers on 22nd February 2019. For their part, some regions and cities already have urban agendas and action plans. This is the case of the Autonomous Communities of Andalusia, Catalonia, Extremadura, and the Basque Country; for its part, since 2014 the Eixo Atlántico has been working on an Urban Agenda for the Urban System of the Euroregion. In the Valencian Autonomous Region, the Valencian Government has been developing the Valencian Urban Agenda. At a local level, cities such as A Coruña, Castellón de la Plana, Málaga, Murcia, and Sant Feliu de Llobregat already have approved their Urban Agenda. Others, such as Alfaro (La Rioja), Algeciras (Cádiz), and Aranda de Duero (Burgos)… have started it; or are on the way to doing it, as the small municipality of Rincón de la Victoria (in the province of Málaga) (for more details see http://www.aue.gob.es/).

Due growing disconnection between population will and political decisions in territorial and urban matters, together with the perception of the persistence of a model based on corruption, various citizen protest movements arose (such as the 15 M Movement, of May 2011), territorial conflicts (NIMBY style) and 'save' movements. All of them have been profusely studied (see Subirats 2011). The evolution from simple protest to proactive proposals (from "not here" to "here, like this, yes") gave rise to a new urbanism of a more locally participated nature in some Spanish cities (Barcelona, Madrid, Valencia). This new urbanism has been associated with a change in the local government in these cities, led by new parties brought to power by these citizen movements (see Montiel and Guillén 2019); respectively 'Barcelona en Comú', 'Más Madrid' or 'Compromís' in Valencia. These new parties took the initiative and helped to break the old bipartisanship hegemony of the more traditional parties. This has promoted changes in the design and implementation of urban policies; although in some cases with a tour limited to a single legislature (from 2015 to 2019) in the case of Madrid (city and regional government), in Zaragoza, and in the three Galician cities of Santiago de Compostela, Ferrol, and A Coruña.

The rise of bottom-up initiatives is going beyond the current established institutional framework, producing a change both in the role and in the relations between local administrations and citizens. Involving citizens in public decision-making processes can promote a greater understanding of the problem and co-responsibility for the solution among the participants, increasing the legitimacy of public policies (Subirats and Parés 2014). The emergence of these initiatives has been determined by various factors (vid. Lloret and Farinós 2018) as: political decisions are far from the citizen's demands; lack of synchrony between the administrative procedures regarding citizen needs; discontent on 'top-down' procedures; the uncomfortable way in which participatory processes are carried out; change in the mentality of many citizens motivated by the context of economic austerity that leads them to act (do-ocracy) instead of waiting for the public power to solve their problems (for example in the face of abandonment and paralysis of many plans, generating many empty and unused urban spaces).

The new participatory governance processes in the city and territory can not only generate greater legitimacy and create new decision-making routines but also promote changes in the values themselves that regulate the relationships between stakeholders and economic and territorial trends (Gallego and Pitxer 2019). This type of relationship is more frequent and possible, and also more conflictive, in those spaces where exchanges and pressures derived from density, diversity, and competition of uses are greater. This happens in large cities and their metropolitan areas or regions, sometimes exceeding not only provincial but also regional borders (vid. Lois and Piñeira 2017; Rando 2019), which leads to thinking about new forms and practices of territorial governance and cooperation (vid. Farinós 2019; Trillo and Paul 2019).

22.3 Is a New Urban Planning and Management Possible? Outstanding Issues

Rules and routines designed to provide security, sometimes end up hindering innovations for better management from which to solve new challenges by developing new forms of governance. The rigidity of the procedures (the time factor is crucial) ends up affecting the possibilities of adequate policy implementation in the city. The participation and empowerment of the population and decision-makers then become the way to advance in resolving this bottleneck.

The complexity of the planning and implementation processes, as well as the laborious and complicated process of their final approval, mean that updating spatial and urban plans is seen as a waste of time and resources (vid. Benabent 2019). Also as a source of uncertainty: unsuspected formal defects in its processing (given the high number of required mandatory reports) may mean the cancellation of the plan in its entirety after long years of proceedings (this is the fundamental reason for the recent bill (on 05/04/24) that seeks to modify the text of Law of 2015). Under these conditions, and given the difficulties of adequate inter-administrative coordination to have the mandatory sectoral reports on time, more and more administrations are opting for: the simplification of procedures, the indefinite term of plans, or replacement plans with projects instead. Some legislative initiatives, such as those carried out in Andalusia and Extremadura, tend to shorten deadlines by simplifying procedures and better intersectoral and interadministrative coordination. The composition and operation of the collegiate bodies (both advisory and decision-making) are reconsidered, increasing their participation before plan processing starts (instead of doing it in its final phases); also new figures such as the Commissions for Provincial Urban Coordination (in the Andalusian case) or the Intersectoral Coordination Commission (in Extremadura) are established. (Farinós et al. 2019).

In the Valencian case, the modification of the Spatial Planning, Urbanism, and Landscape Law (LOTUP) in December 2015 streamlined and simplified the processing and approval of detailed planning (of exclusive municipal competence), except in the case land was not subject to the requirement of environmental assessment (due it should meet the requirement). Later Legislative Decree 1/2021, of June 18, of the Council for approval of the revised text of the Law on Spatial Planning, Urbanism and Landscape (since modification of the previous LOTUP of 2019), tries to make inter-administrative coordination more efficient (art. 55.4 and art. 57.4).[4]

22.4 From the City-Island to the Urban Region; the Pending Challenge of Metropolitan Governance

Since the beginning of the 21st Century, metropolitan areas largely explain the organization of the Spanish territory (Lois and Zúñiga 2017). They represent a particular territorial phenomenon, with specific concerns and challenges demanding attention. Their delimitation, particularly in Spain, does not have a great tradition in the proper treatment and management of the metropolitan phenomenon (Manero 2010). A series of advantages are recognized for them: creative and innovative population, accessibility and connectivity, international visibility, more efficient use of land and energy, space for new markets, social innovation and participation, cooperation, complementarity, and even identity (METREX 2014).

Despite Metropolitan Areas (MAs) have already been considered from an urban point of view since the 1940s, little progress has been made in terms of metropolitan governance. This is a crucial aspect, both in terms of political management (Hildenbrand 2017) as well as for relationships of the metropolis with other cities and with rural areas of the metropolitan region (GOLD IV, 2016). Post-war experimental urban planning, which would lead to the first Spanish Land Law of 1956 (LS56), was based on the figure of the Municipal Urban Master Plan (Plan General de Ordenación

[4] Law 1/2019, of February 5, narrows deadlines for issuing the mandatory sectoral reports (when the sector regulations do not specify their own deadlines). After a time limit of 45 days since the request for the sectoral reports (or the one established by the respective sectoral regulations) the procedure will continue.

For the purposes of collecting the mandatory reports according to the regulations but which have not been issued on time, prior to the provisional approval of the planning instrument, the city council may require the Valencian regional government to realibly convene to attend administrations with affected powers by the plan to attend to an Information Commission of Coordination of the General Directorate of Urbanism. The necessary documentation to issue the report will be attached to the call. This report may be resolved verbally at the same coordination meeting (art. 53.4 LOTUP of 2019). If the report is not issued in writing within the legally established term, nor verbally in the Information Commission of Coordination, it will be considered positive (favorable to the content of the plan) for all purposes. This will be reliably notified to the administrations that have not made an express report, which may oppose the approval through the requirements between administrations provided for in art. 44 of Law 29/1998, of July 13, on contentious-administrative jurisdiction. (Almenar and Antequera 2020).

Urbana) and in the organization of the growth of the central city over its adjacent municipalities.[5]

Two of the objectives of that urban-metropolitan planning were the reinforcement of the administrative capacities of the supra-municipal institution, and the necessary inter-municipal cooperation. Everything would change with the new socioeconomic reality of the Stabilization Plan and the technocratic approaches of the Opus Dei Spanish Government at this time (1956), getting back to the municipality and municipal plans as fundamental instruments. This deeply will penetrate Spanish Town Planning and still persist in current MAs. The metropolitan phenomenon is a challenge for all large cities since it requires mutual deference between governments and the affected territorial levels, which makes metropolitan management a complex and conflictive process.[6]

The legal-administrative formulas decided for MAs in Spain are diverse, but they have in common the participation and influence in different intensity of the higher levels of administrations especially regional ones (on whom MAs institutionalization and financing depends). Since its objective is the decentralization of some regional powers and the efficient concentration of other municipal ones, this has ended up hindering their constitution as a new political subject. Mainly because regional, provincial, and municipal governments oppose them (Lefèvre, 2010).[7] Added to this is the low interest of civil society in the political construction of the metropolitan phenomenon; especially if their demands and needs for services and equipment are satisfied by already existing figures (as commonwealths and, chiefly, consortiums) (Farinós et al., 2018).

Toscano (2011a: 117) distinguishes two types of metropolitan areas: territorial and institutional. The territorial ones have different functions and can be considered a local government with a vocation for general purposes and powers. However, in Spain is not a usual formula. It has only been explored in the MAs of Barcelona (which, with its ups and downs, persists and works) and Valencia and Vigo (where it no longer exists) (see Farinós et al. 2018). The institutional ones are created with

[5] In some cases, such as Madrid, through its assimilation by the main municipality; but not in the cases of Barcelona, Valencia or Bilbao. Madrid (1941) and later (once the work of the Reconstruction Board was completed there) Bilbao (1943), Valencia (1946) and Barcelona (1952), were the only four cities in which urban planning was promoted at metropolitan scale.

[6] Metropolitan Areas are local entities recognized in art. 43 of the Local Government Bases Law. They can be institutionalized in three major possible ways: metropolitan governments, vertical coordination by hierarchical levels or through sectoral agencies, and horizontal cooperation between municipalities. Regarding their organization, the administrative structures and institutional devices, metropolitan reality can be faced from three perspectives (Tomás 2009: 171–172): a) reformist (which proposes maximum institutionalization, with the creation of autonomous and strong metropolitan governments); rational choice (which support new institutions should not be created because the appropriate sphere is the municipal one; inter-municipal cooperation must be voluntary -fragmented model-); c) new regionalism, as halfway between the two (it proposes flexible metropolitan coordination mechanisms with little bureaucratic structures).

[7] The misgivings and problems of mistrust on these intermediate spaces of power are evident; because they imply a change in the power relations between those preexistent levels who already have their powers clearly established. The cases of Barcelona and Vigo are very good examples.

a specific function, clearly delimited, for which it has the necessary powers. For this reason, they have a certain specialized nature; transportation, water supply, and waste treatment are the three most widespread types.

It is not unusual for the metropolitan functions to be assumed by the higher levels if their respective geographical perimeters are coincident. The clearest exponent is the case of the Autonomous Community of Madrid. Consortiums, associative entities governed by public law created to carry out common goals of public interest, can also be set up by administrations of different territorial levels (Autonomous Regions and Municipalities) and include private actors. They are present in Andalusia, Aragon, Catalonia, Valencian Autonomous Region, Madrid, and the Basque Country. Their usual objective is the metropolitan management of a certain area of activity (sectoral), but they could also have a vocation for integrated management and coordination (integral) (Toscano 2011b).[8]

Political factors are a key element that determines the success or failure of the metropolitan experiences and their configuration. MAs must be embedded in a reality in which multiple levels already exist, all of them with power relations in permanent conflict (between the same municipalities of the metropolitan space and between political forces). In this context, it is necessary to think about possibilities and look for alternatives for progress that, although temporary or partial, point in the right direction, facing up to: (a) the blockade caused by misgivings between municipalities and cities; (b) the coordination problems between levels; and (c) the unusualness of a comprehensive vision, as a whole, to which the new metropolitan governments should adjust (because to the sectoral one is already led by the consortiums).

Given this evidence, it is worth looking for alternatives such as an evolution from the concept of the metropolitan area (of a single main city) to that of a polycentric metropolitan region (the Soja post-metropolis, 2011). This implies accepting a soft approach of flexible perimeters based on flows (in a heterarchical network of relationships) between metropolises and between them and their respective environments (as soft spaces). Both the close and frequent ones and the furthest away and occasional ones; in a reconsideration of urban–rural relationships and the resulting territorial matrix. If the redefinition of government levels and institutions to address metropolitan management faces limitations and legal and political limitations and brakes, it is worth expanding, in the meantime, on metropolitan spatial planning instruments with an integrated and strategic approach. Spatial planning appears, also in this case, as the most suitable field of application of the new practices of territorial governance, and we must continue insisting on it.

---

[8] A landscape characterized by plural and flexible metropolitan instruments, that are used simultaneously and in a complementary way ('soft planning' for these 'soft spaces'). Looking for a more coherent and potential management of these spaces (integral), rendering accounts transparently, rather than sectoral (partial) instruments; aspiring to structures, or in this case tools, for their planning and management in an integral way.

References

Almenar M, Antequera E (2020) Acerca del procedimiento de EAE y su incardinación en el procedimiento de elaboración y aprobación de planes. In: Farinós J (ed) Marco legal y procedimental de la ordenación del territorio en España: diagnóstico y balance. Thomson Reuters-Aranzadi, Navarra, pp 59–82

Benabent M (2019) La insoportable rigidez del Plan General Urbanístico. La necesidad de un cambio de modelo. Ciudad y Territorio Estudios Territoriales 201:451–466

Burriel EL (2008) La "década prodigiosa" del urbanismo español (1997–2006). Scripta Nova. Revista Electrónica de Geografía y Ciencias Sociales, vol XII, nº 270 (64). https://www.ub.edu/geocrit/-xcol/383.htm. acceded 14.05.2023

De Gregorio Hurtado S (2016) Is EU urban policy transforming urban regeneration in Spain? Answers from an analysis of the Iniciativa Urbana (2007–2013). Cities 60:402–414

Farinós J (2019) La cooperación horizontal de carácter territorial entre CCAA, un reto para la política de OT y para el modelo de organización del Estado. In: Farinós J, Ojeda J, Trillo JM (eds) España: Geografías para un Estado posmoderno. Madrid/Barcelona, AGE/Geocrítica, pp 187–205. https://www.fundicot.org/_files/ugd/1c299f_12611f439e114a46adbc6e35a931933d.pdf. acceded 14.05.2023

Farinós J, Peiró E, Antequera E (2019) Retos para la planificación y gestión territorial en España: las ineficiencias en el proceso de aprobación de los planes y sus causas. In: Farinós J (ed) Desafíos y oportunidades de un mundo en transición; una interpretación desde la Geografía. Valencia, PUV, pp 601–612

Farinós J, Peiró E, Gomis A (2018) Planificación y gestión territorial de los espacios metropolitanos. Evolución y análisis de la situación en España. In: Baron N, Romero J (eds) Cultura Territorial e Innovación Social. Valencia, IIDL/PUV, Colección 'Estudios y Documentos', vol 26, pp 157–191

Farinós J, Romero J, Sánchez de Madariaga I (2005) Structural problems for the renewal of planning styles: the Spanish case. Eur Plan Stud 13:217–235

Gallego JR, Josep Vicent Pitxer JV (2019) Propuesta de articulación institucional entre la planificación física y la planificación económica: un enfoque evolucionista y territorial. In: IX Congreso Internacional de Ordenación del Territorio: Planificación y gestión integrada. Santander, Fundicot, pp 311–327

González Medina M, De Gregorio Hurtado S (2019) Aproximación al poder de las ciudades globales iberoamericanas: reflexiones para una gestión integrada del territorio. In: IX Congreso Internacional de Ordenación del Territorio: Planificación y gestión integrada. Santander, Fundicot, pp 353–368

GOLD IV-Observatorio Mundial sobre la Democracia Local y la Descentralización- [en línea]: Cuarto Informe Mundial sobre la Descentralización y la Democracia Local. Co-creando el futuro urbano. La agenda delas metrópolis, las ciudades y los territorios. Resumen Ejecutivo. https://www.uclg.org/sites/default/files/gold_iv_resumen_ejecutivo.pdf. acceded 14.05.2023

Gregori Ferrer V, García Jiménez MJ, Hernández Nieto B, Tomás Estellés A (2019) La actividad urbanística en la ciudad consolidada. In: IX Congreso Internacional de Ordenación del Territorio: Planificación y gestión integrada. Santander, Fundicot, pp 169–179

Hildenbrand A (2017) Gobernanza y planificación territorial en las áreas metropolitanas. Análisis comparado de las experiencias recientes en Alemania y de su interés para la práctica en España. Valencia, PUV-IIDL, Colección 'Desarrollo Territorial', vol 18

Lefèvre C (2010) ¿Metrópolis ingobernables? Experiencias europeas. Política y Sociedad 47(3):131–143

Lloret P, Farinós J (2018) La dimensión participativa en el diseño de políticas urbanas. El caso valenciano. GAPP-Gestión y Análisis de Políticas Públicas 21 (Special Issue: 'Gobernanza y políticas de desarrollo urbano: teoría y práctica'), pp 36–52

Lois R, Piñeira MªJ (2017) El modelo territorial y urbano resultante. Un sistema urbano metropolitano y policéntrico más formal que funcional activo. In: Farinós J, Olcina J (eds) Geografía

Regional de España Espacio y comunidades. Bases para una regionalización renovada del territorio español. Valencia, Tirant Lo Blanch-PUV, Serie Humanidades, pp 429–507

Manero F (2010) La cultura territorial metropolitana: de las relaciones de conflicto a la gobernanza supramunicipal. Una aproximación a la experiencia española. Scripta Nova 331(20). http://www.ub.edu/geocrit/sn/sn-331/sn-331-36.htm. acceded 14.05.2023

Méndez R (2016) Renovar economías urbanas en crisis: un debate actual sobre la innovación. Desenvolvimento Regional Em Debate 6(3):4–31

METREX (2014) Metropolitan dimension. Position statement. METREX Secretariat, Glasgow

Montiel A, Guillén G (2019) Acord del Botànic. Balandra Ediciones, Valencia

Rando Burgos E (2019) La extraña relación de la Comunidad de Madrid con las políticas territoriales: un repaso y algunas propuestas. In: IX Congreso Internacional de Ordenación del Territorio: Planificación y gestión integrada. Santander, Fundicot, pp 844–859

Romero J (2005) Capitalismo de casino. El País, 22 de diciembre. https://elpais.com/diario/2005/12/22/opinion/1135206005_850215.html. accedido el 14.05.2023

Subirats J (2011) Otra sociedad ¿Otra política? Del no nos representan a la democracia de lo común. Barcelona, Icaria

Subirats J, Parés M (2014) Cambios sociales y estructuras de poder ¿Nuevas ciudades, nueva ciudadanía? Interdisciplina 2(2):97–118

Tomás M (2009) La governabilitat metropolitana a Europa i a l'Amèrica del Nord. Col·lecció Estudis, Sèrie Govern Local, 13. Barcelona, Diputació de Barcelona

Toscano F (2011a) El área metropolitana en el ordenamiento jurídico español. Cuadernos de derecho local (QDL) 25:112–120

Toscano F (2011b) Los consorcios metropolitanos. Madrid, Fundación Democracia y Gobierno Local, Serie Claves del Gobierno Local

Trillo JM, Paül V (2019) ¿Cooperación territorial alrededor de las fronteras interautonómicas? Algunas preguntas provocativas partiendo de la cooperación territorial europea. In: Farinós J, Ojeda JF, Trillo JM (eds) España: Geografías para un Estado posmoderno. Madrid/Barcelona, AGE/Geocrítica, pp 273–289. https://www.fundicot.org/_files/ugd/1c299f_12611f439e114a46adbc6e35a931933d.pdf. acceded 14.05.2023

Chapter 23
Territorial Governance and Soft Planning in the Region of Lisbon

Cristina Cavaco, Igor Matheus Santana Chaves, and Luís Sanchez Carvalho

Abstract Metropolitan territorial governance has posed significant challenges worldwide, cutting across various states and contexts. Within the ongoing dynamics of governance rescaling that challenge conventional statutory geographies, there has been a notable emphasis on soft planning and soft spaces, impacting both planning practice and theory. In Portugal, these processes have been markedly influenced by the European Cohesion Policy and the corresponding allocation of Structural and Investment Funds. By actively promoting integrated territorial approaches, these policies have played a pivotal role in encouraging non-statutory planning practices. In this context, this chapter delves into the role of soft planning in the metropolitan governance of the Lisbon Metropolitan Area within the framework of inter-municipal cooperation in Portugal. It scrutinizes several strategies and policy documents developed over the last decade, engaging in discussions about how these have fostered a supra-municipal awareness that transcends the mere aggregation of individual and autonomous municipal actions and decision-making.

Keywords Soft planning · Territorial governance · Governance rescaling · Inter-municipal cooperation · Lisbon Metropolitan Area

---

The original version of this chapter was previously published without open access. A correction to this chapter is available at
https://doi.org/10.1007/978-3-031-59679-7_27

C. Cavaco (✉) · L. S. Carvalho
CIAUD, Research Centre for Architecture, Urbanism and Design, Lisbon School of Architecture, Universidade de Lisboa, Lisboa, Portugal
e-mail: ccavaco@fa.ulisboa.pt

L. S. Carvalho
e-mail: lsc@campus.ul.pt

I. Matheus Santana Chaves
Universidade Federal do ABC, São Paulo, Brazil

Lisbon School of Architecture, Universidade de Lisboa, Lisboa, Portugal

I. Matheus Santana Chaves
e-mail: igor.chaves@ufabc.edu.br

R. C. Lois-González and J. A. Rio Fernandes (eds.), *Urban Change in the Iberian Peninsula*, The Urban Book Series, https://doi.org/10.1007/978-3-031-59679-7_23

23.1 Introduction

Over the past decade, soft planning and soft spaces have garnered significant attention in both planning practice and theory (Allmendinger et al. 2015; Cavaco et al. 2023; Purkarthofer and Granqvist 2021). Indeed, these concepts have been propelled by the emergence of innovative planning processes and dynamics of governance rescaling that disrupt the conventional boundaries of administrative jurisdictions and the spatial scales, processes, and instruments associated with hard spatial planning. Two sets of arguments advocate the emergence of this new planning paradigm: firstly, the shift in territorial governance facilitated by the development of more flexible governance arrangements that extend beyond established formal government settlements; and secondly, the ascent of new spaces for planning and political engagement that transcend the scales and boundaries set by traditional administrative-territorial units, having a voluntary and non-statutory character (Haughton and Allmendinger 2007, 2009; Walsh et al. 2012).

As the literature on the subject of soft planning and soft spaces continues to expand, the associated practices and approaches have also broadened. Although the debate initially centered heavily in the UK (Haughton et al. 2010; Stead 2014) and Northern Europe (Purkarthofer 2018), these concepts have transcended boundaries to elucidate similar practices emerging in other geographical contexts, including Southern Europe and notably, Portugal (Cavaco et al. 2021). In fact, the range of approaches within the realm of soft planning is extensive, encompassing not only various scales, from supranational to local community-led initiatives, but also a diverse array of drivers that have been fostering soft planning dynamics in various contexts.

In Portugal, the emergence of soft planning is closely tied to the role of the European Union (EU), particularly in encouraging non-statutory planning practices. This is facilitated not only through non-mandatory policy agendas but especially through the allocation of Cohesion Policy's funds. Despite lacking formal competencies in spatial planning, the EU's Structural and Investment Funds, along with specific European policy initiatives, have been effectively propelling the development of territorial strategies and partnerships. Operating at the periphery of formal planning instruments and processes, essentially on a soft planning basis, these initiatives have played a role in fostering new territorial governance arrangements that ultimately impact the domestic administrative-territorial organization.

Throughout various community programming cycles, the Cohesion Policy has had significant repercussions on Portugal's internal processes of territorial and governance rescaling, with a particular emphasis on the emergence of inter-municipalism. Indeed, the management of community investment funds, grounded in the formulation of a strategic plan for territorial development aligned with the inter-municipal area (NUTS III), has been a central focus of inter-municipal cooperation (Comissão Independente para a Descentralização et al. 2019). Following an unsuccessful attempt at regionalization and in light of diminishing demographic and economic dynamics that deprive many municipalities, particularly those in inland territories—eastern Portugal near Spain, of the critical mass needed for a municipal-scale territorial approach, coupled with the strengthening of metropolitan areas in the national settlement structure and their contribution to the GDP, inter-municipalism—characterized by models of cooperation among municipalities—has become an ever more pervasive and integral aspect of the national political agenda, spanning various governments and party affiliations.

In the absence of administrative regions, inter-municipal cooperation models in Portugal have functioned as a band-aid solution to bolster regional governance and allow for a closer participation of local authorities (municipalities) in the allocation of European funds; when the Cohesion Policy is targeted at a regional scale and therefore traditionally addressed by the central government in a deconcentrated arrangement. However, these models reveal numerous problems and challenges, particularly stemming from the lack of a corresponding administrative and governmental level. Indeed, there is a need to address the sharing of territorial resources and power among cooperating municipalities, while the legitimacy of their intervention relies on the votes cast by local elected officials, thus confined to municipal territories (Comissão Independente para a Descentralização et al. 2019).

Despite the numerous inefficiencies in inter-municipal cooperation and the challenges associated with implementing territorial approaches at a regional scale guided by supra-municipal reasoning—challenges that the instruments of the Cohesion Policy, such as Integrated Territorial Investments (ITI), have been unable to overcome despite their specific design for implementation at the NUTs III level—the Lisbon Metropolitan Area has undergone a substantial shift in recent years, signaling the emergence of a supra-municipal consciousness. Grounded in soft planning processes, particularly related to European funds, climate change and adaptation agendas, socio-spatial inequalities, and food transition, the Lisbon Metropolitan Area, even without formal political powers, has transcended mere inter-municipal logic, paving the way to inter-municipal cooperation capable of voluntarily embracing and visioning the metropolitan space through supra-municipal territorial reasoning. Based on this assumption, this chapter intends to delve into the transformative role of soft planning processes within the Lisbon Metropolitan Area over the past years. It argues that soft planning has served as a catalyst for metropolitan governance, actively contributing to cultivating a metropolitan territorial consciousness while shaping a supra-municipal planning space that transcends the mere inter-municipal association of distinct municipal powers and planning arenas.

This chapter is structured into four distinct sections. In the first section, we delve into the contextualization of inter-municipal cooperation, examining its emergence as a crucial facet of territorial and governance rescaling within the framework of the evolving Portuguese administrative organization. The second section narrows the focus to the Lisbon Metropolitan Area, delving into the intricacies of soft planning. We scrutinize specific planning instruments run between 2014 and the present, employing a lens that encompasses key planning dimensions such as policies, governance, politics, and ethos. The third section engages in a discussion on the impact of soft planning, specifically exploring how it has contributed to the enhancement of metropolitan governance. Finally, the last section provides a comprehensive wrap-up and conclusion.

23.2 Territorial Rescaling and the Rise of Inter-Municipal Cooperation in Portugal

Since the early 1990s, governance and territorial rescaling have drawn substantial interest among scholars in the fields of state theory and planning, closely tied to profound transformations in the statehood and territorial organization of the State. The theme and unfolding dynamics have sparked diverse discussions, ranging from the deconstruction of fordist-keynesian development and accumulation models centered around the nation-state (Jessop 1994) to the dissolution of traditional forms of territorialism (Faludi 2018). This recognizes the impending erosion of national state power, unfolding in a 'hollowing-out' process that contrasts the enduring political sovereignty of the nation-state with the drainage of national powers and the weakening of its borders. It marks the birth of processes involving devolution, decentralization, and the transfer of tasks and responsibilities across various tiers of government and beyond. Simultaneously, it signals the emergence of new governance spaces and institutional configurations, with a particular emphasis on both supranational and subnational scales (Allmendinger and Haughton 2010; Faludi 2013; Jessop 1994; Macleod and Goodwin 1999; Roodbol-Mekkes and Van en Brink 2015; Swyngedouw 2000).

In this context, it is advocated that urban regions, especially metropolitan areas, have been pivotal in the spatial and institutional restructuring of the modern State. The ascendancy of urban and metropolitan regions as novel domains of power and territorial governance was intricately tied to the shift towards territorial competitiveness and economic development over the distributive and equitable rationales of the Welfare State. This dynamic occurred within the context of globalization, where urban and metropolitan regions assume a prominent position in attracting capital investments and promoting economic development (Brenner 2003, 2004; Macleod and Goodwin 1999).

In many European and OECD countries, the ascent of regional and metropolitan power has been solidified through the formalization of a corresponding structure and level of authority, typically manifested in the establishment of a metropolitan government elected through direct and universal suffrage. This model is evident in French and Italian metropolises, following the territorial reforms in 2010 and 2014,[1] respectively, as well as in instances where the metropolitan area functions as a federated state or city-region, as observed in Brussels, Berlin, Vienna or Madrid. In contrast, Portugal maintains a distinct approach. Despite their institutionalized form of inter-municipal cooperation, with a compulsory character, the metropolitan areas of Lisbon and Oporto retain an associative nature, functioning as public associations of local authorities with limited competencies and a significant dependency on the willingness of municipal elected political representatives to collaborate (Freitas do

[1] In the Italian case, although the law (Law No. 56/2014) stipulates the direct election of metropolitan governance bodies, the prefect of the capital city of the former provinces took on the role of metropolitan prefect. Following the establishment of the 10 metropolitan cities in 2015, the metropolitan bodies, at least in their initial phase, were endowed with indirect representation.

Amaral and Coutinho 2019; OECD 2020). According to the OECD's report, "*[t]he Metropolitan Areas (MAs) and Inter-municipal Communities (IMCs) still play a marginal role in the public sector, notably due to modest level of decentralisation in general. While this may change in the coming years, as more responsibilities will be decentralised to municipalities, the challenge is to enlarge the role of the inter-municipal organisations and to guarantee their financial means*" (OECD 2020: 9).

Indeed, in the context of OECD countries, Portugal remains a strongly centralized unitary state (Fernandes et al. 2023; OECD 2020). It is true that in recent years, Portugal's subnational governments have experienced an improvement in their legal and administrative standing. However, despite this growing trend towards decentralization, "*the Portuguese multi-level governance system appears quite centralised compared with other OECD and European countries*" (OECD 2020: 7).

While the Constitution of the Portuguese Republic (CRP) has envisaged the decentralization of powers to subnational levels since its first version in 1976, the process of decentralization through regionalization has faced numerous setbacks. After the constitutional revision of 1997, which introduced a mandatory referendum for the establishment of the administrative regions outlined in the CRP, regionalization faced rejection by over 60% of voters the following year (Melo 2022). As a result, Portugal maintained an administrative-territorial organization that lacks any intermediate governance scale between the central and local levels in the continental mainland. The country's administrative structure is, indeed, quite complex, given the diverse array of entities within it and the specific statutory characteristics associated with each.

After the rejection of regionalization in 1997, inter-municipal cooperation stepped into the role that regionalization was supposed to fulfill, driving the enhancement of inter-municipal governance alongside the decentralization process. Following an initial push for decentralization,[2] the momentum gained further traction with the reform of local administration which established the framework for inter-municipal entities[3] and the subsequent transfer of competencies to the local level in 2018[4] (Fernandes et al. 2023). Efforts were also made to align NUTS III (NUTS 2013) with the territorial delineation of inter-municipal entities (Instituto Nacional de Estatística 2015), enhancing the coherence of public policies, especially those highly dependent on community funding, with the territorial organization of the State.

In the specific context of metropolitan areas, their legal framework has undergone multiple alterations since their establishment in 1991.[5] Ferreira (2022: 146)

[2] Law No. 159/1999, dated September 14, and Decree-Law No. 144/2008, dated July 28.

[3] Law No. 75/2013, dated September 12. According to the law, there are two types of inter-municipal entities: metropolitan areas and inter-municipal communities.

[4] Law No. 50/2018, dated August 16.

[5] The creation of Metropolitan Areas was formalized by Law No. 44/91, dated August 2. Subsequent amendments to the legal framework occurred in 2003 (Law No. 10/2003, dated May 13), 2008 (Law No. 46/2008, dated August 27), and 2013 (Law No. 75/2013, dated September 12), establishing the current regulatory framework.

advocates that these diverse frameworks represent three distinct phases in the formation of metropolitan inter-municipal entities: the initial stage, acknowledging the metropolitan area as a novel territorial scale; the subsequent phase, addressing the need to confer scale and critical mass to municipalities, namely through associativism; and the last phase, specifically recognizing the escalating demands related to the allocation of Cohesion Policy funds.

Nevertheless, inter-municipal cooperation is fraught with various limitations. Operating without political and administrative autonomy, this model of governance hinges on decisions made by locally elected officials from each municipality, accountable solely to their respective electorates. The decentralization of competencies from the central administration to the inter-municipal level, in areas such as education, health, civil protection, social action, and European funding, as mandated by the framework law in 2018, is contingent upon the prior agreement of the municipalities involved (Fernandes et al. 2023). Moreover, the inter-municipal deliberative body being chaired by one of the mayors from the participating municipalities introduces a conflict of representation, pitting the collective interest against the specific interests of individual municipalities (Comissão Independente para a Descentralização et al. 2019: 92).

Besides these considerations, the specific case of the Lisbon Metropolitan Area (LMA) introduces additional complexities that possess a unique character. On the one hand, LMA encompasses both a NUTS II and NUTS III territory. On the other hand, regarding Cohesion Policy and the allocation of community funds, it coincidentally functions as the operational area for both the LMA—acting as an inter-municipal entity within the local administration framework and the Coordination Commission for Regional Development (CCDR-LVT), which operates as a central administration body with deconcentrated powers at the regional level. CCDR-LVT holds responsibilities for both spatial planning and regional development. Furthermore, it also assumes the role of Management Authority for Lisbon Region's Operational Program, granting it additional responsibilities and control over the allocation of regional funds and the financing of projects.

Despite the inherent difficulties and challenges in a governance structure built upon a federation of municipalities and tension between the central and local administrations, the LMA has, in recent years, achieved significant progress in strengthening metropolitan governance. This chapter contends that soft planning, largely instigated and funded by European funds, has played a pivotal role in this process.

23.3 Soft Planning in the Lisbon Metropolitan Area

Soft planning refers to non-statutory and non-binding planning processes aimed at the strategic territorial development of specific areas and regions, building on "*the principles and flagships of strategic spatial planning (e.g., strategy-making; policy integration; collaborative governance; long-term visioning; development-led action)*" (Cavaco et al. 2023: 18). In this framework, these processes operate independently and parallel to the formal procedures outlined in the current spatial planning and territorial management system, adopting a voluntary nature. Centered on the efficient implementation of policy measures in often intricate and challenging contexts (such as the case of a metropolitan area), soft planning processes are grounded in versatile, multidimensional approaches. These approaches emphasize the need for horizontal integration and coordination across diverse sectoral policies, vertical coordination among different government levels, and active engagement of various stakeholders and territorial actors pivotal to the development of the targeted territory. As such, soft planning also introduces several other emerging trends such as the rise of informal planning, territorial rescaling, and the blurring and entanglement of territorial scales (Cavaco et al. 2023).

In recent years, under the umbrella of European Structural and Investment Funds (ESIF), and notably within the latest Community Support Frameworks, the European Commission has introduced a series of policy instruments to catalyze integrated place-based approaches and empower governance at a sub-national level, targeting functional areas. Among these are: Integrated Territorial Investments (ITI), Community-led Local Developments (CLLD), and Integrated Sustainable Urban Development (ISUD) (Cavaco et al. 2021: 118–121). Although non-binding, these instruments are designed to foster the emergence of soft planning processes and spaces that run parallel to or complement formal spatial planning instruments. While the term 'soft planning' may not be explicitly stated, the rationale behind it was to: "*push the emergence of new territorial spaces acting as drivers for spatial rescaling; assist multi-level governance and promote cooperation and coordination between the different levels of government (…); stimulate multidimensional and cross-sectoral interventions; carry out tailor-made and place- or area-based development strategies (…); and promote the capacity building of institutions and territorial actors, as well as the empowerment, involvement and participation of local authorities, local communities and the civil society*" (Cavaco and Costa 2019: 90).

With very limited formal and statutory spatial planning competencies, dependent on the power and decisions made by municipal representatives, as much as on the top-down decision-making of the central administration,[6] the LMA has, in recent years, employed soft planning approaches to shape the planning of the metropolitan region. This process was significantly stimulated in the 2014–2020 community programming cycle, particularly through the establishment of ITI.[7]

[6] Regarding statutory spatial planning instruments, the legislative changes introduced in the legal framework for spatial planning and urbanism in 2014/2015 (Decree-Law No. 80/2015, dated May 15), as well as the new legal regime for local authorities (Law No. 75/2013, September 12), brought significant innovations by explicitly allowing inter-municipal entities to develop inter-municipal spatial programs and plans. In the case of metropolitan areas, when such a program or plan encompasses the entire metropolitan area, the responsibility for development lies with the Metropolitan Executive Commission, subject to approval by the Metropolitan Council. However, to date, there have been no processes in this direction, and the exercise of associated power always remains within the purview of the municipalities, considering the management and governance models of metropolitan areas.

[7] According to the European Commission's factsheet on ITI, "*[t]his approach is multi-dimensional, tailored to place-specific features and outcomes, which may mean going beyond traditional administrative boundaries, and may require greater willingness from different levels of government to*

Despite the Commission's goal to direct ITIs toward more function-oriented (soft spaces) rather than administratively-bounded territories, Portugal ultimately embraced an extensive ITI model applicable to the whole territory at the NUTS III level.[8] Through ITIs, the contracted management of EU funds, involving the development of an Integrated Territorial Development Strategy—EIDT 2014–2020 (AML 2015a) and the signing of a Pact for Territorial Development and Cohesion—PDCT (AML 2015b), has become a central focus within the scope of inter-municipal cooperation.

Despite the tensions and shortcomings identified in the formulation and implementation of the EIDT 2014–2020, pertaining to the bureaucratic nature of the process, the limited autonomy for local and sub-regional actors, and the ongoing tension between central and local administration (Ferreira 2022), the subsequent years saw notable progressions. Following EIDT 2014–2020, the LMA implemented a range of strategic documents and soft planning approaches, elevating metropolitan governance through collaborative efforts among its constituent municipalities and several regional actors, including the CCDR-LVT. Together, they crafted a shared strategic framework addressing critical issues like climate change adaptation, spatial and socio-economic inclusion of deprived communities, and the transition to sustainable food practices.

In this context, the catalyst not only encompasses funds from the Cohesion Policy but also includes the additional post-COVID-19 funding package. This temporary financing, situated within the framework of the Recovery and Resilience Facility of the NextGenerationEU programme, played a pivotal role, particularly in sectors such as housing[9] and interventions in socially and economically disadvantaged communities and territories.

Additionally, the role of the LMA's Executive Commission is noteworthy. Endowed with responsibilities for cooperation and coordination among various municipalities and between them and the central administration, particularly in terms of planning and territorial development, it assumed a crucial role in promoting, energizing, and advancing these diverse metropolitan policy agendas.

In this section, our analysis is centered on a selection of examples that have been developed and approved since the framework provided by ITI and the 2014–2020 community programming cycle. Alongside the EIDT 2014–2020, we delve into

---

co-operate and co-ordinate actions in order to achieve shared goals. (…) Any geographical area with particular territorial features can be the subject of an ITI, ranging from specific urban neighbourhoods with multiple deprivations to the urban, metropolitan, urban–rural, sub-regional, or inter-regional levels. An ITI can also deliver integrated actions in detached geographical units with similar characteristics within a region (e.g. a network of small or medium-sized cities). It is not compulsory for an ITI to cover the whole territory of an administrative unit" (European Commission 2014).

[8] Since January 2015, NUTS III have become territorially coincident with the administrative boundaries of inter-municipal entities.

[9] Housing was not included in this analysis. Despite substantial inter-municipal cooperation concerning housing policies, it did not become the field for the development of a metropolitan strategy. Instead, municipalities chose to formulate Housing Local Strategies at the municipal level.

the Lisbon Regional Strategy 2030 (AML and CCDR LVT 2020), the Metropolitan Plan for Climate Change Adaptation—PMAAC-AML (AML 2019), the FoodLink—Network for Food Transition in the Lisbon Metropolitan Area (CCDR LVT et al. 2022), and the Metropolitan Plan to Support Disadvantaged Communities in the LMA-PMACD-LMA (AML, 2021). This selection is not exhaustive. Despite exclusively focused on non-statutory instruments, therefore voluntary in nature, it aims to cover both instruments and approaches that directly result from European requirements associated with accessing EU funding—as is the case with the EIDT 2014–2020 and PDCT, the Lisbon Regional Strategy 2030, and the PMACD-LMA—and others that stem from both the LMA's and bottom-up initiatives, in response to emergent societal challenges, such as the PMAAC-LMA and FoodLink, respectively.[10]

The analysis of these instruments,[11] outlined in the following table, delves into four pivotal components or dimensions of planning—*policies*; *governance*; *politics*; and *ethos*. Building upon prior reflections (Cavaco et al. 2023), our aim is to advance an approach that encompasses a substantial array of structural aspects essential for comprehending the strategies and processes involved in soft planning within the LMA (Table 23.1).

- In *policies*, we contemplate the policy instruments underpinning these approaches, scrutinizing aspects such as the nature or type of the instrument, its objectives, the associated funding sources, and the process of operationalization.
- In *governance*, our focus is primarily on scrutinizing the governance model intertwined with these soft planning approaches. This involves examining the actors involved and their typology, the organizational model, partnership and/or contractual arrangements, as well as the intricacies of policy implementation and strategy.
- In *politics*, our primary emphasis is on processes related to power dynamics and decision-making, with reflections confined to the instances of the approval of the instruments.
- Lastly, in *ethos*, we seek to delve into the values and motivations driving each of these approaches. Our interpretation encompasses not only the motivations but also the underlying policy agendas and guiding principles.

---

[10] Although the PAMUS (Action Plan for Sustainable Urban Mobility), developed under the umbrella of the EIDT 2014–2020, is considered a form of soft planning, we have chosen not to include it in the selection. This decision is due to transport and mobility becoming formal competencies of the Metropolitan Area of Lisbon since 2015. This shift resulted from the dissolution, through Law no. 52/2015 dated June 9, of the previous Lisbon Metropolitan Transport Authority, empowering the metropolitan area as the new authority for metropolitan public transportations—the Metropolitan Transports of Lisbon—with the exception of railway and metro transportation.

[11] The analysis presented here is based on a thorough review of diverse policy instruments, encompassing consultations and examinations of relevant documentation, including reports, news, and other materials available on the internet concerning the respective processes. Additionally, academic documents featuring research on some of the instruments under consideration were also referenced.

Table 23.1 Instruments and policies

Dimensions	Instrument	Integrated Territorial Development Strategy (EIDT 2014–2020)	Metropolitan Plan for Climate Change Adaptation (PMAAC-LMA)	Lisbon Regional Strategy 2030	Metropolitan Plan to Support AML's Disadvantaged Communities (PMACD-LMA)	FoodLink—Network for Food Transition in the Lisbon Metropolitan Area
		2014	2019	2020	2021	2022
Policies	Type of Instrument	Strategic integrated place-based plan	Strategic plan	Strategic integrated place-based plan	Strategic integrated place-based plan	Network and strategy (Strategy for Food Transition under elaboration)
	Main Goal	Developing an integrated territorial strategy to frame, at the sub-regional level, the development and funding of integrated place-based approaches and respective instruments (ITI, CLLD, ISUD)	Establishing the strategic and operational framework for adapting the territories of the LMA to climate change	Establishing an integrated regional development strategy for the region of Lisbon, to frame the execution of Cohesion Policy Funding in the 2021–2027 period, and promote a multilevel governance	Promoting a sustained, integrated, and long-term intervention in socially and economically disadvantaged communities and territories, with the goal of combating socio-territorial vulnerabilites	Fostering food transition in the Lisbon metropolitan region and support the operationalization of sustainable, competitive, and resilient food systems in a place-based manner

(continued)

Table 23.1 (continued)

Dimensions	Instrument	Integrated Territorial Development Strategy (EIDT 2014–2020)	Metropolitan Plan for Climate Change Adaptation (PMAAC-LMA)	Lisbon Regional Strategy 2030	Metropolitan Plan to Support AML's Disadvantaged Communities (PMACD-LMA)	FoodLink—Network for Food Transition in the Lisbon Metropolitan Area
		2014	2019	2020	2021	2022
	Funding Source	EU Cohesion Policy Structural and Investment Funds (EFDR, CF and ESF) through the Regional Operational Programme (OP) and POSEUR + National co-funding	EU Cohesion Policy Structural and Investment Funds (CF) through the Operational Program for Sustainability and Resource Efficiency (POSEUR) + National co-funding Note: Implementation may be support by other funding sources (e.g. EEA Grants—CLIMA.AML project)	EU Cohesion Policy Structural and Investment Funds (EFDR, CF, ESF, and FEAMPA) through the Regional Operational Programme (PORL 2021–2027) and PACS2030 + National co-funding	*Until 2026:* NextGenerationEU -Recovery and Resilience Facility (RRF) C3—Social Answers National Recovery and Resilience Plan-PRR *Between 2016 and 2030:* EU Cohesion Policy Structural and Investment Funds through the Regional Operational Programme (PORL 2021–2027) + National Funds	FoodLink does not have legal personality, which hinders access to funding. This can only be secured by individual partners

(continued)

Table 23.1 (continued)

Dimensions	Instrument	Integrated Territorial Development Strategy (EIDT 2014–2020)	Metropolitan Plan for Climate Change Adaptation (PMAAC-LMA)	Lisbon Regional Strategy 2030	Metropolitan Plan to Support AML's Disadvantaged Communities (PMACD-LMA)	FoodLink—Network for Food Transition in the Lisbon Metropolitan Area
		2014	2019	2020	2021	2022
	Operationalization	1. Pact for Territorial Development and Cohesion (PTDC—ITI)—intermunicipal level 2. Pacts for Community-led Local Developments (CLLD)—local level 3. Strategic Urban Development Plans (PEDU—ISUD)—municipal level 4. Respective actions and projects	1. Monitoring and management + Metropolitan strategic projects—metropolitan level 2. Integration of climate change adaptations measures within municipal policies and spatial plans—municipal level 3. Development of sector-wide plans and strategies—sectoral level 4. Project Clima.AML—Meteorological Monitoring and Alert Network	1. ITI LMA and respective Action Plan, including actions and projects 2. Community-led Local Developments (CLLD funded by FEAMPA) 3. Other territorial approaches and pacts foreseen in the governance model of the European Funds	Integrated Local Operations based on Local Partnerships, and respective measures, actions, and interventions	Priority Actions / Flagship projects: 1. Prepare a Strategy for Food Transition 2. Territorial mapping 3. Identification of potential actors 4. Sensitization, dissemination and capacity-building

(continued)

Table 23.1 (continued)

Dimensions	Instrument	Integrated Territorial Development Strategy (EIDT 2014–2020)	Metropolitan Plan for Climate Change Adaptation (PMAAC-LMA)	Lisbon Regional Strategy 2030	Metropolitan Plan to Support AML's Disadvantaged Communities (PMACD-LMA)	FoodLink—Network for Food Transition in the Lisbon Metropolitan Area
		2014	2019	2020	2021	2022
Governance	Responsibility for Elaboration and Coordination	LMA (Executive Commission—executive body) supported by a consultancy firm	LMA (Executive Commission—executive body) supported by a consortium and technical team involving the university and consultancy firms	LMA (Executive Commission—executive body) + CCDR-LVT (Central Administration's deconcentrated agency for regional development and spatial planning) supported by external consultants	LMA (Executive Commission—executive body) supported by external consultants	LMA + CCDR-LVT (Central Administration's deconcentrated agency for regional development and spatial planning) + University—Institute of Social Sciences

(continued)

Table 23.1 (continued)

Dimensions	Instrument	Integrated Territorial Development Strategy (EIDT 2014–2020)	Metropolitan Plan for Climate Change Adaptation (PMAAC-LMA)	Lisbon Regional Strategy 2030	Metropolitan Plan to Support AML's Disadvantaged Communities (PMACD-LMA)	FoodLink—Network for Food Transition in the Lisbon Metropolitan Area
		2014	2019	2020	2021	2022
	Actors Involved in Elaboration	1. LMA's representatives 2. Representatives from the municipalities 3. CCDR-LVT and OP Management Authority 4. Different sectors of the central administration 5. Academics and experts 6. Professional associations representatives	1. Technicians from municipalities 2. Public entities across various sectors 3. Public and private stakeholders, encompassing economic, social, environmental and cultural issues 4. Academics and experts 5. Local community and civil society	1. Metropolitan Working Group for Community Funding 2. Metropolitan Council 3. Municipalities 4. Academics and experts in several domains 5. Regional stakeholders in several domains 6. Different sectors of the central administration	1. LMA 2. Municipalities 3. Academics and experts	Working groups: 1. LMA 2. CCDR-LVT 3. Municipalities and other local authorities 4. Universities 5. Regional and local stakeholders 6. NGOs 6. Citizens associations

(continued)

Table 23.1 (continued)

Dimensions	Instrument	Integrated Territorial Development Strategy (EIDT 2014–2020)	Metropolitan Plan for Climate Change Adaptation (PMAAC-LMA)	Lisbon Regional Strategy 2030	Metropolitan Plan to Support AML's Disadvantaged Communities (PMACD-LMA)	FoodLink—Network for Food Transition in the Lisbon Metropolitan Area
		2014	2019	2020	2021	2022
	Governance Model	Estrutura Metropolitana de Governação Territorial 2020: 1. LMA's deliberative, executive and consultive bodies; 2. Technical secretariat; 3. LMA's representative in OP's Direction; 4. Metropolitan working groups	No dedicated governance structure	Two levels of coordination: 1. Executive level—articulation between the LMA's Executive Commission and the CCDR LVT Presidency 2. Technical level—Focal points of both AML and CCDR LVT	Three-level model: 1. *Metropolitan*: LMA's Coordination + Steering Group (AML, Ministries, Municipalities, OP Management Authority) 2. *Sub-regional*: Local Technical Units (based on a partnership agreement) + Monitoring Committee 3. *Parish level*: Local Partnership (Municipality, Parish, Local and Neighborhood Associations, Sectors) + Leadership (Municipal council)	Collaborative platform: 1. Researchers and academics 2. LMA and Municipalities 3. Central administration representatives 4. Local Development Associations 5. Entrepeneurial and logistics stakeholders 6. NGO and Citizens Associations

(continued)

Table 23.1 (continued)

Dimensions	Instrument	Integrated Territorial Development Strategy (EIDT 2014–2020)	Metropolitan Plan for Climate Change Adaptation (PMAAC-LMA)	Lisbon Regional Strategy 2030	Metropolitan Plan to Support AML's Disadvantaged Communities (PMACD-LMA)	FoodLink—Network for Food Transition in the Lisbon Metropolitan Area
		2014	2019	2020	2021	2022
	Responsibility for Implementation	1. LMA as an Intermediate Body with delegated responsibilities to approve projects' applications, under contract with the OP's Management Authorities	1. LMA—Monitoring and follow-up + Strategic management of supra-municipal actions 2. Municipalities 3. Sectorial entities	1. LMA with delegated responsibilities, under a Contract for Territorial Development and Cohesion with the OP's Management Authorities (Dec 2023)	1. LMA's as coordinator 2. Local Technical Units—Municipalities 3. Beneficiaries and partners	FoodLink partners
Politics	Responsibility for Approval	1. Approval by the Metropolitan Council (Municipal Mayors) 2. Evaluation and recognition by the central administration 3. Dispatch by the Secretary of State for Regional Development	Signing of a Commitment on Climate Change Adaptation by the 18 municipalities of the LMA (2019)	Approval by the Metropolitan Council (Municipal Mayors) + CCDR LVT Presidency	1. Approval of the Plan by the LMA's Executive Commission 2. Signing of an Agreement between the LMA and Local Technical Units	Signing of a Charter of Principles and Commitments by the FoodLink partners

(continued)

Table 23.1 (continued)

Dimensions	Instrument	Integrated Territorial Development Strategy (EIDT 2014–2020)	Metropolitan Plan for Climate Change Adaptation (PMAAC-LMA)	Lisbon Regional Strategy 2030	Metropolitan Plan to Support AML's Disadvantaged Communities (PMACD-LMA)	FoodLink—Network for Food Transition in the Lisbon Metropolitan Area
		2014	2019	2020	2021	2022
Ethos	Underlying Motivations	1. Access funding in response to the 2014–2020 programming cycle; 2. Serve as the basis for the contractualization of various instruments (ITI/PDCT; CLLD; ISUD); 3. Strengthen the metropolitan scale and ensure the strategic coherence of municipal action; 4. Enhance cooperation among various territorial actors	1. Contribute to achieving the objectives outlined in international agreements on climate action; 2. Reduce the vulnerability of the metropolitan area to climate change; 3. Raise awareness and build the capacity of decision-makers, technicians, and local stakeholders; 4. Enhance the technical-scientific knowledge base related to climate scenario planning	1. Access funding in response to the 2021–2027 programming cycle; 2. Enhance cooperation between central and local administration and strengthen the region's public governance capacity 3. Strengthen the metropolitan scale and ensure the strategic coherence of municipal action; 4. Enhance cooperation among various territorial actors	1. Access funding in response to the Recovery and Resilient Facility 2. Fight against socio-spatial vulnerabilities and exclusion in an integrated way 3. Technical and institutional capacity-building on socio-territorial development 4. Develop integrated place-based approaches at the local level for targeted disadvantaged communities and territories	1. Strengthen the resilience of natural, social, economic, and infrastructural systems 2. Prepare the territory to enhance food security and sustainability 3. Capacity-building on food transition 4. Promote socio-spatial cohesion 5. Contribute to the further establishment of a metropolitan food policy

(continued)

Table 23.1 (continued)

Dimensions	Instrument	Integrated Territorial Development Strategy (EIDT 2014–2020)	Metropolitan Plan for Climate Change Adaptation (PMAAC-LMA)	Lisbon Regional Strategy 2030	Metropolitan Plan to Support AML's Disadvantaged Communities (PMACD-LMA)	FoodLink—Network for Food Transition in the Lisbon Metropolitan Area
		2014	2019	2020	2021	2022
	Underlying Agendas	*European*: 1. Europe 2020 Strategy 2. Territorial Agenda of the European Union 2020 3. Partnership Agreement Portugal 2020	*International*: 1. United Nations Framework Convention on Climate Change (UNFCCC) 2. COP21 and Paris Agreement *National*: 1. National Strategy for Climate Change Adaptation (ENAAC 2020) 2. National Spatial Planning Policy Programme (PNPOT)	*International*: 1. 2030 Agenda *European*: 1. European Green Deal 2. Towards a Sustainable Europe by 20,303, reflection paper 3. Territorial Agenda 2030 *National*: 1. Partnership Agreement—Portugal 2030 2. Strategy Portugal 2030 3. National Spatial Planning Policy Programme (PNPOT)	*European*: 1. NextGenerationEU *National:* 1. Strategic vision for the economic recovery plan of Portugal 2020–2030	*International*: 1. 2030 Agenda *European*: 1. Milan Urban Food Policy Pact (MUFPP), 2015 2. Glasgow Food and Climate Declaration, 2020 3. EU's 'From Farm to Fork' Strategy, 2020 + European Green Deal 4. Biodiversity Strategy for 2030 *National*: 1. Innovation Agenda for Agriculture 2030—Terra Futura 2. PMAAC-AML 3. National Spatial Planning Policy Programme (PNPOT) 4. Lisbon Regional Strategy 2030 5. RIS3 2030

(continued)

Table 23.1 (continued)

Dimensions	Instrument	Integrated Territorial Development Strategy (EIDT 2014–2020)	Metropolitan Plan for Climate Change Adaptation (PMAAC-LMA)	Lisbon Regional Strategy 2030	Metropolitan Plan to Support AML's Disadvantaged Communities (PMACD-LMA)	FoodLink—Network for Food Transition in the Lisbon Metropolitan Area
		2014	2019	2020	2021	2022
	Guidiing Principles	1. Sustainable development 2. Place-based approaches 3. Policy integration and coordination 4. Territorial integration in functional regions	1. Sustainability 2. Territorial equity 3. Social and spatial justice 4. Intermunicipal Governance 5. Partnership 6. Local action	1. Sustainable development 2. Place-based approaches 3. Multilevel governance 4. Intra and Interregional Cohesion	1. Place-based approaches 2. Multi-level governance 3. Policy articulation and complementarity 4. Socio-spatial equity 5. Partnership 6. Local action	1. Ecological, social and economic resilience 2. Sustainability 3. Proximity 4. Urban–rural relationships 5. Circular economy 6. Fair trade

23.4 Soft Planning as a Driver of Metropolitan Governance

The examination of the selected instruments yields insights into the pivotal role and significance of soft planning approaches and policy instruments. These contribute not only to the consolidation of a metropolitan planning scale but also play a crucial role in strengthening forms of territorial governance. This influence extends across various dimensions, encompassing inter-municipal cooperation, vertical coordination between central and local administration, horizontal coordination among diverse sectors, and fostering partnerships among multiple territorial actors.

From a governance standpoint, notable advancements have been observed during the transition from the 2014–2020 to the 2021–2027 European Union programming period, specifically in the consolidation of metropolitan territorial governance, particularly regarding multilevel coordination. Despite being spearheaded by the LMA, the EIDT 2014–2020 underwent rigorous oversight from the central administration and the Managing Authority of the Regional Operational Program. This scrutiny, marked by assessment, recognition, and evaluation, underscores the limited autonomy granted to local and sub-regional actors. Concurrently, throughout the same programming period, the Regional Action Plan for Lisbon (PARL) was also elaborated and handled, covering the same territorial area as the EIDT. It delineated a development strategy and action plan for the distribution of community funds, albeit in a top-down direction orchestrated by the central administration entity (CCDR-LVT).

In transitioning to the subsequent programming period 2021–2027, Integrated Territorial Investments (ITI) at the NUTS III level continued as a foundational instrument for sub-regional integrated territorial approaches, formalized through the signing of a Contract for Development and Territorial Cohesion (CDCT). A substantial paradigm shift occurred though, with the revision of the strategic framework, initiated at the minister's behest. LMA and CCDR-LVT collaborated to synchronize positions and strategies, presenting a unified vision—the Lisbon Regional Strategy 2030. LMA retained delegated responsibilities for ITI management following the CDCT, highlighting a newfound inter-institutional alignment and collaboration between the two administrative levels.

This multilevel collaboration goes beyond the instruments directly addressing Brussels' requirements and subsequent partnership agreements. Through FoodLink network, CCDR-LVT and LMA synergize additional efforts in a collaborative partnership and strategy, this time centered around the theme of food transition. The academic sector, represented by the Institute of Social Sciences at the University of Lisbon, joins in coordinating this initiative, which encompasses a growing network of diverse partners and sub-regional/local stakeholders.

In the realm of inter-municipal and multi-actor collaboration, the influence of soft planning approaches on formalizing and strengthening partnership structures is evident. These approaches facilitate cooperation among diverse local authorities, enabling them to collectively address shared concerns, and coordinate municipal policies and interventions. Notably, these instruments are guided by an integrated, place-based logic—an orientation emphasized by the European Commission when shaping some of its policy initiatives, instruments, and funding. Since the 2007–2013 Community Framework, particularly following the Barca Report 2009, integrated territorial approaches have gained prominence within Cohesion Policy. These approaches significantly contributed to sub-regional planning scales, holding the potential to shape future spatial planning and territory-focused endeavors. Despite the 2014–2020 programming cycle being characterized by fragmented municipal interventions, particularly in sustainable urban development in response to Article 7 of ERDR (ISUD) (Carvalho et al. 2022), positive dynamics are emerging, and increased cooperation and coordination among municipalities and sub-regional stakeholders are anticipated for the upcoming cycle.

Considering the policy-making and the motivations behind it, funding emerges as a primary catalyst, having a pivotal role in stimulating soft planning processes. Indeed, strategies explicitly designed to meet European requirements (such as EIDT 2014–2020, Lisbon Regional Strategy 2030, PMACD-LMA) hold a clear instrumental character. They serve as a fundamental prerequisite to secure funding access for municipalities and other beneficiaries. This has a two-sided effect. On the one hand, this utilitarian and operational dimension acts as a catalyst for inter-institutional strategic cooperation and the formation of multi-actor partnerships, potentially fostering a transition to a more collaborative and cross-cutting organizational and planning culture. However, on the other hand, shortcomings and latent contradictions also come up, notably the strongly hierarchical and technocratic nature intrinsic to funding allocation that often restricts bottom-up responsiveness, posing a risk of strategic standardization conflicting with place-based and tailor-made aspirations.

However, the case of LMA and the ongoing non-statutory planning process conducted at the inter-municipal level vividly demonstrates installed capacity and initiative that far surpasses the requirements for accessing major European funding packages. The examples of PMAAC-LMA and FoodLink are clear illustrations of this voluntarism and proactive approach, addressing emerging themes and agendas at both European and international levels. The voluntary commitment, whether through the Commitment on Climate Change Adaptation embraced by all 18 municipalities' mayors or the Charter of Principles and Commitments endorsed by FoodLink partners, serves as a non-binding bond revealing the partners' dedication to the initiatives. All this contributes not only to institutional innovation and sub-regional capacity-building but also to the enhancement of metropolitan governance towards the construction of a supra-municipal space.

23.5 Conclusions

Examining the case of the AML and inter-municipal cooperation within the context of Portugal's territorial administrative structure, this chapter endeavors to delve into the dynamics and impact of soft planning approaches on the reinforcement of territorial rescaling and the establishment of metropolitan-scale governance. It acknowledges the influential role played by Cohesion Policy and the allocation of European Funds in propelling these dynamics, particularly by invigorating and providing financial incentives for instruments focused on integrated territorial approaches.

The scrutinized cases, spanning from the inception of the initial ITI and its corresponding strategy to the present day, unveil substantial progress made in recent years. Beyond fortifying the Metropolitan Area from an institutional and administrative point of view, these instances also contribute to the gradual development of a 'supra-municipal awareness,' transcending the mere aggregation of individual and autonomous municipal intentions—an ordinary record in Portugal's inter-municipal cooperation.

The LMA case additionally supports the argument for the positive impact of soft planning. This counters the prevailing interpretation in the literature, which often depicts the shift to soft planning as a 'malefic' and pernicious process tied to the neoliberal hijacking of planning by obscure private interests (Cavaco et al. 2023). In contrast to these perspectives, and drawing from the observed experience in the Lisbon region, we contend that soft planning presents an opportunity for introducing institutional and planning innovation. It plays a role in strengthening multi-level governance and fostering genuine forms of cooperation and partnership, serving as effective instruments for place-based policies and planning.

Acknowledgements Authors thank the Portuguese Fundação para a Ciência e a Tecnologia (FCT), for funding the Strategic Project UIDB/04008/2020 e UIDP/04008/2020, as well as the research project SOFTPLAN—From Soft Planning to Territorial Design: Practices and Prospects (Ref. no PTDC/GES-URB/29170/2017). Igor Matheus Santana Chaves also acknowledges the financial support of the Fundacão de Amparo a Pesquisa do Estado de São Paulo (FAPESP), Brazil, through the PhD Grant Ref. n.º 2021/09660-0, n.º 2022/16640-8, and n.º 2022/08402-0.

References

Allmendinger P, Haughton G (2009) Soft spaces, fuzzy boundaries, and metagovernance: the new spatial planning in the Thames gateway. Environ Plann A: Econ Space 41(3):617–633. https://doi.org/10.1068/a40208

Allmendinger P, Haughton G (2010) Spatial planning, devolution, and new planning spaces 28:803–819. https://doi.org/10.1068/c09163

Allmendinger P, Haughton G, Knieling J, et al (eds) (2015) Soft spaces in Europe. Re-negotiating Governance boundaries and borders. Routledge, New York. Available at: https://bookshelf.vitalsource.com/books/9781317666332

AML (2015a) Estratégia Integrada de Desenvolvimento Territorial da Área Metropolitana de Lisboa 2014–2020. Documento de Política. Área Metropolitana de Lisboa, Lisboa

AML (2015b) Pacto para o Desenvolvimento e Coesão Territorial. Área Metropolitana de LIsboa. Área Metropolitana de Lisboa, Lisboa

AML (2019) Plano Metropolitano de Adaptação às Alterações Climáticas. Resumo. Área Metropolitana de Lisboa, Lisboa

AML (2021) Plano Metropolitano de Apoio às Comunidades Desfavorecidas. Área Metropolitana de Lisboa, Lisboa

AML and CCDR LVT (2020) Lisboa 2030: Estratégia Regional de Lisboa. AML. LIsboa: Área Metropolitana de Lisboa e Comissão de Coordenação e Desenvolvimento Regional de Lisboa e Vale do Tejo (CCDR LVT). Available at: www.aml.pt

Brenner N (2003) Metropolitan institutional reform and the rescaling of state space in contemporary western Europe. Eur Urban Reg Stud 10(4):297–324. https://doi.org/10.1177/09697764030104002

Brenner N (2004) New state spaces: urban governance and the rescaling of statehood. Oxford University Press, New York. https://doi.org/10.1093/acprof:oso/9780199270057.001.0001

Carvalho L, Pereira A, Pagliuso A et al (2022) Instrumentos de Planeamento Suave na Região de Lisboa. Relatório. Academia de Escolas de Arquitetura e Urbanismo de Língua Portuguesa, Lisboa

Cavaco C, Costa JP (2019) Administrative organisation and spatial planning in Portugal: a push towards soft planning spaces in Europe? In: Shaping regional futures. designing and visioning in governing rescaling. Springer, Cham, pp 87–101. https://doi.org/10.1007/978-3-030-23573-4

Cavaco C, Mourato J, Costa JP et al (2021) Spatial planning and regional development in Portugal. Direção-Geral do Território, Lisboa. Available at: https://www.forumdascidades.pt/content/publicacao-spatial-planning-and-regional-development-portugal

Cavaco C, Mourato J, Costa JP et al (2023) Beyond soft planning: towards a soft turn in planning theory and practice? Plan Theor 22(1):3–26. https://doi.org/10.1177/14730952221087389

CCDR LVT, ICS – ULisboa and AML (2022) FoodLink - Rede para a Transição Alimentar na AML - Enquadramento Estratégico. CCDR-LVT; Instituto de Ciências Sociais – U Lisboa, Lisboa, AML

Comissão Independente para a Descentralização, Cravinho J, Ferrão J et al (2019) Avaliação independente sobre a organização e funções do Estado. Assembleia da República, Lisboa. Available at: https://www.parlamento.pt/Documents/2019/julho/descentralizacao/Relatorio-Final-descentralizacao.pdf

European Commission (2014) Integrated Territorial Investment. Cohesion Policy 2014–2020. European Commission. https://doi.org/10.2776/56347

Faludi A (2013) Territorial cohesion, territorialism, territoriality, and soft planning: a critical review. Environ Plan A 45(6):1302–1317. https://doi.org/10.1068/a45299

Faludi A (2018) The poverty of territorialism. A neo-medieval view of Europe and European spatial planning. Edward Elgar Publishing, Cheltenham

Fernandes HC, Martins LC, Santos A et al (2023) O processo de transferência de competências para os Municípios – Lei n.º 50/2018, de 16 de agosto. Tribunal de Contas, Lisboa. Available at: https://www.tcontas.pt/pt-pt/ProdutosTC/Relatorios/relatorios-oac/Documents/2023/rel-oac004-2023-2s.pdf

Ferreira F (2022) A Governança Multinível: Proposta de um modelo de avaliação de instrumentos de política pública territorial. Universidade de Lisboa. Available at: https://www.ulisboa.pt/prova-academica/governanca-multinivel-proposta-de-um-modelo-de-avaliacao-de-instrumentos-de-politica

Freitas do Amaral D, Coutinho JF (2019) Áreas Metropolitanas e Comunidades Intermunicipais. Assembleia da República, Lisboa. Available at: https://www.parlamento.pt/Paginas/2019/julho/relatorio-comissao-independente-descentralizacao.aspx

Haughton G, Allmendinger P (2007) 'Soft spaces' in planning. Town and country planning 76 (the quarterly review of the Town and Country Planning Association), pp 306–308

Haughton G, Allmendinger P, Counsell D et al (2010) The new spatial planning. Territorial management with soft spaces and fuzzy boundaries. Routledge, New York

Instituto Nacional de Estatística (2015) Nuts 2013 | As Novas Unidades Territoriais Para Fins Estatísticos. INE, Lisboa. Available at: https://www.ine.pt/xurl/pub/230205992

Jessop B (1994) The transition to post-Fordism and the Schumpeterian workfare state. In: Burrows R, Loader B (eds) Towards a Post-Fordist welfare state. Routledge, London and New York, pp 13–37

Macleod G, Goodwin M (1999) Space, scale and state strategy: rethinking urban and regional governance. Prog Hum Geogr 4(23):503–527. https://doi.org/10.1191/030913299669861026
Melo MCFX (2022) O Processo de Descentralização em Portugal: Uma análise à escolha dos Municípios. Universidade do Minho. Available at: https://repositorium.sdum.uminho.pt/handle/1822/81583
OECD (2020) Decentralisation and regionalisation in Portugal: what reform scenarios? OECD Multi-level Governance Studies. OECD Publications
Purkarthofer E (2018) Diminishing borders and conflating spaces : a storyline to promote soft planning scales. Eur Plann Stud 26(5):1008–1027. https://doi.org/10.1080/09654313.2018.1430750
Purkarthofer E, Granqvist K (2021) Soft spaces as a traveling planning idea: uncovering the origin and development of an academic concept on the rise. J Plan Lit 36(3):312–327. https://doi.org/10.1177/0885412221992287
Roodbol-Mekkes P, Van en Brink A (2015) Rescaling spatial planning: spatial planning reforms in Denmark, England, and the Netherlands 33:184–198. https://doi.org/10.1068/c12134
Stead D (2014) European integration and spatial rescaling in the Baltic Region: soft spaces, soft planning and soft security. Eur Plan Stud 22(4):680–693. https://doi.org/10.1080/09654313.2013.772731
Swyngedouw E (2000) Authoritarian governance, power, and the politics of rescaling. Environ Plann D Soc Space 18:63–76. https://doi.org/10.1068/d9s
Walsh C, Jacuniak-suda M, Knieling J et al (2012) Soft spaces in spatial planning and governance: theoretical reflections and definitional issues. Available at: https://www.researchgate.net/publication/256461649_SOFT_SPACES_IN_SPATIAL_PLANNING_AND_GOVERNANCE_THEORETICAL_REFLECTIONS_AND_DEFINITIONAL_ISSUES

Chapter 24
Decentralization and Governance in Portugal

Filipe Teles

Abstract This chapter discusses subnational governance in Portugal, within an historical context of centralization, and the problems it creates for suitable ways of addressing contemporary urban challenges. The argument presented here is based on a general model of successful subnational governance that is dependent on the capacity and competences of multiple tiers of governmental agencies and institutions. This is, ultimately, indissociable from a clear institutional design and from opportunities for effective engagement in policy coordination, which has practical implications for the definition of competences of each of the intervening actors in urban governance and policies. Portugal's challenges related to subnational governance, namely its illusive two-tier system, highlight problems of scale and capacity of regional and local authorities. The imbalances between municipalities, with strong asymmetries in institutional capacity and governance tools, amplify these problems and limit urban potential. This chapter's discussion about the need for unambiguous subnational governance for urban policies underlines the relevance of an appropriate design of territorial arrangements and effective multilevel mechanisms.

Keywords Decentralization · Governance · Portugal · Urban policies · Municipalities

24.1 Introduction

Two friends walk into a bar. This is how any attempt to explain the state of the art of the political debate in Portugal about decentralization should start. The consolidation of the democratic institutions in the country, a process which started after the revolution in 1974, has walked hand in hand with promises of reforms of the subnational governance architecture of the country. Democracy and decentralization

F. Teles (✉)
University of Aveiro, Aveiro, Portugal
e-mail: filipe.teles@ua.pt

R. C. Lois-González and J. A. Rio Fernandes (eds.), *Urban Change in the Iberian Peninsula*, The Urban Book Series, https://doi.org/10.1007/978-3-031-59679-7_24

are closely tied in the Constitution of 1976 and have been kept together, since then, in most of the political rhetoric and parties' electoral manifestos. However, a closer look reveals how one of the partners has been left behind.

This chapter discusses Portugal's subnational governance, within an historical context of centralization, and the problems it creates for suitable ways of addressing contemporary urban challenges. The approach to urban territories has changed from a structural focus on the development of plans and regulations, together with local public investments, to one where strategic planning, multilevel governance and urban–rural cohesion start to play a key role. Urban policies require a much more complex dialogue and negotiation between tiers of government and different types of actors. The broader picture, useful to provide a clearer understanding of this context, is the one of an evolution from direct action and policies, to the required multiagent and multilevel governance focused on programmes and policies, rather than plans and infrastructures.

The evolution of urban policies from a consequence of political inputs and leadership to an answer to major challenges, namely related to liveability, demography, housing, sustainability and economic development, has put significant pressure over its subnational governance architecture, explicitly highlighting important aspects related to decentralization. In addition, despite the apparent consensus and acknowledgement of needed significant changes regarding urban policies in Portugal, with specific programmes and tools, as well as the development of capacity building at the local level, the need to question the existence of a tangible national urban policy (e.g. Ferrão 2018) is obvious. The result is a problematic situation, where relevant questions need be tackled regarding the competences of different governance tiers, the capacity, and efficiency of policy arrangements, and the overall coordination between different policy strategies in Portuguese cities.

24.2 The Task of Subnational Governance

Policy issues that are inherently complex in the number of actors involved, and in scope and scale, require a diversity of resources and information that need to addressed through efficient subnational governance arrangements (Pope and Lewis 2008; Teles 2016). The growth of a significant number and diversity of such arrangements has prompted the need to assess the choice of the architecture (Teles 2012), the capacity of such collaborative schemes (Silva et al. 2018), and the advantages and weaknesses of the available repertoire of policy and governance tools (Silva et al. 2016). To navigate in such a complex chart, with multiple options, and to ensure its efficiency is, in terms of coordination, an endeavor which political actors and decision-makers struggle to achieve. Additionally, it raises new problems related to political accountability and to the capacity of local authorities to perform as steering agents within the network, as the demand to operate beyond traditional policy borders require much more than the usual 'governing by the book' approach.

As suggested previously by Heinelt and Kubler (2005), particularly regarding urban governance, success depends on multiple dimensions: the organizational culture, external incentives, and leadership. Available scholarship and research in the field has, however, largely disregarded the impact of subnational governance architecture, crucial in forging links and networks between multiple actors (Nelles 2013). Policy coordination is, at this level, mostly related to the alliances established between stakeholders, where clear partnerships between empowered actors are more likely to mobilize required resources and capacity (Silva et al. 2016). This hard-to-reach equilibrium for successful coordination requires working across institutional boundaries and the capacity to promote consensus. Unclear structures of decision-making, limited capacity and autonomy of local actors, and centralized governance processes undermine efficiency and the ability to function as expected. Thus, a particular important aspect to consider regards the subnational political actors' autonomy and legitimacy to manage diverging interests, while securing agreement on solutions to specific territorial challenges (Teles 2016).

To operate in a weakened and complex subnational governance world it creates excessive tensions between role attribution and role demands and undermines policy efficiency. The following sections of this chapter address the Portuguese subnational governance context, underlining these same context dilemmas, resulting from its centralization and complexity.

24.3 Understanding Centralization in Portugal

Decentralization is often mentioned as a political priority in Portugal, as a consequence of its administrative and political culture that favors centralized control and decision making. European countries provide very few comparable examples where such a relevant disproportion between local and central government's roles is recognizable, independently of the indicators used to measure it. Metrics of public expenditure, policy discretion, revenues or total number of public servants have kept Portugal significantly below the average of decentralized countries in Europe.

The transition to democracy in Portugal, despite the significant political change, was clearly insufficient to transform its administrative culture (Teles 2020). The parallel processes of democratization and decentralization, often observed in the wider European context, has seen less progress in a country where, with democracy, the growth in the state's functions came to generate new opportunities and sources of power for central bureaucracy (Tavares 2019). After 1986, the year in which Portugal entered the—then—European Economic Community, public administration reform is placed high in the political agenda. The strategy was highly demanding and aimed at changing organisational aspects, processes, and practices, rooted in decades of administrative culture. However, the course of the transformation over the following years resulted in a set of incremental and sectoral changes, rather than a mobilising reform and strategic orientation regarding subnational governance.

Within this context, reforms addressing the country's centralisation have been partially implemented and have not kept pace with the European trend, even that of the countries that have most recently introduced local autonomy in their constitutions, after decades of non-democratic regimes (Teles 2022). The considerable differences in the autonomy of European municipalities (Ladner et al. 2019) and the extent to which it translates into the democratic capacity of local communities to influence policies, show how dependent are subnational political entities on the institutional central mechanisms and government agencies (Kersting and Vetter 2003). In the case of Portugal, there is an important difference between the legal framework for local functions, and the actual discretionary capacities of the municipalities. The wide range of policy responsibilities fall short of their legal status. The erroneous assumption resulting from a mere legalist understanding of the role of subnational institutions is easily refuted in several indices of decentralisation. Portugal's consistent ranking as one of the most centralized countries in the EU over the past decades, despite the growth in competences, highlights the significant limitations regarding organizational, political and financial means, which are clearly not extensive (Magone 2011).

Despite this wide range of competencies, local governments have been traditionally weak in terms of capacity, particularly given their low share of national public revenue and spending, operating in a recognizable centralized setting (Teles 2021). Current local budget and fiscal framework resort to the period of the financial crisis and the concomitant Troika bailout in 2011 (Raudla et al. 2023), where municipalities were challenged by an intense fiscal supervision, decline in transfers, and strict debt limits. Their spending in Portugal constitutes 13.4% of general government expenditure (Eurostat), significantly lower than the EU average of 21.9%. The subnational governance architecture reinforces such centralized perspective and mitigates the formal constitutional autonomy of municipalities (Silva 2017).

This important recentralization trend regarding government control, structure, staffing and fiscal autonomy is reflected on the way local share of general government revenue shrank significantly after the crisis, with local taxes accounting—after 2012—for a larger share than grants. This limited room for maneuver results, regarding investments, in a high dependency of extraordinary transfers from competitive EU regional development and cohesion funds. The share of local public expenditure is also mirrored in the kind of competences assigned to municipalities, with a strong role of central administration in service provision and policy control. The three most relevant categories of subnational expenditure in the EU are education and welfare, while in Portugal these comprise of general services and economic affairs (Teles 2021). Similar features only prevail in other centralised countries.

In addition, available resources for local investment are largely resulting from European cohesion and regional development funds, which induce deceptive conclusions as they have positive impact over decentralisation metrics. However, it reflects an exceptionally complex form of centralisation, as it induces policy and strategic adaptation as a way to follow the funding opportunities, which are to a large extent centrally determined.

Finally, despite the relevant heterogeneity (in population size per municipality, internal migrations, employment rate, economic profile) across Portuguese municipalities, the competences and governance tools are the same, regardless of the territory. Usually, one of the purposes of decentralization is the need to meet differences in territorial, demographic, and economic circumstances across subnational jurisdictions. Unlike other examples pursuing asymmetric decentralization strategies, the Portuguese subnational governance system is still an outlier bearing the burden of its distinctive centralization.

24.4 An Illusive Two-Tier Structure

Another aspect of subnational governance in Portugal requires particular attention, since it has an important impact on the complexity of governance. A closer inspection to its architecture reveals that its two constitutional tiers of government, national and local, is far from depicting the reality.

Several adjustments, in recent years, have added to this complexity. The reinforcement of intermediate levels of public administration, through deconcentration and allocation of coordination and planning responsibilities to regional public agencies, with the rise in the number of competencies of the Regional Coordination and Development Commissions, acting also as Managing Authorities for European Regional Development Funds, is a clear example of such efforts.

In addition, the adjustment to the map of the civil parishes (Tavares and Teles 2018; Callanan et al. 2023) and the incentives for intermunicipal cooperation, have put significant pressure over local governments' institutional capacity. More technical, financial, and human resources are needed, and supra-municipal coordination, together with new mechanisms of inter-institutional and multi-level governance are expected to be in place. In addition, regional development programs and competitiveness and innovation strategies that municipalities are required to implement, demand a swift adaptation and the adoption of new policy agendas. These changes are almost always beyond the scope of local jurisdictions and have highlighted the need of a consolidated supra-municipal and regional scale of governance (Silva et al. 2016). The implementation model of European Regional Development Funds initiatives has also created strong incentives toward collaborative arrangements and led local authorities to engage in new policy implementation procedures, meeting standards in a number of areas requiring new skills and capabilities by municipalities (Silva et al. 2016).

However, an institutional framework to support and answer to such policy demands cannot be solely based on local political commitment, supplemented by regional technical support and the coordination capacity from central administration. In fact, this has resulted in unbalanced approaches across the country (Tavares et al. 2018). Against this demanding context of governance, it is necessary to ensure that policy decision and implementation instruments are clear and in place at the different scales. This is only possible with a clear and well-defined subnational governance architecture (Teles 2021).

This is why presenting Portugal as having an unambiguous tow-tier governance system is evidently illusive. Notwithstanding the expansion and rationalization of the country's deconcentrated public administration, the enlargement of local competencies and institutional additions to its multilevel governance, namely though the establishment of intermunicipal communities, which had a significant impact on subnational governance, Portugal is far from coping with the challenges it faces. Working across this complex multilevel system, going beyond the existing institutional boundaries, coordinating with a significant number of public agencies with multiple territorial schemes, requires from local authorities the tools and capacity to navigate and negotiate with a set of public agents.

Local policies and policymakers often struggle to steer within the existing complex set of institutions involved in the policy process and face substantial coordination challenges, menacing their governance capacity (Teles 2016). Beyond the obvious challenges related to political accountability, there are also limitations in the current governance system to adapt and answer to particular policy challenges which are significantly place-based.

Taken together, both centralization and the complexity of subnational governance have generated a growing interest in the choice of architecture and intensity of collaborative and multilevel governance, which ultimately underline the relevance of decentralisation in Portugal (Silva et al. 2018).

24.5 Urban Policies Require Unambiguous Subnational Governance

The overall picture of the Portuguese context of subnational governance, presented earlier in this chapter, helps in explaining why it generates concerns about its instruments to meet new policy agendas. The prevailing narrative is that of reviewing the role of local politics and its functioning, rather than addressing the overall architecture (de Sousa et al. 2015). As suggested by Heinelt and Kubler (2005), organizational and administrative culture are required to be successful in complex subnational governance settings. In fact, existing research tends to be focused on the role of local institutions, local governance arrangements, elected politicians and the impact of political leadership (Teles 2014), largely disregarding the overall institutional aspects of the phenomenon, critical in promoting multi-level coordination and building alliances between actors (Nelles 2013; Teles 2016).

The challenges of urban governance emphasize these scale and capacity problems. In urban policy, local authorities are more than just service providers and require an adequate design of effective territorial arrangements, which entails obvious pressures over multilevel policy coordination and requires a wide range of policy mechanisms.

Moreover, the period under consideration in this book is precisely the one where these structures of subnational governance were under significant change. However, the aim to maximize its efficiency did not meet with an adequate institutional architecture, as most urban authorities were under pressure to cope with demands for governance capacity and were under strong incentives to strengthen their tools to face new competencies and autonomy.

Weak governance capacity, unbalanced structures and policy instability are the main features of the national broader picture. It underlines relevant obstacles for efficient multilevel governance in urban policies and may help in explaining most of the shortcomings and policy inefficiencies presented in this volume. The argument in favour of a multilevel approach to urban policies underlines the needed relationship between tiers of government, which has practical implications for the actors engaged in and for the institutional design itself. It requires a stable multilevel arrangement with a clear assignment of competencies to the different scales. However, these are precisely the most relevant challenges met in Portugal during the last decades, where sub-national entities are occupied with unpredictable, complex and volatile contexts.

In a country that lacks political legitimacy resulting from elected bodies at the regional level, urban policies have been understood as a product of national incentives and regulations. Despite the degree of policy discretion at the local level in Portugal (Ladner et al. 2019), municipal policy action requires a negotiated consensus between governance actors. The delivery of urban policies is, thus, highly dependent on multilevel governance capacity, where the talent of political actors to select the relevant tools and to negotiate policies amongst tiers of government is required (Pierre and Peters 2000). Evidently, these processes of alliance building are tied with their perceived legitimacy, and the consistency of such arrangements are highly dependent on their stability and on the commitment between actors.

To claim that policy efficiency requires a rather stable and well institutionalized subnational governance landscape, is a recognition of the dynamism of the policy process that requires from actors in the broader context (regional and local authorities, national urban policies, EU programmes and incentives) the tools to avoid scale-traps, information and decision costs, inter-institutional competition and blurred boundaries of competencies. These would lead each of them to a biased autonomous approach to their role in urban policies. Without effective decentralization in Portugal and the absence of a clear, empowered and democratically legitimized subnational governance arrangement, key aspects for the success of urban policies related to efficiency and stability remain unanswered. And if policies are unlikely to persist, it will, ultimately, also affect trust in institutions.

24.5.1 Evidences of Inefficiency

Three main characteristics of the Portuguese subnational governance architecture provide clear evidence of the abovementioned challenges: the ambiguous map of regional public administration, the fragility of the intermunicipal tier, and the unconsolidated regional level. In the first case, the enduring complexity of the deconcentrated public administration bodies (despite the recent reforms aimed at their concentration) results in unnecessary overlaps and unclear jurisdictional territories. The consequent redundancies and source of inefficiencies, translate into odd cases of territorial misalignment. Some agencies are coherent with the existing Regional Coordination and Development Commissions scale, which follows the NUTS2 maps. Others are aligned with the current Intermunicipal Associations territories, accordingly to NUTS3, but not all. There are even particular cases of regional administration with responsibilities over distinct regions or sub-regions. To add to this bizarre picture, one could even refer to a set of subnational structures of public administration that are still organized following to the extinct districts (over a decade ago). The intricate and costly matching and negotiation between local authorities, public agencies, and—especially—citizens add severely to the abovementioned complexity of multilevel governance.

The second case is a simple outcome of the fragility of intermunicipal cooperation arrangements. Despite the importance of this scale in urban policies, namely given the polycentric nature of most of the Portuguese urban landscape, it remains institutionally underdeveloped and politically contested (Teles 2016). Portugal sought to strengthen these mechanisms through the creation of twenty-one Intermunicipal Communities and two Metropolitan Areas, to which tasks have been assigned to address policy issues that go beyond the borders of single municipalities. The rules of the recent European funding frameworks are very clear in valuing municipal cooperation as condition for accessing the funds. Nevertheless, their role remains limited, as only a fraction of local expenditure is allocated to their activities and still have a limited capacity to collect their own revenues, playing a marginal role in the public sector and the intensity of cooperation between member municipalities varies significantly (Camões et al. 2021). These are governed through a council of Mayors, which accentuates accountability problems related to the fact that all governance structures between the local and central government in Portugal are indirectly elected or appointed by the government.

The third and final evidence is also related to appropriate scales of governance in a multilevel arrangement, and results from the fact that the current tow-tier model is not designed to efficiently address problems at the regional level (Teles 2022). The existing Regional Coordination and Development Commissions, as public bodies of deconcentrated central administration, lack the legitimacy that democratically elected structures could have. Their current institutional setting and competencies convert regional policy issues in central government responsibilities. Despite the recent election in 2020, for the first time, as part of the Executive Board of the Regional Coordination and Development Commissions, this has not significantly changed the previous appointment character of such bodies. And, ultimately, it does not reflect a subnational system that allows for regional policy coordination to be addressed from a subsidiarity principle approach.

24.5.2 Decentralization and Urban Policies

Governance capacity, coordination of policy networks, multilevel governance, and accountability, sum up the most prominent aspects related to Portuguese subnational governance architecture that require a place in a future reform agenda.

A decentralization strategy needs to address these issues at their different scales (local, subregional and regional), recognizing the undeniable role of clear competence attribution and multilevel governance capacity as indispensable for effective urban policies. The sole focus on the local level or on the national policy formulation is clearly insufficient when information and coordination costs are clearly at stake. This integrated approach to decentralization—and not just a sum of policies or scattered legislative initiatives—is among the critical requirements for a successful reform process. In addition, recognizing territorial urban asymmetries, needs institutional tools to address their challenges accordingly to their specific requirements and implementing policies that better meet local needs. Thus, a simple translation of the subsidiarity principle must apply to its architecture.

In the Portuguese context, the formulation and steering of urban policies implies good governance and articulation of municipalities with civil parishes, intermunicipal structures, deconcentrated regional administration, and other public agencies and networks. The design of appropriate territorial arrangements isn't, therefore, independent from the way policy collaborative mechanisms efficiently operate.

24.6 Conclusion

The urge to adapt the architecture of subnational governance arrangements in Portugal (de Sousa 2015; Teles 2021) is not new, but the picture of a centralized country coping with complexity, ambiguity and uncertainty presents clear challenges.

The argument presented in this chapter is based on a general model of successful subnational governance that is dependent on the capacity and competences of multiple tiers of governmental agencies and institutions. The good functioning of an interdependent networked model of policy implementation depends on the mechanisms available for institutions of government. This is, ultimately, indissociable from a clear institutional design and from opportunities for effective engagement in policy coordination, and have practical implications for the definition of competences of each of the intervening actors in urban governance and policies.

The country's challenges related to subnational governance highlight the problems of scale and capacity of regional and local authorities. The imbalances between municipalities, with strong asymmetries in institutional capacity and governance tools, amplify these problems. Urban policies require more than local politicians and institutions able to steer and provide services. The design of appropriate territorial arrangements and effective multilevel mechanisms underlines the need to implement an appropriate approach to service restructuring, modernization agendas, and public administration reforms (Teles 2021).

The consolidation of democratic institutions and decentralization need to go hand in hand, as the two friends that walked into a bar.

References

Callanan M, Houlberg K, Raudla R, Teles F (2023) "Top-Down" local government mergers: political and institutional factors facilitating radical amalgamation reforms. J Urban Aff

Camões PJ, Tavares A, Teles F (2021) Assessing the intensity of cooperation: a study of joint delegation of municipal functions to inter-municipal associations. Local Gov Stud 47(4):593–615. https://doi.org/10.1080/03003930.2020.1857245

de Sousa L, Cruz N, Tavares AF, Jorge S (eds) (2015) A reforma do poder local em debate. ICS, Lisboa

Ferrão J (2018) Novas Agendas Urbanas: Que Desafios para o Ordenamento do Território? Congresso Ibérico Direito do Urbanismo, "O Direito e a prática urbanística na Península Ibérica: Desafios da Nova Agenda Urbana da União Europeia", Faculdade de Direito da Universidade de Lisboa, Lisboa, 16-03-2018

Heinelt H, Kübler D (eds) (2005) Metropolitan governance. Capacity, democracy and the dynamics of place. Routledge, London/New York

Kersting N, Agelika V (eds) (2003) Reforming local government in Europe. Closing the gap between democracy and efficiency. Leske + Budrich, Opladen

Ladner A, Keuffer N, Baldersheim H, Hlepas N, Swianiewicz P, Steyvers K, Navarro C (2019) Patterns of local autonomy in Europe. Palgrave, London

Magone JM (2011) Portugal: local democracy in a small centralized Republic. In: Loughlin J, Hendriks F, Lidström A (eds) The Oxford handbook of local and regional democracy in Europe. Oxford University Press, Oxford, pp 384–409

Nelles J (2013) Cooperation and capacity? Exploring the sources and limits of city-region governance partnerships. Int J Urban Reg Res 37(4):1349–1367

Pierre J, Peters GB (2000) Governance, politics and the state. Macmillan, Houndmills

Pope J, Lewis JM (2008) Improving partnership governance: using a network approach to evaluate partnerships in Victoria: research and evaluation. Aust J Public Adm 67(4):443–456

Raudla R, Callanan M, Houlberg K, Teles F (2023) Municipal finances in unitary systems: the effects of crises on financial autonomy in four European countries. In: Johnson C, Moldogaziev T, Ross J (eds) Research handbook on city and municipal finance. Edward Elgar Publishing, Cheltenham

Silva CN (2017) Political and administrative decentralisation in Portugal: four decades of democratic local government. In: Silva CN, Bucwek J (eds) Local government and urban governance in Europe. Springer, Basel, pp 9–32

Silva P, Teles F, Rosa Pires A (2016) Paving the (hard) way for regional partnerships: evidences from Portugal. Reg Federal Stud 26(4):449–474

Silva P, Teles F (2018) The biggest loser? Local Public Services under austerity measures in Portugal. In: Lippi A, Tsekos T (eds) Local public services in times of austerity across Mediterranean Europe. Palgrave, Basingstoke

Tavares A (2019) Administração Pública Portuguesa. Fundação Francisco Manuel dos Santos, Lisboa

Tavares A, Teles F (2018) Deeply rooted but still striving for a role: the Portuguese Freguesias under reform. In: Hlepas N, Kersting N, Kuhlman S, Swianiewicz P, Teles F (eds) Submunicipal governance in Europe: in search of expressive and responsive communities? Palgrave, Basingstoke

Teles F (2012) Beyond paternalism towards social capital: local governance reform in Portugal. Int J Public Adm 35(13):864–872

Teles F (2014) Local government and the bailout: reform singularities in Portugal. Eur Urban Reg Stud 23(3):455–467

Teles F (2016) Local governance and inter-municipal cooperation. Palgrave, Basingstoke

Teles F (2020) Public administration in Portugal. In: Bouckaert G, Jann W (eds) European perspectives for public administration: the way forward. Leuven University Press, Leuven, pp 439–454

Teles F (2021) Descentralização e podel local em Portugal [Decentralization and Local Government in Portugal], Fundação Francisco Manuel dos Santos, Lisboa

Teles F (2022) Mind the gap: the hiddden challenges of intra-state governments in Portugal. In: Copus C, Kerley R, Jones A (eds) A modern guide to local and regional politics. Edward Elgar Publishing, Cheltenham

Chapter 25
Governance and Urbanism in the Revitalization of the City Centre of Albacete

Francisco Cebrián Abellán and Irene Sánchez Ondoño

Abstract The current context is one marked by globalized ultraliberalism (Méndez in Tiempos críticos para el capitalismo global. Una perspectiva geoeconómica, 2023), planetary urbanization—the cosmopolis—(Beumer in Sustainability 9:845, 2017) and economies of agglomeration and density as justifications for the opposition between densified urban spaces and the depopulation of many rural areas (Rodríguez-Pose in Camb J Reg Econ Soc 11:189–209, 2018). This partly explains the changes in urbanization dynamics and dominant planning paradigms that have been replicated in different sociocultural settings. Over the past two decades, there has been a progressive shift in interest from more or less distant peripheries, which have been the focus of investments during periods of expansive urbanism, to renewed attention on city centers. The latter feature a unique functional mixture that combines residential, commercial, leisure and specialized service uses, lending them a high social and symbolic value. Furthermore is the heritage aspect, which in some cases fosters tourism use. The intensity of the transformation process is noteworthy in cities whose traditional centers are lacking in significant historical and cultural value. This paradigm shift in planning has occurred against a background of the need to adapt to changes in the economic cycle, following the collapse of urban development and housing production activity after the 2008 crisis. It is also driven by a growing awareness of the need to address environmental and social issues more efficiently. The process has been underpinned by the deployment of governance strategies, with urban policies incorporating more participatory mechanisms into the decision-making process. It also involves debate and consensus between civil society and technical services for city mobility design. The goal is, among others, to promote the social use of public spaces, encourage more sustainable modes of transportation, and support economic activities linked to services, commerce, and leisure, without

F. Cebrián Abellán (✉)
Department of Geography and Town and Country Planning, Castilla-La Mancha Universtity, Albacete, Spain
e-mail: Francisco.Cebrian@uclm.es

I. Sánchez Ondoño
Department of Geography, University of Salamanca, Salamanca, Spain
e-mail: Irene.SanchezOndono@usal.es

R. C. Lois-González and J. A. Rio Fernandes (eds.), *Urban Change in the Iberian Peninsula*, The Urban Book Series, https://doi.org/10.1007/978-3-031-59679-7_25

undermining the traditional residential vocation of the city center. This text focuses on strategies designed to foster a more urban planning approach and to return the city to its citizens. It is based on the analysis of strategies deployed in an intermediate city located in the southeastern quadrant of inland Spain, Albacete, which has opted to incorporate functional reactivation strategies for its traditional center, accompanied by a change in the urban mobility model. The chapter is organized in three parts. The first is a contextual introduction presenting ideas on the role of governance in our cities. The second aims to capture the general characteristics of the city, while the third focuses on the process of pedestrianizing the city center.

Keywords Governance · Urban planning · Sustainable mobility · Urban change · Albacete

25.1 Paradigm Shifts in Urban Planning

The term governance has become progressively more widespread and consolidated in Europe and Spain since the end of the last century. At its core is a willingness to address and solve problems through collaboration and consensus between administration and civil society, employing, to this end, more democratic processes of participation. Its incorporation into the urban sphere is part of the paradigm shift from government to governance for public administration, public policies, and the functioning of collective life (EC 2003). Numerous territorial and urban governance initiatives have emerged, affecting decision-making processes in social, territorial, and city policies. This has been achieved through citizen participation, coordination, and cooperation at different levels (Farinós 2015: 301–304). In essence, it aims to involve society and incorporate collective interests into urban planning decisions.

The increasing prominence of governance must be understood as a logical reaction to the dominant planning model in the 1990s and 2000s, with the process being delegated to the large real estate companies. Planning during this period was characterized by the dominance of functionalist, economic and unsustainable urbanism. This model weakened urban compactness, leading to extensive consumption and sealing of new land at the city edges and more or less distant peripheries, causing significant landscape impacts and social segregation (González 2018). It was heavily influenced by urban design patterns associated with dispersed cities, where physical and social isolation was constant, undermining the role of public space. Local and regional administrations, responsible for planning, have been necessary accomplices in urban territorial fragmentation. Motivated by evidently unrealistic expectations of demographic and residential growth, they prioritized maximizing profits at the expense of massive land consumption, delegating the city construction process to private urban developers (Gaja 2008; Romero 2010; Burriel 2014; Naredo 2015). This also occurred alongside a model where housing transitioned into an investment asset. All this happened in the context of an increasingly financialized housing market (Méndez 2019). During these years, the focus shifted towards creating low-density residential units outside the city or on the urban edges. Commercial and leisure areas were built, physically separated and connected by private vehicles. This was followed by the trend of urban megaprojects based on urban marketing, with high economic costs but minimal or no social profitability.

However, the interest in peripheries led to a neglect of the traditional consolidated city, which generated less capital gain and thus residual interest. The early years of the twenty-first century were a prodigious decade for Spanish urban planning (Burriel 2008), but also one of territorial mismanagement (Romero 2010). There had never been so much regulation, yet there was less control over urban dynamics and the expansion of urban sprawl (Calderón and García 2017). In these years, the financial sector immobilized capital surpluses in the second circuit of capital, in the forms of developable land, developed land, and housing (Harvey 2005; Lois et al. 2016; Méndez 2019). The large construction firms, banks and administrations went hand in hand, with the effects derived from the land and housing bubble and the price per square meter of urbanized land and construction.

The effects of the 2008 financial crisis, with undoubted cause-effect links to the real estate sector (Méndez 2019), disrupted this prior neoliberal urbanism framework and contributed to a reassessment of criteria that has favored a certain territorial resilience. Since the second half of the 2010s, interest in urban centers has recovered. The revaluation and pursuit of social interaction in public spaces are bolstering the function of streets and squares as meeting places. Sectors loaded with historical and symbolic values, traditional commerce, services and leisure have aroused greater interest. The focus has returned to the compact city, the mixture of uses, building density, and urban complexity; in short, to the essence of the city.

In parallel, strategies that work towards economic and environmental sustainability and social cohesion have been promoted, but also others towards a more collective and participatory social construction. Contributions from the scientific realm, citizen movements, and political debate, emerging as reactions to the previous period, have contributed to the change. The result is an urban planning practice that combines the traditional technocratic and administrative model with a more democratic one, characterized by greater social, environmental, cultural, and economic awareness. From an institutional perspective, this is reflected in the theoretical contributions of urban ecology and new urbanism. This embodies the idea of the "negotiated city" (Theurillat et al. 2016), incorporating new actors into the process of urban value generation (academic sphere, citizen movements, social groups, environmentalist groups), in addition to those that traditionally led the process and worked closely together (real estate developers, builders, financial sector, administration, and political authorities).

A reflection of this change can be seen in the new less hierarchical and technocratic agendas, aimed at creating a more human (Gelh 2014) but also a more urban urbanism, given the interest in the compact city (Cebrián 2020). Habitat III, the European Urban Agenda, which lays out its priority areas of action (from the Leipzig Declaration -2007 to the Bucharest Declaration 2019, including the 2016 Amsterdam Pact and the Spanish Urban Agenda of 2019) bolsters innovation in working methods and adjustment (European Sustainable Urban Develpment Strategy

2023; Ministry of Transport, Mobility and Urban Agenda 2023a, c). It places a strong focus on decision-making processes, grounded in the principles of multilevel governance, urban policy management and approaches to city planning (white paper on governance, U.E. 2001). These principles encompass the importance of participation, transparency, responsibility, accountability, an integrated approach to effectiveness, inter-administrative collaboration and cooperation, as well as subsidiarity at the most appropriate level (Pemán-Gavín 2019: 703–705).

Specifically, the Spanish Urban Agenda features a commitment to promoting a more rational use of land, controlling urban sprawl, preventing climate change, sustainable resource management, fostering social cohesion, supporting urban economy, ensuring access to housing, embracing digital innovation and utilizing governance tools. This framework also includes policies dedicated to promoting proximity and sustainable mobility. The goal is for cities to allocate more comfortable and attractive higher-quality public space for people's use and movement. Reduced car use is encouraged, with the goal of more sustainable transportation to decrease pollution, noise and accidents. Intervention and governance tools play a role in achieving these objectives. The European Union's Sustainable and Integrated Urban Development Strategies (EDUSI) work in this direction, shaping actions taken by some cities.

These principles have generated recent changes in many European cities, with Spanish cities also engaging in these transformative dynamics. The following section focuses on the strategies implemented and commitments made to promote these types of actions in the urban center of Albacete, an intermediate city of inland Spain.

25.2 Albacete, a Service City with a Transformed but Functional Historic Center

25.2.1 The Transformation Process of the City of Albacete

Albacete is a city located in the center of the southeastern quadrant of the Iberian Peninsula (Fig. 25.2). Its original location is on small promontories within an extensive, almost perfectly horizontal plain, flanked by minor mountainous reliefs to the south and east, featuring a rich underground aquifer. Its origin dates back to the medieval period, with documented evidence of its existence in texts from the early twelfth century (Pretel 2010: 13–14), although some authors trace it back to the ninth century (Panadero 1976). Its subsequent development is associated with a series of factors and events that, until the early nineteenth century, kept it in a state of relative stagnation. Its population was less than 5000 inhabitants (1200households) in the sixteenth century, although this decreased in the seventeenth century, only to grow again and surpass 7,000 inhabitants at the beginning of the nineteenth century (Panadero 1976: 258–261). The first population census conducted in Spain in 1857 indicated a population of 7,875. Until then, its economic base relied on agriculture (dryland cereal farming and primarily sheep farming) and commercial activities (boosted by the granting of a fair by Felipe V in 1710).

Since the mid-nineteenth century, the role of infrastructure and administration has favored productive diversification and connectivity, driving its demographic growth. In 1833, it became a provincial capital, and in 1834, it became the seat of the Territorial Court. Its position in the national communication network, both on major roads until the beginning of the industrial revolution and later in the railway network (in 1855, it became a stop on the second national railway line), helped bolster its role as a privileged location for communication, storage, intermediation, and distribution. This improvement in infrastructure allowed for the accelerated incorporation of new urban facilities and services in the transition from the 19th to the twentieth century, including electric power in 1888, sewerage in 1908–1910, and running water in 1901–1905. All of this facilitated the emergence of the first agro-industrial factories. In parallel, there was an increase in population, due to both new employment demands and the initiation of rural-to-urban migration processes that would accelerate from the 1930s onward. The city had 18,958 inhabitants in 1877. Since the beginning of the twentieth century, its population has been multiplied by eight: 21,512 inhabitants in 1900, 32,163 in 1920, 41,885 in 1930, 64,222 in 1940, 74,417 in 1960, 117,126 in 1981, 148,934 in 2001, and 172,359 in 2022 (Fig. 25.1).

The most significant change has occurred since the 1960s, as in most Spanish cities. In these years, there has been a considerable demographic increase due to an intense national socio-economic transformation associated with delayed industrialization, and rapid tertiarization fueled by a significant rural exodus, which was

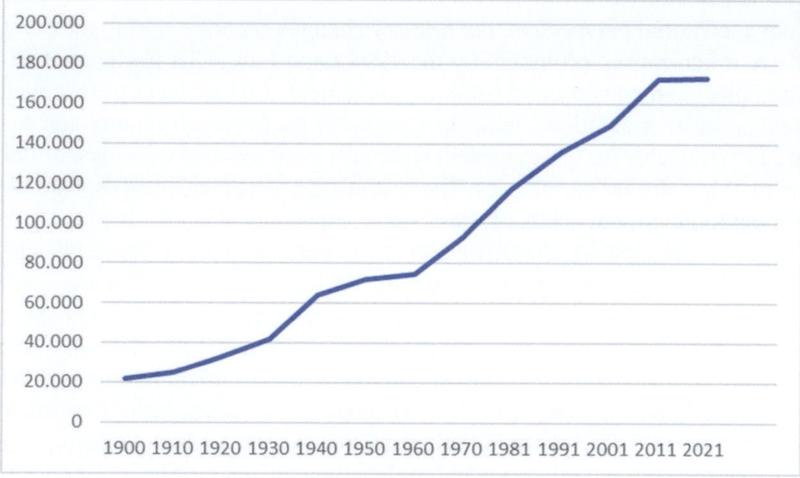

Fig. 25.1 Evolution of the population in the municipality of Albacete (1900–2021). *Source* National Institute of Statistics, Government of Spain, 2022. Own preparation

particularly intense in the 1960s and 1970s. Its demographic size (172,359 inhabitants as of January 1, 2022, according to the INE; Fig. 25.1) places it in the first position within the regional urban system, although its growth has stabilized over the past decade, lagging behind other cities in the Autonomous Community of Castilla-La Mancha. It is the provincial capital, the seat of the Territorial Court, and it offers services and commercial areas that encompass most of the provincial territory and extend into the neighboring province of Cuenca. The city has several industrial estates in its vicinity, employing nearly 20,000 workers in industries of various sizes. Among them, a new aeronautical industrial hub is emerging near the air force logistics center (*La Maestranza*) and the Los Llanos airbase. Albacete has specialized healthcare facilities, a campus of the University of Castilla-La Mancha, and is well integrated in the national communication network, both in terms of road and rail (High-Speed Rail—AVE—has been operational since 2010).

These processes significantly impacted the city, reflected in the extensive destruction and reconstruction of the real estate market, especially in the central sectors of the city. As a result, the city remains well defined historically, with a somewhat chaotic verticalization in certain sectors built during the decades from the 1950s to the 1970s. Until the late 1980s, the urban morphology was characterized by urban compactness and territorial continuity. In fact, there were well-defined limits set by the first ring road generated in the thirties based on the road connecting Madrid to Valencia, which served as the urban edge well into the last century, now transformed into an avenue that channels a significant portion of the city's peripheral traffic. The second ring road is more recent, from the early twenty-first century, and delineates the new axes of growth. It is also essential to note the creation of a significant stock of second homes in the vicinity of the city in the 1970s and 1980s, especially towards the west and south.

From a territorial perspective, the primary changes occurred during the period of expansive urbanization, promoted by the 1998 Land Law. This law helped modify morphological and constructive patterns, implemented under local governments of different political orientations. In the last decade of the twentieth century and the first of the 21st, strategies for urban growth were promoted towards the edges of the city and more or less distant peripheries. These strategies were conceived as responses to the demographic growth needs of those years, especially since the beginning of the century (Albacete grew by 25,000 inhabitants in the first decade). These changes are reflected in Fig. 25.2, where it is worth noting that the oldest residential areas are now located in the areas of urban expansion from the mid-twentieth century. There are still some groups of houses in the city center that are more than 100 years old, although there are not so many and these correspond to the few historical buildings that have been preserved and are now protected. Meanwhile, there has been a considerable process of renewal in the urban building stock in the center, where tradition coexists with modernity, both in real estate and in the dominant activities and functions.

Meanwhile, newly developed neighborhoods have emerged in new sectors, sometimes consisting of attached or detached single-family homes, and in other cases, under the model of multi-family blocks with four or more storeys in new, aseptic areas, constructed rapidly without connections to the community. These have been

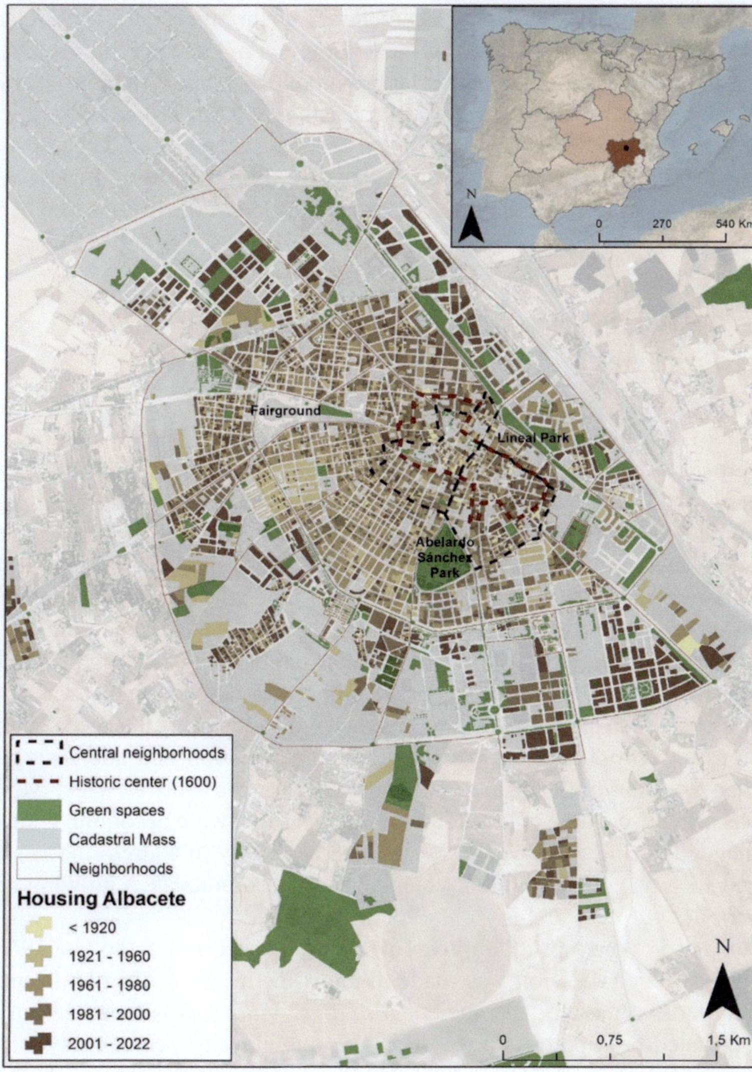

Fig. 25.2 Year of construction of housing in Albacete. *Source* CAT file (Cadastral Electronic Site). Own preparation

primarily erected to the south and north (Fig. 25.2). This is a consequence of the real estate boom that occurred at the beginning of the century, fueled by demographic growth, the need for housing, government responses, and initiatives from a small number of influential urban developers. The model relies on the creation of a significant housing stock for the free market, including a number of blocks of social housing. Ambitious projects for the construction of distant urban islands also appeared in the last years of the real estate boom in neighboring municipalities, at distances ranging from ten to twenty kilometers from the city center. Their fates have varied. There are failed examples of urbanization of a marked speculative character and questionable economic, social, or environmental viability (a clear example is found in the projects of El Cañaveral or La Losilla, in Chinchilla de Montearagón) (Cebrián et al. 2022). Others have settled and become new residential neighborhoods of detached single-family homes, such as Aguasol, from the end of the last century, or La Felipa, 14 km from the city, since the first decade of the twenty-first century.

25.2.2 The Adaptation of the Urban Center to Recent Processes of Change

The origin of the city is situated in the current urban center. The explanatory reason for the fan-shaped growth is that there are no significant topographical features that have conditioned its urban expansion towards the south, north, and west. For this reason, processes of urban and social transformation have been intense and critical in some sectors of the old urban center. The most significant transformations occurred between the 1960s and 1980s. It was a time when dominant architectural criteria were based on an eclectic style, which aimed to replace old buildings, some with certain heritage value (palatial houses, convent buildings, or simple residences of pre-industrial types of construction), with others of greater volume. It was a context of functionalist urban planning, more concerned with addressing the needs of housing production to accommodate a rapidly growing population and addressing social issues than with the preservation and/or recovery of the built heritage. More real estate was needed, capable of providing the services that were becoming widespread and demanded during those years. It was also necessary to address the needs of a motorized society, which required wider roads to accommodate a growing number of automobiles.

Possibly one of the most disruptive interventions occurred in the late 1960s and early 1970s, when a decision was made to intervene in one of the oldest and most traditional parts of the city, the ancient *Alto de la Villa*. It was located on a small, slightly elevated promontory above the plain, retaining its medieval layout, originally surrounded by an Almohad wall. By the mid-twentieth century, the housing was in a significant state of deterioration and the neighborhood was one of considerable social vulnerability. To address the problem, it was decided to implement a disruptive innovation operation. The solution was to demolish the historic center, empty the small promontory, and replace it with an artificial architectural solution articulated around a central open space, with standardized multi-family residential blocks, between six and eight stories high, open to the perimeter streets and an interior public space (Fig. 25.3). The result is a space organized around a closed, colonnaded, and elevated square (either to preserve the condition of a small promontory in the collective imagination or purely for speculative reasons—increased building volume

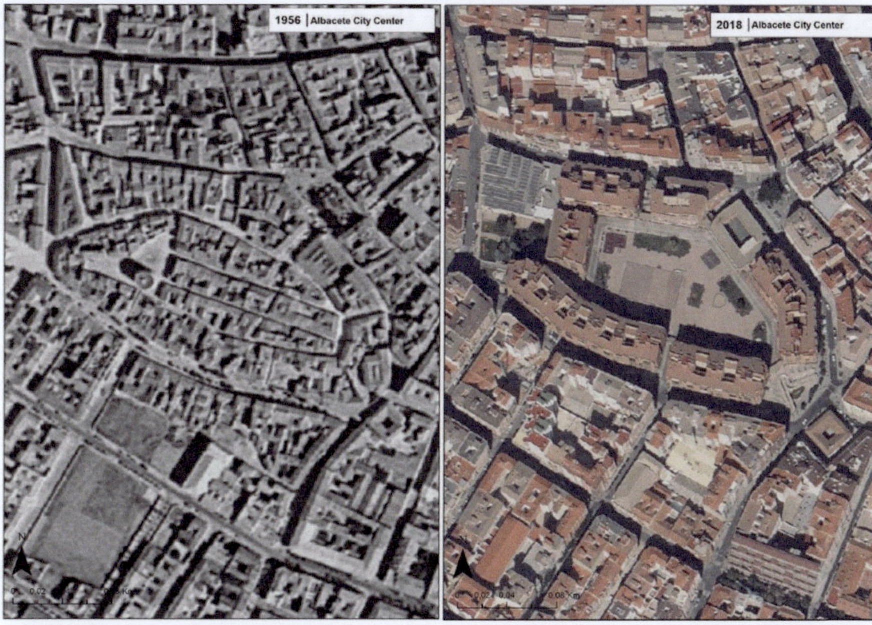

Fig. 25.3 The transition from the medieval Alto de la Villa to the current Villacerrada. *Source* National Geographic Institute. National Aerial Orthophoto Plan 1956–2018. Own preparation

and more underground parking). This solution, regarded as a mistake by many residents of the city, has resulted in a large, enclosed pedestrian interior space, devoid of personality and life, only retaining its residential function. In contrast, there are peripheral streets full of commercial activity and various services.

Meanwhile, the rest of the city center has been affected by a significant renewal of the residential housing stock, but with more specific actions. Generally, the focus has been on the scale of urban plots, replacing traditional stone, rammed earth, and wood constructions of two or three stories with concrete multifamily buildings of six or more floors. Many old buildings are still preserved among the newer ones, now protected by urban regulations highly sensitive to the preservation of the historical built heritage. The result is a certain chaotic verticalization of the center, which preserves the layout and maintains many of the original street names (some named after the guilds that were grouped in them in the past), preserves some old buildings (now converted for cultural or administrative functions), and continues to exhibit its traditional multi-functionality. This explains why, since the early twentieth century, a process of selective pedestrianization has been initiated, conceived in terms of commercial and public use of streets and squares.

25.3 Revitalization Strategies for the City Center

As mentioned in the previous section, much of the historic center retains its personality and urban morphology, although it has undergone a profound renewal of real estate stock. This sector continues to maintain a symbolic component for the citizens, as it is home to the most emblematic buildings. Some of the most representative ones are the old City Hall, the Cathedral, the Convent of the Assumption, the Palace of the Provincial Council, the *Pasaje de Lodares* (a covered commercial gallery), the *Posada del Rosario* (sixteenth century inn), the *Teatro Circo*, and the Fairground. Albacete is a typical example of the urban trends experienced in Spanish intermediate cities in the twenty-first century. The urban growth of the periphery led to a loss of vitality in the historic center, among other things, due to the arrival of large commercial centers. These are areas of commerce and leisure located at the city's edges (with ample parking spaces) that clearly compete with the traditional commercial centers. These new spaces have managed to satisfy and adapt to the new general consumer trends, offering fresh perspectives in commerce and services. Nevertheless, both the City Council and business associations have consistently promoted initiatives to boost local commerce and attract more visitors to the historic commercial center (events and activities, as well as road reforms to improve accessibility). Nonetheless, the center of Albacete has succeeded in maintaining its functional diversification: residential, leisure and hospitality, and traditional commerce (small businesses focused on clothing, footwear, accessories, and food). It also has a considerable offer of specialized services (financial, health, legal, and personal). Most of the commerce and services are concentrated in the Villacerrada neighborhood (Fig. 25.4).

Since the mid-2020s, the municipal government, in coordination with various citizen groups and under the prevailing logic of participatory and sustainable urban planning/renewal, has driven a process of transformation in the city center. Two initiatives stand out. One is the Commercial Revitalization Plan for Albacete (Albacete City Council 2015), which was conceived with the aim of supporting small businesses in the face of growing competition from large commercial areas strategically established on the city's outskirts since the early 1990s. The plan includes, among other objectives, the revival of traditional markets in the city center. It also sets out proposals for improvement, addressing issues such as street interventions, replacement of street furniture, lighting and the maintenance of green areas. It is a comprehensive but generic document, lacking specific actions.

This was the precedent of another initiative on which attention is focused in this chapter. The city of Albacete has, by agreement of the Local City Board of June 18, 2014, a Sustainable Urban Mobility Plan for the city of Albacete (hereinafter SUMP). It was inspired by the Spanish Sustainable Mobility Strategy (approved by the Council of Ministers on April 30, 2009—Ministry of Transport, Mobility and Urban Agenda 2023a, c). The plan was agreed upon in the city in 2016 and has had a greater impact than the previous one (aimed at boosting commercial activity). The SUMP was developed by the Infrastructure and Urban Mobility Service of the City Council and aims to expand the pedestrianized area of part of the urban center,

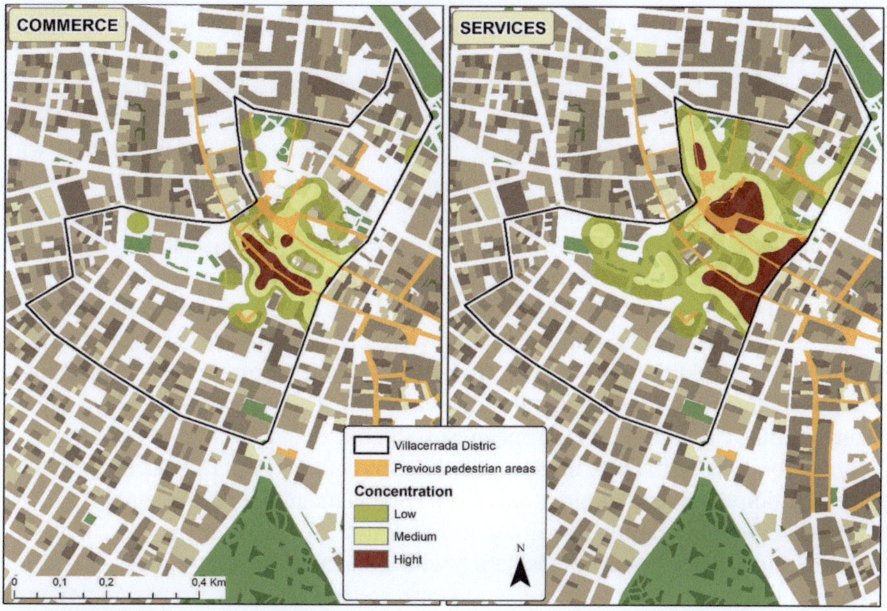

Fig. 25.4 Concentration of commerce and services in the neighborhood of Villacerrada. *Source* Own preparation

whose boundaries were set under criteria that are more functional than historical. The Mobility Commission was created to implement the initiative. The plan is inspired by the principles of urban ecology, as its goals include reducing fuel consumption, reducing environmental and noise pollution, increasing road safety (reducing accidents), more democratic and accessible management for different social groups that inhabit or use the city center, and improving health to contribute to the well-being of the population. The working methodology is based on both technical planning and citizen participation. The SUMP ultimately aims to modify the mobility model in the city center, without substantially altering the traditional residential, commercial, and service functions. It also seeks, however, to promote the social use of public spaces and incorporate sustainable forms of transportation.

The dominant strategies of the SUMP are aimed at pedestrianization and semi-pedestrianization of certain sectors, promoting mobility on foot, by bicycle and public transport. From a methodological perspective, it follows an analytical-propositional model articulated under the following: (a) the analysis of mobility within the city; (b) the definition of a sectoral mobility model; (c) the proposal of sustainable forms of mobility (on foot, by bicycle, or by public transport); (d) the reduction of car traffic by regulating and controlling access of private vehicles to the center; (e) the design of a parking policy for residents and visitors to the center; (f) the organization of loading and unloading spaces and schedules; (g) the proposal of strategies to improve the offer and services of public transport; (h) support for public transport mobility policies for different groups.

It is based on the idea of providing greater availability of more habitable and walkable spaces. In terms of territory, the model relies on topological logic, incorporating a structure of pedestrian axes and nodes (streets and squares/parks) with the aim of promoting the use of the center for citizens. It is organized around a series of radial axes with a periphery-to-center direction for pedestrian and cyclist access to the center. For this purpose, it leverages green infrastructure (Abelardo Sánchez, Lineal, Fiesta del Árbol and Jardinillos parks), certain sectors and iconic buildings (Fairground; urban ensemble around the Plaza del Altozano; Railway Station; Football Stadium) and streets that were previously pedestrianized or semi-pedestrianized. Additionally, circular routes have been designed around the center, acting as connectors for peripheral neighborhoods. The project has been financed through an EDUSI program (European Sustainable Urban Development Strategy), designed for cities with more than twenty thousand inhabitants. Its structure includes citizen participation and administrative capacity as mandatory requirements.

The interventions aim to re-order motor traffic to peripheral routes, pedestrianize certain streets, enhance aesthetics with new street furniture and ornamental elements, create single-level streets, separate pedestrian areas from vehicle zones using bollards, reduce motor vehicle speeds on streets, limit parking areas for vehicles and re-organize parking for motorcycles, loading and unloading, and blue zones. It also includes widening sidewalks, modifying lighting, installing containers for urban solid waste (USW), improving tree planting and vegetation, incorporating bicycle parking, updated signage, regulating motor vehicle access hours, regulating leisure and hospitality activities, and re-opening emblematic buildings and premises located in the city center that are currently unused.

The governance methods employed for its design are based on a working method that combines technical actions, commissions and participation workshops, and numerous working meetings. These formats have given rise to the action plan, dissemination methods, and awareness strategies for citizens. Therefore, consensus from different groups on mobility planning and pedestrian improvement proposals is mandatory. Among the groups involved in the process are municipal technical services, political parties, taxi associations, city-center trading associations, groups of retailers from the streets affected, the urban cyclists' association, the Albacete Neighborhood Federation, the College of Architects, the College of Technical Architects, the "967" collective, associations of people with disabilities, city-center beverage distributors, and the city's business federation. The implementation of the process is summarized in Fig. 25.5, where it can be seen that the majority of the pedestrianized area corresponds to the streets of the 16th-century historic center, on which the two neighborhoods that make up the city center are organized (Villacerrada and Carretas square).

The map synthesizes the actions that have contributed to transforming, expanding and improving pedestrian spaces in the urban center. It is based on the premise that this is a process closely related to the historic center of the city, which retains fully operational secular multi-functionality. The analytics instrument on which the

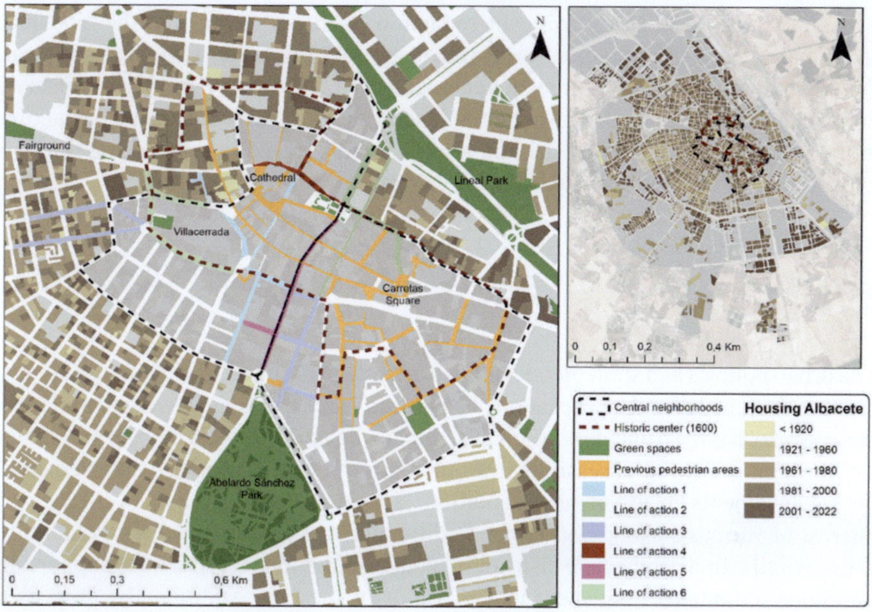

Fig. 25.5 Object of the actions, final results, and age of housing. *Source* Own preparation based on the information generated by the Mobility Service of the Albacete City Council

cartography is based synthesizes the action plan to extend these types of spaces, the works executed and the period in which they were done. The construction period ran from April 2018 to August 2020, organized in three phases between April 2018 and August 2020 (the first from April to October 2018, the second from June to December 2018, and the third from October 2019 to August 2020). The result was the closure of some streets to traffic and pedestrianization, with vehicle access only for residents or those authorized (taxis, ambulances, loading and unloading operations). Another change was the re-orientation of traffic to readjust the flow of vehicles throughout the city. The final investment exceeded 2.4 million euros, coming from EDUSI funding. The main consequences of the street remodeling work are the reduction of vehicle traffic and the expansion and improvement of sidewalks and pedestrian areas. In parallel, changes have been made to supply and distribution schedules for goods, modifications to the forms and times of selective collection of urban solid waste, and the introduction of a bicycle lending service. These measures complement others that have been in progress for some years, such as the creation of an underground public parking network to relieve congestion, operated through concessions to private companies, and aimed at addressing the growing parking limitations and traffic closures in some sectors of the center. The city's bus fleet has also been strengthened and modernized. The cycle lanes have been expanded, and the city has been equipped with a bicycle rental service.

As a result, there is a significant reduction in traffic throughout the city and, specifically, in the urban center. According to available mobility service information, the decrease exceeds 30% in recent years, although it is worth noting the impact of the pandemic over a number of months. One of the critical aspects is the reduction in the vehicle parking surface area on streets and the regulation of parking through paid zoning (blue zone and green zone), except for residents (red zone).

25.4 Conclusions

The urban policies and strategies developed after the 2008 crisis have shifted attention towards the renewal and regeneration of central city areas. The aim has been to address these spaces from a triple dimension. The first is the much-needed recovery of social use through improved accessibility, an increase in pedestrian zones, accompanied by enhancements in urban ornamentation. These actions demonstrate a renewed interest of various stakeholders in the traditional center involved in urban planning. This revitalization perspective also holds economic significance, linked to supporting traditional commerce, prevalent in this sector of the city. Additionally, there is a growing awareness of the necessary energy transition and more sustainable mobility, leading to the inclusion of strategies for limiting or eliminating motorized traffic in certain areas. This convergence of realities arises in a context of different political scenarios and less technocratic and more participatory practices.

The SUMP has focused its attention on actions aimed at the sectors less affected by urban developments generated in the first decade of the century, primarily located on the urban edges, especially towards the south, north, and west of the city. These developments have relied on low building density modalities, with single-family homes to the south and detached multifamily blocks to the north. This is where large commercial areas have been located. In fact, mobility restrictions in the central area seek to address a present and future problem, although they only do so partially and in a very localized way. The city still has an oversized car population, and there is great dependence on private vehicles, especially for short distances. It is also a reality that low-density neighborhoods (townhouses, single-family homes, etc.) have proliferated, with the distance to public services and the lack of commercial facilities making their inhabitants dependent on cars, as public transportation fails to meets the needs of residents. This common reality in our cities requires a rethinking of mobility beyond strategies aimed at city centers, and which is shaped by the city model chosen. This encourages us to think that urban policies need to regain a sense of time and scale. Realistic long-term strategies are needed for demographic growth (our cities grow very little), an urban model based on the multi-functionality of different city sectors (there are dormitory communities without sufficient service provision), and more sustainable mobility, where accessibility is a constant reference, including peri-urban spaces.

The morphology of the city of Albacete has been characterized by territorial continuity, although there has sometimes been a strong temptation to adopt models of dispersed cities. The contradictory strategies that have unfolded in recent decades, aimed at promoting or restricting the car-centric city, require serious review. This necessitates studies and planning processes in terms of the urban area, especially when designing the location of new growth sectors of the city. This is particularly important considering the socioeconomic scenarios in which we are immersed, far removed from the fictitious growth of the early years of this century. In addition to this, there is an increase in the average age of the population (especially in the city center), a slowdown in demographic growth, and a progressive incorporation of information and communication technologies into everyday life. These changing scenarios necessitate a review of the city model to allow for more urban-focused urbanism, which is more attentive to citizens, and with greater involvement of different social groups. In this context, planning, citizen participation, and research on the city, where the University plays a prominent role, can, and should regain, the leading role it has lost.

Acknowledgements This text is part of the research projects "Territorial Reconfiguration Processes and Socio-spatial Fragmentation in Medium-sized Cities of Castilla-La Mancha (2000–2019)," funded by the University of Castilla-La Mancha in the 2022 call (Reference: 2022-GRIN-34425), and "Spatial fragmentation and residential space (housing and public space) in Spanish medium-sized cities and their urban areas" (PID2021-124511NB-C22), funded by the Spanish Ministry for Science and Innovation.

The text is an updated revision of Cebrián, F. (2022). Governance, sustainable mobility, and the use of public spaces. Initiatives for the reactivation of the urban center in a Spanish medium-sized city: Albacete. In Rio Fernandes, J.A., Lois González, R.C., and Beltrao Sposito, M.E. El mundo visto de las ciudades, Tirant Humanidades, Valencia, 781–794. ISBN 978-84-18970-21-4.

References

Ayuntamiento de Albacete (2015) Plan Municipal de Dinamización del Comercio de la Ciudad de Albacete. Ayuntamiento de Albacete

Ayuntamiento de Albacete (2016) Guía del plan de acción. Desarrollo del PMUS. Documento de inicio. Servicio de Infraestructuras y movilidad urbana. Ayuntamiento de Albacete

Beumer C (2017) Sustopia or cosmopolis? A critical reflection on the sustainable city. Sustainability 9:845. https://doi.org/10.3390/su9050845. https://www.mdpi.com/2071-1050/9/5/845

Burriel E (2008) La década prodigiosa del urbanismo español (1997–2006). Scripta Nova XII 270(64) (Universidad de Barcelona)

Burriel E (2014) El estallido de la burbuja inmobiliaria y sus efectos en el territorio. In: Albertos JM, Sánchez JL (eds) Geografía de la crisis económica en España. Universidad de Valencia, Valencia, pp 101–140

Calderón B, García JL (2017) Legislación urbanística y planeamiento urbano en España, 1998–2015. Del despilfarro a la sostenibilidad. GeoCrítica, Scripta Nova, Vol XXI, nº 570

Cebrián F (2020) Dinámicas de urbanización en ciudades medias interiores. ¿Hacia un urbanismo más urbano? Tirant Lo Blanc, Valencia

Cebrián F, Sánchez I, Jover FJ (2022) El proceso de transformación de Albacete: la dialéctica entre la ciudad compacta y el cambio de modelo. In: Bellet C, Cebrián F (eds) Ciudades medias en España. Urbanización y políticas urbanísticas (1979–2019). 40 años de ayuntamientos democráticos. Asociación Española de Geografía, pp 259–292

Comisión Europa (2003) Informe de la Comisión sobre la Gobernanza Europea. Luxemburgo. Servicio de Publicaciones de las Comunidades Europeas. https://ec.europa.eu/governance/docs/comm_rapport_en.pdf

European Sustainable Urban Development Strategy (2023) Available in https://europa.albacete.es/es/fondo-europeo-desarrollo-regional-feder/edusi-albacete. http://www.edusi.es/. Fecha de consulta 15–11–2023

Farinós J (2015) "Gobernanza Territorial" y "Goberanza". In: López L (ed) Diccionario de Geografía aplicada y profesional. Universidad de León, León, pp 301–306

Gaja F (2008) "El "tsunami urbanizador" en el litoral mediterráneo. El ciclo de hiperproducción inmobiliaria 1996–2006, X Coloquio Internacional de Geocrítica. Diez años de cambio en el mundo, en la Geografía y en las Ciencias Sociales, 1999–2008, Barcelona, Universidad de Barcelona

Gelh J (2014) Ciudades para la gente. Ed. Infinito. Buenos Aires. Libro Blanco del Comité de las Regiones sobre la Gobernanza Multinivel (2009/C 211/01). https://eur-lex.europa.eu/LexUriServ/LexUriServ.do?uri=OJ:C:2009:211:0001:0027:ES:PDF

González Pérez JM (2018) Urban inequality: the city after the 2007 crisis. In: González Pérez JM (ed) Urban inequality. MDPI, Basel, Switzerland

Harvey D (2005) A brief history of neoliberalism. Oxford University Press, Oxford

Lois RC, Piñeira MJ, Vives S (2016) El proceso urbanizador en España (1990–2014): una interpretación desde la Geografía y la teoría de los circuitos del capital. Revista Scripta Nova, vol XX, nº 539

Méndez R (2019) Ciudades en venta. Estrategias financieras y nuevo ciclo inmobiliario en España. UPV, Valencia

Méndez R (2023) Tiempos críticos para el capitalismo global. Una perspectiva geoeconómica. REVIVES, Madrid. https://revives.es/publicaciones/tiempos_criticos/

Ministry of Transport, Mobility and Urban Agenda (2023) La Agenda Urbana de la Unión Europea, el Pacto de Ámsterdam (2016) y la Declaración de Bucarest (2019). Available in https://www.mitma.gob.es/arquitectura-vivienda-y-suelo/urbanismo-y-politica-de-suelo/actividad-internacional/union-europea/el-pacto-de-amsterdam-2016-y-declaracion-de-bucarest-2019. Fecha de consulta: 15–11–2023

Ministry of Transport, Mobility and Urban Agenda (2023) Estrategia Española de Movilidad Sostenible. Available in https://www.miteco.gob.es/es/calidad-y-evaluacion-ambiental/participacion-publica/estrategia_esp_movilidad.html. Fecha de consulta: 15–11–2023

Naredo JM (2015) Un episodio relevante: la burbuja especulativa y la crisis inmobiliaria en perspectiva. Cuadernos de Investigación Urbanística 100:77–82 (mayo junio)

Panadero M (1976) La ciudad de Albacete. Caja de Ahorros de Albacete, Albacete

Pemán-Gavín I (2019) La Agenda Urbana en el marco europeo. Experiencias internacionales de políticas urbanas. Ciudad y Territorio, Estudios Territoriales LI(202):701–722

Pretel A (2010) La Villa de Albacete en la baja Edad Media. IEA, Albacete

Rodríguez-Pose A (2018) The revenge of the places that don't matter (and what to do about it). Camb J Reg Econ Soc 11(1):189–209

Romero J (2010) Construcción residencial y gobierno del territorio en España. De la burbuja especulativa a la recesión. Causas y consecuencias. Revista Cuadernos Geográficos 47:17–46

Theurillat T, Vera-Büchel N, Crevoisier O (2016) From capital landing to urban anchoring: the negotiated city. Urban Stud 53(7):1–10

UE (2001) Goberananza Europea. Un libro Blanco, 428, 25 de julio de 2001

Chapter 26
The Power of Neighbourhoods. Bottom-Up Governance and Urban Planning in a Working-Class District of Barcelona, Nou Barris

Asunción Blanco-Romero

Abstract It has been shown that the spaces most threatened by the COVID-19 health crisis were cities, due to their high population density and levels of labour or tourist mobility. However, they are also the places with the highest human, material and social capital and from where cures and alternatives can arise (Nel·lo in La ciudad y la plaga, en Eldiario.es, 2020). The most hopeful scenarios are based on community organisation and on the accumulation of many successful historical experiences (Observatorio Metropolitano de Madrid, Diez and Col·lectiu Víric, L'aposta municipalista. De les bullangues a les lluites actuals pels municipis lliures a Catalunya. Virus Editorial, Barcelona, 2015). This case study of the city of Barcelona conducts a review of: (1) the theoretical aspects referring to democratic radicalism, (2) governance and the government system, and (3) its applicability to urban planning and then (4) it describes how the case study is applied to the neighbourhood scale in Barcelona and specifically (5) the neighbourhood movement of the district of Nou Barris.

Keywords Democratic radicalism · Citizen empowerment · Participative governance · Civic action

26.1 The Theories of Democratic Radicalism

The environmental, social and COVID-19 health crises have given rise to challenges that are difficult to solve. The most individualist alternatives are always at our disposal (Pigem 1994), while state-oriented organisation shows weaknesses in terms of authoritarian, corrupt, productivist and arms biases in state capitalism systems

A. Blanco-Romero (✉)
Geography Department, Universitat Autònoma de Barcelona, Cerdanyola del Vallès-Barcelona, Spain
e-mail: asuncion.blanco@uab.cat

R. C. Lois-González and J. A. Rio Fernandes (eds.), *Urban Change in the Iberian Peninsula*, The Urban Book Series, https://doi.org/10.1007/978-3-031-59679-7_26

(Amin 1974). Somewhere in between, community self-organisation promotes democratic radicalism and governance, with the best results shown through proximity. In this respect, Bookchin (1991) maintains that separation and self-reliance, such as that prevailing in eco villages, go against the principles of social organisation and political counterpower that a transforming structure should assume. This author's thesis of libertarian municipalism proposes a movement linked to society, without becoming isolated from the institutional system or renouncing elections. This implies breaking away from the predominant formalist and liberal view of democracy in which participation is only present in the election of representatives and not in decision-making or ensuring the redistribution of wealth and power (Aguiló 2014). Real democracy would require accepting *demodiversity* (Santos and Avritzer 2003), assuming the coexistence of different democratic models and practices and searching for alternative means other than the usual democratic channels of liberal politics. It requires unlearning and transforming, fighting against the *monoculture* of this globalised liberal democracy and single discourse with projects and campaigns that generate practices of articulation and complementarity between participative, radical and representative democracy (Aguiló 2013).

Citizen empowerment would be based on the concept of *prefigurative* politics with ways of organisation and tactics that anticipate the structure of the desired society, "… those forms of social relations, decision-making, culture, and human experience that are the ultimate goal" creating more radical and participative political practices (Boggs 1977) and generating ways of popular self-governance able to assume the power that is currently in the hands of privileged minorities. Within this context, the new political forms are conceived in spaces of broader and higher quality participation, where the public domain constitutes an element of social change and the capacity of State intervention is widened to politicise an economy oriented towards the common good and transparency.

In recent years, we have seen how the democratic imagination, through the social movements of the *indignados* and the anti-austerity movement of 2011 in Spain, may be deployed in practices that go beyond conventional actions. Political, social or cultural transformation can be reinforced by actions in at least three social dimensions: (1) grassroots and assembly movements; (2) the neighbourhood and association fabric, together with NGOs; and (3) the academic world (Col·lectiu Víric 2015). Their contribution, going beyond the personal commitment, adopts teamwork practices, for which there are many recommendations derived from direct experience (Cembranos and Medina 2003). The establishment of alternative mechanisms of economic organisation have added to the defence of the rights of social citizenship, the reduction of vulnerability and dependence on market mechanisms (García and Pradel 2019). Therefore, the vindications beginning with the denouncement of capitalist privatisation have evolved into the development of self-management models of amenities, neighbourhood assembly networks, community food supply experiences, popular culture, etc., which are starting to resemble governance systems.

26.2 Governance and the Government System

Although the concept of governance can be considered to be as old as human civilisation, it became a widely used concept from the 1990s. Despite its complex and sometimes polysemic nature, according to many authors, in governance it is possible to identify a social form in which the formal government, civil society and business fabric interact to manage public affairs in a different way.

This form of acting is due to the acknowledgement of the existence of a crisis, of certain social transformations and the exhaustion of the organisational principles of social life in general and of the government in particular (García 2015). It is no coincidence that the practical application of governance has come about precisely as a consequence of the crisis of the Fordist model of regulation (Méndez 2018) and of the welfare state in the 1980s, that had begun in the 1970s. The crisis was used for industrial delocalisation and the restructuring of European urban areas, with the emergence of new neoliberal entrepreneurial forms of local regulation that imply structural changes in urban management (Pradel Miquel 2007).

To do this, it is initially necessary to acknowledge, accept and integrate conflict management as an intrinsic element of the political process, so as to subsequently create a government system through the participation of different actors within the framework of plural networks and establish a new position for the public powers in the government processes. In addition, new functions and the use of new management instruments are adopted (Blanco and Gomá 2006). In this way, the government becomes an agent of governance, shaping social spaces in which the rest of the actors interact, normalising the creation of public–private relational structures to direct policies, projects and the management of democratic institutions (Somoza Medina 2013). The undertaking of this function by the government suggests that from its origin in the American neoliberal schools of thought, the concept has a distinct ideological nature and its development is not neutral. Its polysemic nature enables it to be adopted by many different positionings, from the New Public Management of the neoliberals, to the new social state of the neo institutionalists, or the community universalism of civil society and non-government organisations (Farinós Dasí 2008). Being adopted by all institutional levels of the government fulfils the objective of de-ideologising it, reducing it to the mere decision-making structures regarding the public realm (Centelles and Portella 2006). Nevertheless, as indicated by Farinós, it is possible to consider governance from many different dimensions: multilevel governance between different political and administrative systems, governance as horizontal cooperation between sectoral and territorial policies, economic governance through networks of social actors and businesses and citizen participation (Farinós Dasí 2006). Considering governance with respect to its potential in relation to civil society is precisely the objective of a critical analysis that goes beyond simple structures, converting it into a fighting mechanism, driving social creativity against exclusion and poverty. It constitutes a catalyst of democratic opportunities and of obtaining citizens' rights through new forms of more horizontal management.

Among them, civil society plays an important role as a local manager, participating in decision-making through consensus between the different actors involved: the market, the state and civil society (Pradel Miquel 2007). It also acquires a central role within projects of social transformation, democratic expansion and participative governance (Bua et al. 2018; Della Porta and Felicetti 2019; Roth et al. 2019; Blanco et al. 2020). Some authors, such as Swyngedouw, have been indicating for some time that this model has two sides, given that its underlying practices and discourses are often contradictory (Swyngedouw 2005). According to this criticism, the model is based on the assumption that these new regulatory systems will generate a vicious circle, giving rise to economic growth, social cohesion and greater democracy. An example of this are the new and diverse management models that cities seek to develop through so-called governance in order to obtain a competitive position on a global level, without, in theory, putting internal social cohesion at risk. Therefore urban policies change direction in order to generate a proactive, dynamic and entrepreneurial role, seeking opportunities where the public and private sectors can implement joint actions that increase urban competitiveness, with the consequences that this brings about.

26.3 The Governance of Urban Planning

One of the principal instruments of territorial hegemony of capital is an urban planning that converts the space into a good one. Its most effective weapon is its disguise, based on knowledge systems, jargon and techniques that are incomprehensible to ordinary people. This shows us the potential of deploying citizen governance policies. In terms of urban planning, complacent, acritical and menial experts are crucial for power. With their aseptic contribution, a territorial design is created that favours the accumulation of capital, defining the land uses, which spaces will be developable or how the layout of the infrastructures (roads, ports, airports, etc.) will be shaped. The dominant class, capital and the state impose their logic in this way, appropriating the planning instruments through an abstract objectification of the space that is functionalist, quantifiable, Cartesian, cold and repressive.

Urban and territorial planning confers an artificial value to the space. Instead of defending its value of use which should meet the collective needs and common good, it favours its exchange value as good to extract speculative profits. This subjection of the decision of the public administrations to the orders of capital robs the local population of major territorial and urban planning decisions. In other words, the most important issues, such as urban planning or the expansion of infrastructures are decided by the highest institutional levels, where the board of directors of large corporations are often confused with the public governments. In this way, the people are robbed of their power of decision regarding their space and also their future.

From the 1990s, despite the many and different forms of applying citizen participation, urban planning on a global level has been led proactively by governments, with major interventions, promoting private investments through different methods of attraction, such as tax incentives, urban branding or public–private consortia. The renovations of the cities have followed homogeneous patterns, with highly significant results of socio-spatial segregation of the city, a territorial fracture between the globalised urban space and the rest of the physical and social framework (Somoza Medina 2013). Therefore, it could be said that the governance processes applied have not gone beyond the design of exchange structures and an increase in bureaucratisation.

However, if the goal is to establish elements of democracy and agreement in a true political and democratic dimension of citizen participation (Le Gales 2007, cited by Somoza Medina 2013), it would be necessary to favour initiatives of general interest and the principle of subsidiarity, governed through social empowerment (Rojo Salgado 2009; Romero and Farinós 2011). The model of urban governance would provide what Geddes called major political assets: consensus between the actors involved, whatever the field of application of a policy; economic growth together with social cohesion through the right to participate and integration; and the combination of representative democracy with participative democracy with the involvement of the whole of society in the application of policies (Geddes 2006).

The governance of urban planning should aspire to do everything necessary to decipher the territorial discourses which often only seek to overwhelm and demotivate citizens. The ruling class, capital and the state thereby administer the space to organise it so as to favour the accumulation of capital, commercialising every corner of the urban environment, putting well-being and sustainability at risk. Participating in the discussions and administrative processing phases of territorial and urban planning can be highly frustrating due to the hegemony of the desire for profit and the unsustainable growth paradigm. This can only be corrected through the democratic radicalism of knowledge and information, with the academic support of the self-organised and assembly-based neighbourhood fabric.

In the case of large cities such as Barcelona, these urban renovation processes have often been questioned (Capel Sáez 2005; Casellas 2006; Delgado 2007; Borja 2009) as they are sustained by public investments that have principally favoured the interests of capital and real estate business, leading to an increase in prices and the subsequent segregation processes. On the other hand, some authors such as Pradel advocate social innovation as a tool to reinforce territorial-based social relations on a neighbourhood or district level (Pradel Miquel and García Cabeza 2018) and thereby slow the dynamics of territorialised social exclusion. In this way, although the predominant neoliberal conception regards the neighbourhood as a resource for obtaining capital gains through gentrification, segregation and the overproduction of exclusive urban spaces (Davidson 2008; Harvey 1997), depoliticised from all possibility to actively construct and produce the city (Letelier 2018), the neighbourhoods have historically been relational spaces for voicing demands.

26.4 The Neighbourhood Scale in the City of Barcelona

The development of the industrial city in Spain in the 1950s was seen as an opportunity for real estate development in the metropolitan areas through urban renovation projects and the mass construction of new homes (Borja 1975). The reforms adopted between 1969 and 1974 to order urban growth and the use of the land could not prevent the terrible living conditions in the new peripheral suburbs or the dreadful access to amenities, transport, etc. This reality gave rise to the emergence of a neighbourhood movement subject to the Law of Associations of 1964 (Letelier 2018). Through the configuration of association networks, they became a first line political agents (Borja 1975), participating directly both in the processes of urban negotiation and addressing broader social problems in an exercise of critical governance at the city and even regional level (Gail 1980). These actions were mainly developed between 1976 and 1979, a time of an upsurge in the neighbourhood movement, the vindication of amenities and the opposition to large speculative projects (Bonet i Martí 2012).

The consideration of neighbourhoods as exclusively relational spaces on a neighbourhood level at the end of the 1970s, reducing the importance of territorialism meant that the neighbourhood spaces were only considered based on the relations prevailing within them on different levels (Harvey 2012; Massey 2012). This conception of urban fragmentation facilitated the creation of isolated, delimited spaces disconnected from the nearby global environment. Without a critical view of the neighbourhood as a place of global vindication, the neighbourhood struggles are unable to contribute to the structural aspects of the overall planning of the city, abandoning the political role of the neighbourhood existing in the 1960s, in response to the strategies of urban renovation (Letelier 2018). Contemplating the neighbourhood without its radical critical point of view and reducing its scale to the relational dimension was a convenient strategy for applying governance policies in the 1980s, mainly focused on the management of aspects such as the development of the provision of services to the community, preventing the neighbourhood network from influencing the institutions and those responsible for the policies (Katznelson 1982). The administrations absorbed the vindicative and transforming role and integrated this into a new scale of urban governance to combat the urban problems of tension, inequality, segmentation, the fragmentation of the urban landscape, the so-called "quartiers en crise" (Andersson and Musterd 2005).

From 1979 to 1983 there was a period of new relations between the institutions and the neighbourhood associations of Barcelona, embodied in the Special Plans of Interior Reform (PERIs) which the first democratic governments in Barcelona (1979–1983) implemented in areas of special attention. In 1984, Barcelona regulated citizen participation and decentralisation in the districts, which increased the administrative competences of the territories (Borja 2001), but also hindered the introduction of new deliberative components and the involvement of citizens in public affairs (Villasante and Gutiérrez 2000). These measures, principally focused on regeneration and processes of social cohesion, have been seen from a critical perspective as testing laboratories for applying new urban policies (Martín 2003) within liberal democratic governability (Atkinson et al. 2009). After these actions are implemented, a neoliberal agenda is established to convert the city into a source of capital gains through gentrification, segregation and the overproduction of elitised urban spaces (Davidson 2008; Harvey 1997), rooted in the so-called "new localism" (Brenner and Theodore 2002) and the transfer of responsibilities to local areas in a fragmented way. Therefore, the neighbourhood becomes a "communitarian trap" (Harvey 1997), being responsible for finding the solution to global urban problems through the local community, appealing to the ethical content and radical community spirit typical of the neighbourhoods in the 1960s and 1970s (Letelier 2018).

Between 1990 and 2000, the so-called Communitarian Plans were implemented that focused on reinforcing the neighbourhood on a community level in areas where there were greatest shortfalls and creating rationalisation processes of citizen participation (Letelier and Valdosky 2019). In 2004, these processes gave rise to the *Llei de Barris de Catalunya*, and with the initiative "*Els Barris de Barcelona*", to the creation of the Consejo de Barrio (Neighbourhood Council), thereby institutionalising the channels of participation in an action of contention or political expropriation of neighbourhood representation (Leiva et al. 2007). The consequences of the crisis and the unease that it caused in different areas of the city reactivated the role of the neighbourhood movement and its participation in the anti-austerity campaign, again showing that it still had articulation and politicisation potential. A result of this was the renewal of the progressive municipal government, which, in 2015 deployed the so-called Pla de Barris, with social, economic and urban measures for 16 of the most disadvantaged neighbourhoods (Ajuntament de Barcelona 2017). Five of them belong to the district of Nou Barris, included in three of the ten plans to be developed throughout the whole city, due to the special characteristics of the area. As we shall see later, they have a long tradition of neighbourhood vindications and associative movements.

26.5 The Neighbourhood Movement in Nou Barris

With the establishment of the first democratic local governments, in 1984, Barcelona shaped the map of the city based on a new territorial division into 73 neighbourhoods and 10 districts (Fig. 26.1), including the 13 neighbourhoods of the eighth district called Nou Barris, taken from the name of the magazine of the Neighbours and Owners Association of the area in the year 1977 (Fig. 26.2).

The current district has 13 neighbourhoods: Ciutat Meridiana, Vallbona, Torre Baró, les Roquetes, Trinitat Nova, Canyelles, Verdum, Prosperitat, Guineueta, Porta, Vilapiscina i Torre Llobeta, Turó de la Peira, Can Peguera (Fig. 26.3), which account for 7.94% of the 101 km^2 area of the city.

According to the latest available data of 2022, the district has 170,736 inhabitants, 10.4% of the total population of the city, of which 40% are over the age of 85, which represents an over-ageing index higher than that of Barcelona (Ajuntament

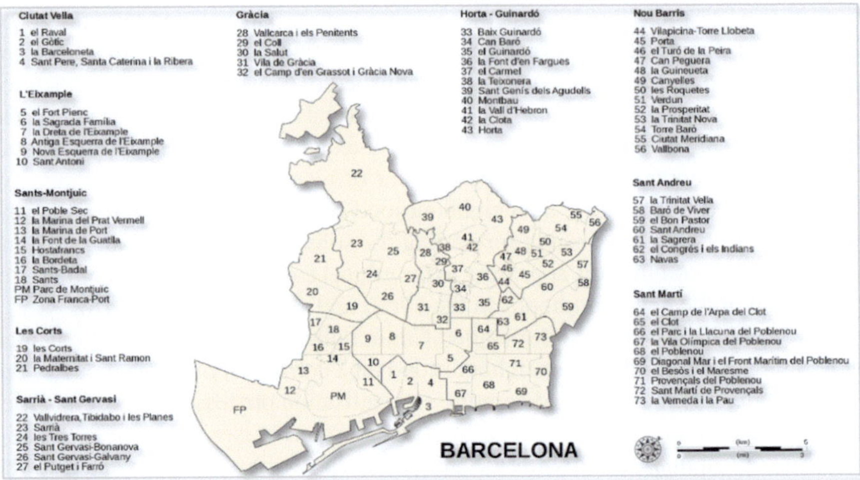

Fig. 26.1 Districts of Barcelona. *Source* © Sémhur/Wikimedia Commons, CC BY-SA 4.0. https://commons.wikimedia.org/w/index.php?curid=3914779

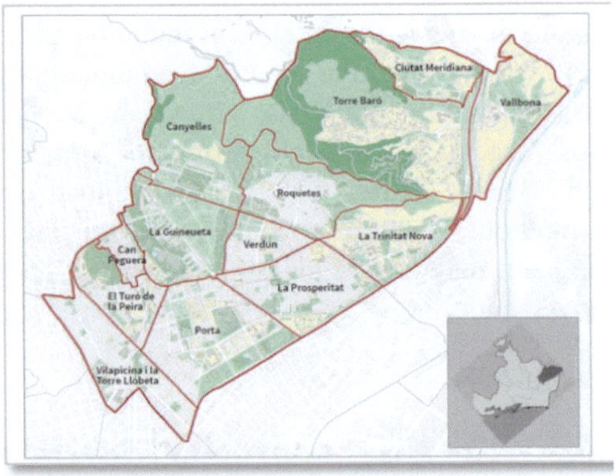

Fig. 26.2 Neighbourhoods of the District of Nou Barris

de Barcelona 2023). The majority of its population have average levels of education, with 3% considered as having no education and only 16.1% with university or higher education training, which is far removed from the city average of 34.6%. The levels of disposable family income are the second lowest in the whole of Barcelona (only behind the district of Ciutat Vella), with some of the poorest neighbourhoods of the city (Torre Baró and Ciutat Meridiana) and the highest unemployment rate of the 10 existing districts (13.5% of the total of Barcelona). Therefore, 25.2% of the population receives social benefits (PIRMI), the highest percentage of all the districts. Between 2001 and 2010, the district experienced the greatest rise in its migrant population. In 2001 it received a little under 5000 people while in 2010 this figure had grown to 27,100. This pace has reduced in recent years. In 2022, the proportion

Fig. 26.3 First issue of the magazine 9 Barrios. *Sources* HansenBCN. Own work, Public domain, https://commons.wikimedia.org/w/index.php?curid=3390681 and http://www.periodicosregalo.com/wp-content/uploads/2015/02/19.jpeg

of the foreign population reached 20%. This is lower than the average of Barcelona, but with large differences between the neighbourhoods, oscillating between 7.2 and 31.3% of migrants in different areas of the district (Ajuntament de Barcelona 2023).

Nou Barris is one of the districts that has undergone major transformations, drastically modifying its structure in less than a century; from large agricultural areas worked by a small number of farmers and small population nuclei to a dense and diverse urban fabric. The first urban nuclei, formed around a medieval hostel and a chapel, dates back to the eighteenth century and had large farmhouses that were active until the second half of the twentieth century; only twenty have survived the real estate speculative process (Pauné 2014), although some of them have given their names to neighbourhoods or emblematic parts of the new district. Between 1875 and 1877, the first actions to improve the area were taken and in 1884 the Local Council of Sant Andreu created the general urban plan for the Vilapiscina sector to prevent the proliferation of small urbanisations and enable harmonisation (Fabre 1991: 34) (Fig. 26.4).

For a long time the area was used for municipal amenities that were too large to build in the urban centre of Sant Andreu, shaping a territory of "container spaces", such as the cemetery, factories, workshops or the first mental health building in Spain (1889), the Instituto Mental de la Santa Creu, located in the current neighbourhood of Guineueta (Fig. 26.5).

The new dynamics and the boom in the residential construction activity in the periphery of Barcelona, occurring between 1897 and 1935, affected the whole area, when, in 1897, the municipality of Sant Andreu de Palomar was incorporated into Barcelona and the western part of the current district of Nou Barris began to grow, although in a slower and more dispersed way. Different urban plans and proposals, such as those of Ricard Alsina (1899), the Pla Jaussely (1905), or the Ciudad Jardín planned for the Torre Baró and Vallbona (1904), contemplated, to a greater or lesser extent, the development of the urban network in the most mountainous areas of the district. Although none of them were fully implemented, they served as a base for subsequent urban plans (Pujol 2003: 69).

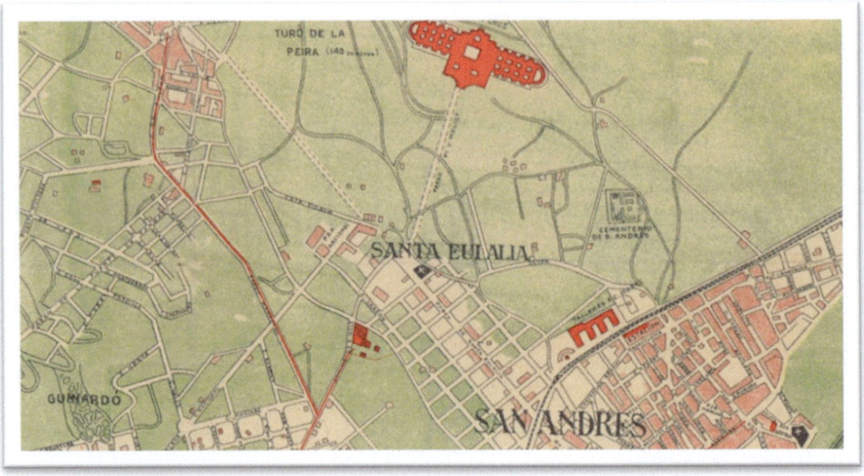

Fig. 26.4 Ordered growth of the suburbs from the Santa Eulàlia promenade (1916). *Source* Cartoteca Digital ICGC. http://cartotecadigital.icc.cat/cdm/singleitem/collection/catalunya/id/1335/rec/252

Fig. 26.5 Instituto Mental de la Santa Creu i Sant Pau. *Source* Noubarris.net, http://www.noubarris.net/web40/?p=63606

The 1920s marked the beginning of the urbanisation of the land of different neighbourhoods, such as Verdum or Prosperitat (with a modest and economic version of the garden city), Roquetes or Vilapiscina, through initiatives of owners or cooperatives; and the reservation by the City Council of land for the construction of two roads which would subsequently become two main arteries of the district and the city, the Vía Júlia and the current Ronda de Dalt.

From the 1940s, the district of Nou Barris grew significantly, giving rise to its current configuration. The years of the dictatorship and its management and the administration of the territory profoundly marked the idiosyncrasy of the district and its development. Between 1950 and 1970, the accelerated construction of poor quality real estate predominated, in which public actions were combined with private actions and self-construction projects, accompanied by a chronic deficit of services and facilities, with no connection with the rest of the city. This gave rise to a precarious urban plan that did not respond to the basic needs of the inhabitants (Borja 1975; Domingo i Clota and Bonet i Casas 1998). A working class population settled there, derived from the mass migratory population from more depressed areas on a state level and attracted by the dynamism of the development that was occurring in the metropolitan area of Barcelona. The Plan Comarcal (District Plan) of 1953, attempted to provide an overall plan for the area of influence of Barcelona, but beyond facilitating the reservation of land for constructing the large fast metropolitan roads, it did not contemplate areas such as Nour Barris. This reality became the cornerstone of the many neighbourhood demands and vindications of the 1960s and 1970s, such as the planning and construction of a sewage network by the neighbours themselves in 1964, from which the phrase "urbanisation on Sundays" was coined (Fig. 26.6).

The period which began in 1970 and lasted until well into the 1990s was characterised by a strong social and neighbourhood awareness, beginning with the founding of the Neighbourhood Association "9 Barris" and the fight against the Plan Parcial Torre-Baró-Vallbona-Trinidad Nueva (1969) (Naya 2023). This plan prioritised major roads and the liberalisation of land, which gave rise to a greater volume of construction being placed in the hands of real estate companies, favoured by the direct interests of the city council (Valdosky 2017). The need to transform the urban precariousness and the political characteristics of the aftermath of the Franco dictatorship constituted a catalyst for demonstrations, collaborative and neighbourhood organisation (Andreu 2015) and trade union, political and church movements.

From this moment, the neighbourhood demands and struggles became a phenomenon that we could currently call democratic radicality. Governance understood as such did not exist and the different demands were channelled through direct actions of protest on the streets. Unlike the direct involvement in the construction of basic services of 1964, the action was impregnated with a political character, adding to the demands of the resolution of basic needs the questioning of the political actors regarding planning processes, decision-making and the steps taken.

There were many vindications of very different types. In addition to the demands for improved access to housing (emblematic in Canyelles and Guineueta affected by the expropriations carried out due to the construction of the Ronda de Dalt) (Fig. 26.7), there were demands for the better planning of public spaces, as in the

Fig. 26.6 Urbanizar en domingo. *Source* Regillant Nou Barris. Arxiu Històric de les Roquetes – Nou Barris. http://www.noubarris.net/relligantnb/?p=2500

case of the Plaza Sóller in the Porta neighbourhood, in 1975–1976; the demand for the refurbishment of the Ronda de Dalt to be covered in its sections passing through the affected neighbourhoods; the construction of green spaces such as the Parque de la Guineueta (opened in 1971) or opposition to speculation and swindling carried out by the real estate sector in the Prosperitat neighbourhood.

On 19 July 1976, the Corporación Metropolitana de Barcelona (legal entity of the local government, created in 1974 to review and update the District Plan of 1953) approved the Plan General Metropolitano (General Metropolitan Plan), a new urban planning tool, which, among its objectives, sought to palliate the lack of facilities suffered by the Nou Barris district (Fernández 2010). The implementation of this plan in the district was unequal and caused, for example, the direct rejection by 90% of the population of the Can Peguera neighbourhood. Protests were presented against the Plan due to the proposal to demolish the whole neighbourhood. Even today, there are remains of the consequences of this plan in a series of constructions classified as unique in the Passatge de Santa Eulàlia (Vilapiscina) and the neighbours still have unsurfaced roads.

On the other hand, the range of demands was broad, particularly referring to the need for a minimum provision of educational, sports, cultural and health facilities, embodied in the demands of the neighbours of Guineueta and Verdum for the Professional Training Centre; the lock-in in 1974 in the parish of Santa Engracia for the construction of the Sant Andreu Baccalaureate School which opened in 1984 or the continuance of a public kindergarten; the demands to Social Security for the installation of a Health Centre in Roquetas or the fight for the Mental Health Institute.

Fig. 26.7 Vindication of decent housing. *Source* Regillant Nou Barris. Arxiu Històric de les Roquetes—Nou Barris. http://www.noubarris.net/relligantnb/?p=2500

Given the characteristics of the orography of the neighbourhoods of the northern area (Roquetes, Torre Baró, Ciutat Meridiana, Vallbona) and its urban fragmentation, one of the greatest deficits was its connection via public transport. The government used the poor state of the roads to justify this deficit and the network was reduced to the bus service which did not run through most of the neighbourhoods with the narrowest and steepest streets. For this reason, throughout 1974, the neighbours of Roquetes seized the number 11 bus eleven times to demonstrate that it was possible to reach the neighbourhood. In the same way, a driver of line 47 (Plaza Cataluña to Guineueta), Manuel Vital, a neighbour of Torre Baró, seized the vehicle with which he worked in 1978 and drove it through the steep and winding streets, accompanied by other neighbours who joined his cause. This led to the expansion of the route six months later to the Canyelles neighbourhood (Figs. 26.8 and 26.9).

Following the tradition of the area when it formed part of the municipality of Sant Andreu in the 1970s, Nou Barris continued to house large facilities (container spaces), some that were even industrial and polluting. This was the case of an asphalt plant installed in the neighbourhood of Trinitat Nova in 1976, conflicting with the legal provisions that stipulated a minimum distance of two kilometres from inhabited areas. Built for producing the asphalt necessary for paving the II Cinturón (the current Ronda de Dalt), it was only active for a few days as the indignation of the neighbours regarding the smoke and contaminating fumes that it generated was such that an assembly was called and it was paralysed. In January 1977, five hundred people dismantled the chimney and staged a sit-in in the factory, which, from this moment was managed by the 9 Barris Pro-Athenaeum Organising Committee (CPA) (Ateneu

Fig. 26.8 Bus seized to reach Torre Baró. *Source* Fernández (2010). https://eltranvia48.blogspot.com/2010/09/manuel-vital-para-siempre.html?m=1

Fig. 26.9 Bus seized to reach Torre Baró. *Source* Fernández (2010). https://eltranvia48.blogspot.com/2010/09/manuel-vital-para-siempre.html?m=1

popular de 9 Barris 2018) (Figs. 26.10 and 26.11). The initiative became a self-managed project for the development and democratisation of culture, currently a reference as a social project related to the circus on an international level.

In the same respect, there were companies that installed their factory installations in the district, mainly in Prosperitat, taking advantage of the proximity of the Meridiana road as an access way and the remodelling of the Passeig Valldaura as an industrial estate. This gave rise to an increase in the demands of workers, such as those of Harry Walker, a car engine and accessories company (Pujol 2003), among others.

This small sample of the many actions carried out by the neighbours of the district, from the lowest level of the urbanisation of the streets to the social amenities, shows that almost everything was obtained, thanks to the fight of its inhabitants and almost

Fig. 26.10 Asphalt plant close to the residential housing (1976). *Source* Arxiu Ateneu Popular Nou Barris

Fig. 26.11 Panoramic view of the Ateneu Popular Nou Barris during the 30 h Festival (4 and 5 June 1977). *Source* Arxiu Ateneu Popular Nou Barris

always after the construction of housing. This is contrary to what a process of governance and urban planning would have looked like. When they existed the plans were conceived and designed without contemplating the territory as a whole but only their immediate scope. New urban planning models and different types of plan were imposed, ignoring the shape and natural elements of the territory. This gave rise to the juxtaposition of different urban layouts and a greater disconnection and marginality of the neighbourhoods between one another and in relation to the rest of the city (Navarro 1997). The rapid execution of the plans multiplied the defects of their design, with the construction of very low quality housing with serious aluminosis problems and no collective services or amenities. In order to mitigate part of these dysfunctions, from the 1980s, the local governments sought ways to plan the empty spaces within the densification of the neighbourhoods. Converting the discontinuities in the layout into poles of attraction and points of civic communication constituted a challenge with respect to previous periods. Therefore, the first democratic local governments (1979) reinforced the neighbourhood initiatives that vindicated and fought to obtain all the basic amenities and services that they lacked. No other part of the city was as belligerent as that of Nou Barris, making it a historical reference.

During the 1980s, the district benefited from the local welfare system of Barcelona, characterised by the collaboration between civil society and the local government. The left-wing parties sought to develop a social and democratising agenda within a context of economic crisis and major social needs. The neighbourhood associations supported the city project and the association fabric of the neighbourhoods became consolidated and acquired a great decision-making capacity which enabled it to request a redistributive agenda and the promotion of profound urban reforms, particularly in the working-class neighbourhoods (Cano Hila and Pradel Miquel 2018).

The promotion of social cohesion was also sought through the establishment of citizen participation mechanisms and the reinforcement of community life in the neighbourhoods. Despite the recentralisation promoted from the Catalan government, more market-oriented and with less focus on social policies (Nel·lo 2013), from 1984, with the decentralisation of the districts, long-term projects were developed and civic centres and neighbourhood centres were opened (such as the Casal de Joves de Roquetes in 1980 or the Casal de Barri de Prosperitat in 1988), understood as spaces of free citizen and cultural expression.

In the same way, the Special Plans of Interior Reform (PERIs) were implemented, such as those of Roquetes and Torre Baró (1985), the latter barely applied or that of Prosperitat (1989). The Integral Plans accompanied the existing processes in the neighbourhoods, supporting grassroots experiences that responded to the objectives of the municipal government of urban renovation but also of participation and the consolidation of the association fabric in the management of facilities. Therefore, the business incubators brought about initiatives but also cultural associations such as the Nou Barris 9 or project or the Associació Juvenil Sobreàtic de Dinamitzacio Socio-Cultural, which, in 1992, began to self-manage the Casal de joves de Roquetes as a first experience of a self-reliant youth facility. This was followed a little later by Bidó Nou Barris, which, from 1994, began to manage the Ateneu Popular or the Coordinadora d'Associacions i Entitats de Nou Barris.

During the different socialist governments (1979–2011), the neighbourhood associations had a close relationship with the local government and constituted representatives of the neighbourhood demands, obtaining funds for the organisation of local festivals and other cultural activities. Little by little, the debts with the inhabitants of Nou Barris were paid off and the district began to integrate with Barcelona as another space of the city and was no longer seen as a suburban and marginal space. In this respect, an attempt was made to change the traditional concept of the working-class periphery (Fernández 2010). However, at the same time, there was an institutionalisation of the demands. In this scenario the participative dynamics are enriched but civil society becomes fragmented; the demands are captured and the activist leaders are domesticated and the dynamics turn against one another at the heart of the social movements.

From the second half of the 1990s, Barcelona implemented the Community Development Plans as global projects that contemplated specific inclusion and local development actions. In the case of Nou Barris, after significant pressure from the local entities, community management became widespread (traditional in different neighbourhoods of the district) as a mechanism to manage the public cultural centres by the entities in collaboration with the local government.

A paradigmatic case is the Centro Cultural Ton i Guida, located in the building of the former school with the same name after many demands from the inhabitants of the Roquetes neighbourhood. The project, which was created in 1993, emerged as a result of the neighbourhood struggle and its self-management (which was achieved in 2002) was conducted through a platform of entities of the neighbourhood within the framework of civic management. Even today, it is an atypical facility which implements three levels of action with a wide diversity of entities and neighbourhood activities (assuming the function of a civic centre) and it is also open to the rest of the city and is based on a type of co-management between the municipality and the citizen initiative.

In parallel, the city of Barcelona has participated in large mega projects and mega events that seem to follow a completely different line. Actions such as the refurbishment of the waterfront for the Olympic Games, the gentrification process of Raval or the Forum of Cultures drastically changed participation which shifted from being a right to an obligation, leading to the disaffection of the citizens who seemed to have lost control of the future of the city (Pradel Miquel 2020). The post-industrial city was much more fragmented with an economic model based on tourism and which did not respond to the needs of the residents or reduce inequalities. The initial sensation of working towards a fairer city began to crumble and the population of Nou Barris, despite the urban and environmental improvements, did not feel like a participant in this city project or included in the so-called pride of belonging. Between the mid-1990s and the beginning of the 2000s, social mobilisation reduced and the entities of each neighbourhood began to focus more on their own individual realities (Cano 2017). A certain level of coordination between the neighbourhood associations was maintained (Coordinadora d'entitats de Nou Barris) in aspects such as health, education, housing, employment and culture, but this did not translate into any significant joint action.

Two of the main war horses were, and continue to be, the scarcity of public housing policies that respond to the real situations of the citizens and immigration, for which new measures of self-management are being sought. Therefore, initiatives have arisen, such as the association for public and affordable rent (Associació 500 × 20); or the Red 9 Barris Acull, created in 2001 and currently made up of eighty entities, which work in a coordinated way to promote conviviality and facilitate the incorporation of new immigrant neighbours into the neighbourhoods and which, as an emblem, celebrates a famous international festival, The World Soup Festival (https://9bacull.org/es/).

The crisis of 2008 revitalised the decentralisation of the movements and the creation of neighbourhood social assemblies oriented towards political transformations, in the same way and driven by the anti-austerity movements. Initiatives to promote employment, the insertion and promotion of the sharing economy, such as the *Asamblea de parados y paradas de Nou barris*; community-based network initiatives for production, distribution and consumption according to ethical, democratic, ecological and solidarity criteria, such as the *Mercat ImPorta* (Social Market of Porta); the *Alianza contra la pobreza energética (Alliance against energy poverty)* (APE), created in 2014, to demand a new energy and natural resource management model in light of the crisis, to denounce the deprivation of the population of access to basic resources (water, electricity, gas) and to modify the legal framework that guarantees them; the platform *Salvemos las pensiones 9 Barris*. All of these were agglutinated from 2012, when the demands were channelled into a general appeal formulated through a public manifest of demands (http://9bcabrejada.blogspot.com/p/manifest.html), bringing together more than 100 entities and their demands under the slogan: *Nou Barris Cabrejada*, and which, in 2014, published the report "*No es pobreza, es injusticia*" (Baquero 2014; Nou Barris cabrejada 2014) (Fig. 26.12). Thanks to this, based on an extraordinary plan, the government of the district elaborated the "Plan de Acción para la cohesión y los derechos sociales de Nou Barris 2016–2019" [Action plan for cohesion and social rights of Nou Barris 2016–2019] (Ajuntament de Barcelona 2016).

The election of Barcelona en Comú in 2015 as the leader of the local government brought about the emergence of more transversal social policies, with the development of citizen participation proposals which became sovereign and placed the

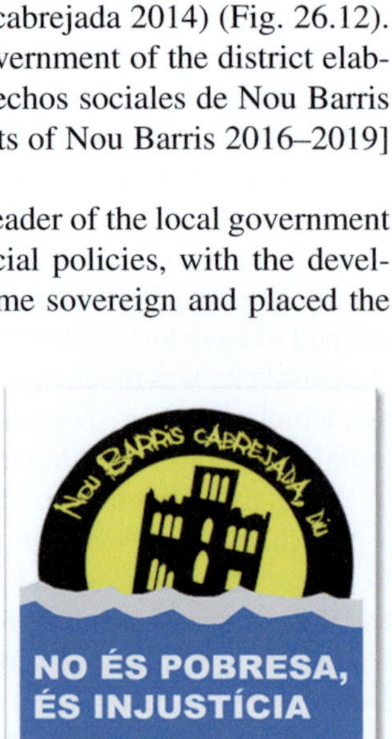

Fig. 26.12 Nou Barris cabreada. "No es pobreza, es injusticia" (2014). *Source* http://9bcabrejada.blogspot.com/

discourse in the sphere of citizen rights (right to housing office, elimination of energy poverty, etc.). During this phase many new initiatives arose inspired by the mechanisms of the 1980s, this time creating new forms of association, management and relations with the government and the so-called grassroots governance spread from Nou Barris to other districts of the city.

From 2016, the implementation of the "Decidim" process by the city council brought together proposals to collectively elaborate the municipal action plan (2016–2019) with the slogan: *"Construyamos una sociedad más abierta, transparente y colaborativa. Únete, participa y decide". [Building a more open, transparent and collaborative society. Join, participate and decide.]* (https://www.decidim.barcelona/?locale=es). This has involved the execution of different levels of action, including redevelopment and action plans for the district, plans for the protection of heritage, education plans, the elaboration of a low emissions plan, youth facilities, cultural and green spaces, special retail planning projects, the regulation of terraces, urban mobility (PMU), accessibility, the use of the public space, modifications to the General Metropolitan Plan in terms of the urban regulation of housing, gender equality, etc. In the case of Nou Barris, between 2016 and 2019, as well as the city projects, four specific plans were implemented: the facilities plan (2018), "Repensemos Trinitat Nova" (2017–2018), the redevelopment of La Meridiana (2016–2018) and the modification of the General Metropolitan Plan for the neighbourhood of Ciudad Meridiana (2018–2019).

In the period 2020–2023, there are plans to take a step further in the participative process of the Municipal Action Programme (PAM), promoting participative budgets for the first time for the whole of the city, reserving 75 million euros of the municipal budget for the neighbours to decide how to invest it in their districts (Fig. 26.13).

Although this could initially appear to be an appropriate initiative and one of participative democracy, the neighbourhood associations, such as that of Prosperitat believe it to be a "bad idea" as this action does not seem to respect the agreements reached in the neighbourhood assemblies. In 2019, social intervention plans were made by the residents addressing the needs and priorities of each neighbourhood for four years; the implementation of new processes by the City Council is perceived as a form of discrediting these assemblies, delegitimising them and "turning the neighbours and neighbourhoods against each other" (Elia Herranz, cited in López, 12/04/2020). The idea of a fair city for the residents goes a step further and once again places the institutional processes of tenocratisised democracy in conflict with a vindication of bottom-up democratic radical innovation. The spatiality and articulated form of neighbourhood action of the Nou Barris organisations enables them to act on a district scale, which bestows then with a political voice to communicate with the institutions (Bonet i Martí 2012; Letelier and Valdosky 2019).

Fig. 26.13 Informative campaign of participative budgets. Decidim Barcelona. *Source* Ajuntament de Barcelona, 2020. https://www.decidim.barcelona/processes/PressupostosParticipatius?locale=es

26.6 Final Reflection

Between 1950 and 1960, cities took shape in many different ways. The peripheral territories of the large cities, such as Barcelona, sprang up based on speculation or self-construction processes. The people settling in the peripheries such as Nou Barris knew nothing about partial plans, but necessity led them to organise themselves so as to construct marginal spaces, with no planning, equipping the neighbourhood with basic services and "urbanisation on Sundays".

The neighbourhoods of the 1960s and 1970s did not use the term governance, but they were deprived of rights and suffered the consequences of a lack of social justice. With their demands, they sought a fair and equal city, going beyond the basic recognised rights of citizenship, exercising what has been called real and radical democracy. Despite a kind of "domestication" and capture of the activist initiatives by the public administrations in the 1980s and 1990s, in neighbourhoods such as those referred to this this study, the associations were characterised by relevant and innovative elements, such as collective self-organisation sometimes combined with collaboration with the institutions, where the city council has been an economic and technical support partner but without leading the initiatives.

Repeated situations of uncertainty and abandonment caused by the economic crisis of 2008 or the COVID-19 crisis have led to the resurgence of participative processes that never completely disappeared. In these processes, voluntary reciprocity prevails through decision-making mechanisms, involving those affected by the dynamics of exclusion, empowerment and collective organisation, where the pertinent professionals and experts are a complement, generating innovative ways of organisation. In short, they are bottom-up participative actions of radical democratic innovation which take the tecnocratisation of democracy a step further so as to continue working towards a truly fairer city. Some have called it the era of citizen tecnopolitical emancipation. New forms of organisation, coordination and civic action, with new ethics and new methodologies are enabled by new tools, spaces and actors, which require democratic institutions able to adapt, innovate and respond. As suggested by Blanco, these new informal practices and narratives formed the basis of the rebirth of participative governance associated with the government of Barcelona en Comú (2015–2019), as the dynamics of participative governance depends on the relationship between the rules, practices and narratives (Blanco et al. 2022). This could generate a very different type of dynamics and diverging paths of governance, as shown by Bua and Bussu in their "Locating Governance-driven democratization (GDD) and Democracy-driven governance (DDG) spaces of democratization in dynamic relationship" in the case of Barcelona or Nantes (Bua and Bussu 2021; Bua et al. 2023).

Therefore, in the second decade of the twenty-first century, it is equally necessary to conserve and expand "the democratic common senses of the masses arising from the popular struggles in the streets and squares" (Aguiló 2014: 81). An emancipating "new culture" needs to be constructed based on radical policy (Gramsci 2018), which, in reality, is not so new but has inherited the "social footprint" left by popular movements (Calle 2005) through demands, democratic complementarity and citizen empowerment, which already existed in the self-management of popular neighbourhoods in districts such as Nou Barris. The different platforms have contributed, particularly in the city of Barcelona, to politicising neighbourhood action, vindicating social-spatial inequalities related not only to the local context but also to global processes (Massey 2012). They weave strategic territorial relations on a complementary scale going beyond fragmentations and define a political character for the neighbourhood relations, constituting grassroots instruments and participants.

References

Aguiló AJ (2013) Una cartografía abismal de la democracia: representación política y democracia de baja intensidad. En Instituto Internacional Casa de Mateus (org.) Representação | Representation. IICM, Vila Real (Portugal), pp 54–79

Aguiló AJ (2014) Radicalidad democrática y nuevos movimientos populares: las luchas por otras democracias. Agora 1(2):65–86. https://doi.org/10.6035/Kult-ur.2014.1.2.3. ISSN 2386-5458

Ajuntament de Barcelona (2016) Mesura de Govern. Pla d'Acció per la Cohesió i els Drets Socials de Nou Barris 2016–2019. Consell del Districte de Nou Barris. Sessió extraordinària 19 de maig de 2016. Recuperado de http://ajuntament.barcelona.cat/premsa/wp-content/uploads/2016/05/mesuragovernnoubarris.pdf

Ajuntament de Barcelona (2017) Pla de barris, Barcelona (2016–2020) / El plan de los barrios de Barcelona. Recuperado de https://pladebarris.barcelona/es

Ajuntament de Barcelona (2023) Nou Barris. Estadística i difusió de dades. Observatori de Districtes. Recuperado de: https://ajuntament.barcelona.cat/estadistica/catala/Estadistiques_per_territori/Documents/districtes/08_NouBarris_2023.pdf
Amin S (1974) El desarrollo desigual: ensayo sobre las formaciones sociales del capitalismo periférico. Editorial Fontanella, Barcelona
Andersson R, Musterd S (2005) Area-based policies: a critical appraisal. Tijdschr Econ Soc Geogr 96(4):377–389. https://doi.org/10.1111/j.1467-9663.2005.00470.x
Andreu M (2015) Barris, veïns i democràcia: El moviment ciutadà i la reconstrucció de Barcelona (1968–1986). L'Avenç, Barcelona
Atkinson R, Dowling R, Mcguirk P (2009) Home/neighbourhood/city/+. Environ Plan 41(12):2816–2822. https://doi.org/10.1068/a42110
Baquero C (2014) El memorial de agravios de Nou Barris. El País. Recuperado de https://elpais.com/ccaa/2014/11/20/catalunya/1416511908_286112.html
Blanco I, Gomá R (2006) Del gobierno a la gobernanza: oportunidades y retos de un nuevo paradigma. Politika. Revista de Ciencias Sociales, nº 2/Diciembre, 11–27
Blanco I, Salazar Y, Bianchi I (2020) Urban governance and political change under a radical left government: the case of Barcelona. J Urban Aff 42(1):1–21. https://doi.org/10.1080/07352166.2018.1559648
Blanco I, Lowndes V, Salazar Y (2022) Understanding institutional dynamics in participatory governance: how rules, practices and narratives combine to produce stability or diverge to create conditions for change. Crit Policy Stud 16(2):204–223. https://doi.org/10.1080/19460171.2021.198426
Boggs C (1977) Marxism, Prefigurative Communism, and the problem of workers. Publicado en Libcom.org el 23 de septiembre de 2010. Disponible en: https://libcom.org/library/marxism-prefigurative-communism-problem-workers-control-carl-boggs
Bonet i Martí, J. (2012). El territorio como espacio de radicalización democrática. Una aproximación crítica a los procesos de participación ciudadana en las políticas urbanas de Madrid y Barcelona. Athenea DigitalK 12(1):15–28. Recuperado de https://dialnet.unirioja.es/servlet/articulo?codigo=4149752
Bookchin M (1991) Libertarian municipalism: an overview. Green Perspectives. Disponible online: https://theanarchistlibrary.org/library/murray-bookchin-libertarian-municipalism-an-overview.a4.pdf
Borja J (2009) Luces y sombras del urbanismo de Barcelona. UOC, Barcelona
Borja J (1975) Movimientos sociales urbanos. Siap-Planteos, Buenos Aires
Borja J (2001) Ciutadans i participació. Participació ciutadana. Ponencia presentada en el 2º Congreso de Municipios de Catalunya, Barcelona, España
Brenner N, Theodore N (2002) Preface: from the "new localism" to the spaces of neoliberalism. Antipode 34(3):341–347. http://citeseerx.ist.psu.edu/viewdoc/download?10.1.1.475.9527&rep=rep1&type=pdf
Bua A, Bussu S (2021) Between governance-driven democratisation and democracy-driven governance: explaining changes in participatory governance in the case of Barcelona. Eur J Polit Res 60:716–737. https://doi.org/10.1111/1475-6765.12421
Bua A, Davies J, Blanco I, Chorianopoulos I, Cortina-Oriol M, Feandeiro A, Gaynor N, Griggs S, Howarth D, Salazar Y (2018) The urban governance of austerity in Europe. In: Kerley R, Liddle J, Dunning P (eds) The handbook of international local government. Routledge, Abingdon, UK, pp 280–295. https://doi.org/10.4324/9781315306278
Bua A, Bussu S, Davies J (2023) Democracy-driven governance and governance-driven democratisation in Barcelona and Nantes. In: Bua A, Bussu S (ed) Reclaiming participatory governance. Social movements and the reinvention of democratic innovation. Routledge, Abingdon, UK. https://doi.org/10.4324/9781003218517
Calle A (2005) Nuevos movimientos globales. Hacia la radicalidad democrática. Edit. Popular

Cano Hila AB, Pradel Miquel M (2018) Barcelona. ¿Hacia nuevas formas de institucionalización? In: Pradel M, García Cabeza M (coord.) El momento de la ciudadanía: innovación social y gobernanza urbana, pp 101–122. ISBN 978-84-9097-391-2

Cano B (2017) Malestar social y tiempos de populismo: crisis y desafección política en Nou Barris. Trabajo de fin de grado, Departament de Mitjans, Comunicació i Cultura, Universidad Autónoma de Barcelona

Capel Sáez H (2005) El Modelo Barcelona: un examen crítico. Barcelona. Ed. del Serbal

Casellas A (2006) Las limitaciones del "modelo Barcelona". Una lectura desde Urban Regime Análisis. Documents D'anàlisi Geogràfica 48:61–81

Cembranos F, Medina JA (2003) Grupos inteligentes. Teoría y práctica del trabajo en equipo. Editorial Popular, Madrid

Centelles I, Portella J (2006) El buen gobierno de la ciudad. IIG, Barcelona

Col·lectiu Víric (2015) Present del municipalisme a Catalunya. In: Observatorio Metropolitano de Madrid, Xavier Diez y Col·lectiu Víric. L'aposta municipalista. De les bullangues a les lluites actuals pels municipis lliures a Catalunya. Virus editorial, Barcelona

Davidson M (2008) Spoiled mixture: where does state-led 'positive' gentrification end? Urban Stud 45(12):2385–2405. https://doi.org/10.1177/0042098008097105

Delgado M (2007) La ciudad mentirosa: fraude y miseria del «modelo Barcelona. Los Libros de la Catarata, Barcelona

Della Porta D, Felicetti A (2019) Innovating democracy against democratic stress in Europe: social movements and democratic experiments. Representation. https://doi.org/10.1080/00344893.2019.1624600

Domingo i Clota M, Bonet i Casas RM (1998) Barcelona i els moviments socials urbans. Barcelona: Mediterrània. Recuperado de https://www.fbofill.cat/sites/default/files/1034.pdf

Fabre J (1991) Nou Barris. La penúltima Barcelona. Ajuntament de Barcelona

Farinós Dasí J (2006) Gobernanza territorial de ámbito metropolitano. En Feria Toribio JM (coord) Los procesos Metropolitanos: materiales para una aproximación inicial. Sevilla, Junta de Andalucía, pp 155–166

Farinós Dasí J (2008) Gobernanza territorial para el desarrollo sostenible: estado de la cuestión y agenda. Boletín de la Asociación de Geógrafos Españoles 46:11–32. https://bage.age-geografia.es/ojs/index.php/bage/article/view/668

Fernández R (2010) Breve historia de Nou Barris VIII: los "Nueve Barrios" en lucha (1970–1979)- El Tranvía 48. Blog. Disponible en: https://eltranvia48.blogspot.com/2010/08/breve-historia-de-nou-barris-viii-los.html

Gail A (1980) Crecimiento urbano y participación vecinal. Madrid: Centro de Investigaciones Sociológicas. Colección Monografías, n.33

García S, Pradel M (2019) Innovació social: diferents respostes de la ciutadania organitzada. Barcelona Metròpolis 112:38–43. Ajuntament de Barcelona. http://hdl.handle.net/2445/154508. https://www.barcelona.cat/metropolis/ca/continguts/la-innovacio-social-diferents-respostes-de-la-ciutadania-organitzada

García S (2015) Un cuestionamiento de los supuestos que vertebran la sociedad moderna en búsqueda de nuevos modelos de gobernanza: el gobierno, la sociedad civil y la empresa nadando en un mar de complejidad. Cuadernos de Gobierno y Administración Pública 2(2):187–211. https://doi.org/10.5209/rev_CGAP.2015.v2.n2.51572

Geddes M (2006) Partnership and the limits to local governance in England: institutionalist analysis and neoliberalism. Int J Urban Reg Res 30(1)

Gramsci A (2018) Pasado y presente: Cuadernos de la cárcel. Gedisa Editorial, Barcelona

Harvey D (1997) The new urbanism and the communitarian trap. Harvard Design Magazine (1). http://www.harvarddesignmagazine.org/issues/1/the-new-urbanism-and-the-communitarian-trap

Harvey D (2012) La geografía como oportunidad política de resistencia y construcción de alternativas. Revista de Geografía Espacios 2(4):9–26. https://doi.org/10.25074/07197209.4.339

Katznelson I (1982) City trenches: urban politics and the patterning of class in the United States. University of Chicago Press, Chicago, IL
Le Gales P (2007) Gobernanza y cohesión social. En: Hall P (dir.) Congreso Regiones Capitales. Comunidad de Madrid, Madrid, pp 327–347
Leiva E, Miró I, Urbano X (2007) De la protesta al contra poder: nous protagonismes socials en la Barcelona metropolitana. Virus, Barcelona
Letelier F, Valdosky F (2019) La acción vecinal más allá del barrio: el caso del distrito Nou Barris en Barcelona. Rev Urban 40:1–16. https://doi.org/10.5354/0717-5051.2019.53618
Letelier L (2018) El barrio en cuestión: fragmentación y despolitización de lo vecinal en la era neoliberal. Scripta Nova. Universitat de Barcelona 22(602). https://doi.org/10.1344/sn2018.22.21518
López H (2020) La peleona Prospe, un siglo defendiéndose en comunidad. Érase una vez en el barrio (21). elPeriodico. 12 April 2020. Recuperado de: https://www.elperiodico.com/es/barcelona/20200412/la-peleona-prosperitat-un-siglo-defendiendose-en-comunidad-7920895
Martin D (2003) Enacting neighborhood 1. Urban Geogr 24(5):361–385. https://doi.org/10.2747/0272-3638.24.5.361
Massey D (2012) Un sentido global del lugar. En: Albet A, Benach N (eds) Doreen Massey. Un sentido global del lugal. Icaria, Barcelona, pp 112–228
Méndez R (2018) La telaraña financiera. Una geografía de la financiarización y su crisis. Instituto de Estudios Urbanos y Territoriales, Universidad Católica de Chile, Santiago de Chile
Navarro F (1997) Un espai urbà desordenat i fragmentari. En: AA.VV. (coord.) Nou Barris, 25 anys. Kim Manresa. Document gràfic. Ajuntament de Barcelona
Naya A (2023) 50 anys fent barri. Cronologia de les lluites veïnals a la Prosperitat. 1972–2022. Edit. Comissió de Memòria i Patrimoni de l'AV Prosperitat. ISBN 978-84-09-47848-4
Nel·lo O (2020) La ciudad y la plaga, en Eldiario.es. 31 March 2020. Madrid. https://www.eldiario.es/tribunaabierta/ciudad-plaga_6_1011808823.html
Nel·lo O (2013) Barcelona y Cataluña: las raíces del debate sobre el policentrismo del sistema urbano catalán. Ciudad y territorio: Estudios territoriales. Ministerio de Vivienda
Nou Barris Cabrejada diu prou! | Barcelona (2014) No és pobresa, és injusticia. Recuperado de https://jaumeuab.files.wordpress.com/2015/01/documentpobresa_definitiu-1.pdf
Pauné M (2014) Sant Andreu de Palomar ha perdido el 87% de sus masías. La Vanguardia. https://www.lavanguardia.com/local/barcelona/20140327/54404454542/sant-andreu-perdido-masias.html
Pigem J (1994) La odisea de Occidente. Modernidad y ecosofia. Kairós, Barcelona
Pradel Miquel M (2007) Análisis crítico del discurso de la gobernanza urbana. In: Conference: IX Congreso Español De Sociología, Barcelona (Spain). Aproximación crítica al discurso de la gobernanza urbana contemporánea. https://changingcities.files.wordpress.com/2007/10/aproximacion-critica-al-discurso-de-la-gobernanza.pdf
Pradel Miquel M (2020) Analysing the role of citizens in urban regeneration: bottom-linked initiatives in Barcelona. Urban Res Pract, 1–18. https://doi.org/10.1080/17535069.2020.1737725
Pradel Miquel M, García Cabeza S (eds) (2018) El Momento de la ciudadanía: Innovación Social y Gobernanza Urbana. La Catarata, Madrid
Pujol F (2003) Relligant Nou Barris: recull d'articles d'història publicats a la revista Rotllana (1988–2002). Ajuntament de Barcelona, Barcelona. ISBN 978-84-7609-731-1
Rojo Salgado A (2009) Una aproximación al concepto de gobernanza. En: Rodríguez González R (dir.) Ordenación y gobernanza de las áreas urbanas gallegas. A Coruña, Netbiblo, pp 3–27
Romero J, Farinós J (2011) Redescubriendo la gobernanza más allá del buen gobierno. Democracia como base, desarrollo territorial como resultado. Boletín de la Asociación de Geógrafos Españoles 56:295–319. https://bage.age-geografia.es/ojs/index.php/bage/article/view/1354
Roth L, Monterde A, Calleja López A (2019) Ciudades Democraticas. La Revuelta Municipalista en el Ciclo post-15M. Icaria, Barcelona, Spain

Santos BS, Avritzer L (2003) Introdução: para ampliar o cânone democrático. En Santos BS (org.) Democratizar a democracia: os caminhos da democracia participativa, Afrontamento, Porto, pp 35–66.

Somoza Medina J (2013) Gobernanza urbana, competitividad y procesos de renovación en las ciudades medias españolas. Boletín de la Asociación de Geógrafos Españoles 0(61). https://doi.org/10.21138/bage.1535

Swyngedouw E (2005) Governance, innovation and the citizen: the Janus face of governance beyond-the-state. Urban Stud

Valdosky F (2017) Das ações de resistência à consciência da expropriação. En: Alessandri Carlos AF, Alves G, Faleiros de Padua R (orgs.) Justiça Espacial e o direito a cidade. Editora Contexto, São Paulo

Villasante T, Gutiérrez V (2000) El movimiento vecinal: trayectoria y perspectivas. En: Grau E, Ibarra Güell P (coords.) Anuario de Movimientos Sociales 2000. Participando en la Red. Icaria – betiko Fundazioa, Barcelona, pp 70–86. Recuperado de http://fundacionbetiko.org/wp-content/uploads/2012/11/el-movimiento-vecinal-trayectoria-y-perspectivas.pdf

Correction to: Territorial Governance and Soft Planning in the Region of Lisbon

Cristina Cavaco, Igor Matheus Santana Chaves, and Luís Sanchez Carvalho

Corection to: Chapter 23 in: R. C. Lois-González and J. A. Rio Fernandes (eds.), *Urban Change in the Iberian Peninsula*, The Urban Book Series, https://doi.org/10.1007/978-3-031-59679-7_23

The original version of the chapter "**Territorial Governance and Soft Planning in the Region of Lisbon**" was previously published as non-open access, which has now been changed to open access under the terms of the Creative Commons Attribution 4.0 International License (https://creativecommons.org/licenses/by/4.0/) and the copyright holder has been updated to 'The Author(s)'. The chapter and the book have been updated with the changes.

The updated version of this chapter can be found at
https://doi.org/10.1007/978-3-031-59679-7_23

R. C. Lois-González and J. A. Rio Fernandes (eds.), *Urban Change in the Iberian Peninsula*, The Urban Book Series, https://doi.org/10.1007/978-3-031-59679-7_27

MIX
Papier aus verantwortungsvollen Quellen
Paper from responsible sources
FSC® C105338

If you have any concerns about our products,
you can contact us on
ProductSafety@springernature.com

In case Publisher is established outside the EU,
the EU authorized representative is:
Springer Nature Customer Service Center GmbH
Europaplatz 3, 69115 Heidelberg, Germany

Printed by Libri Plureos GmbH
in Hamburg, Germany